T0186622

Dynamic and Mobile GIS

Investigating Changes in Space and Time

INNOVATIONS IN GIS

SERIES EDITORS

Jane Drummond
University of Glasgow, Glasgow, Scotland

Bruce Gittings
University of Edinburgh, Edinburgh, Scotland

Elsa João
University of Strathclyde, Glasgow, Scotland

Dynamic and Mobile GIS: Investigating Changes in Space and Time
Edited by Jane Drummond, Roland Billen, Elsa João, and David Forrest

Dynamic and Mobile GIS

Investigating Changes in Space and Time

Edited by

Jane Drummond
Roland Billen
Elsa João
David Forrest

CRC Press
Taylor & Francis Group
Boca Raton   London   New York

CRC Press is an imprint of the
Taylor & Francis Group, an **informa** business

CRC Press
Taylor & Francis Group
6000 Broken Sound Parkway NW, Suite 300
Boca Raton, FL 33487-2742

First issued in paperback 2019

ISBN-13: 978-0-8493-9092-0 (hbk)
ISBN-13: 978-0-367-38993-2 (pbk)

Library of Congress Cataloging-in-Publication Data

Dynamic and mobile GIS : investigating changes in space and time / edited by Jane Drummond and Roland Billen.
    p. cm. -- (Innovations in GIS)
    Includes bibliographical references (p.).
  ISBN 0-8493-9092-3
  1. Geographic information systems. 2. Mobile communication systems. 3. Space and time. I. Drummond, Jane, 1950- II. Billen, Roland. III. Title. IV. Series.

G70.212.D96 2007
910.285--dc22   2006050478

Visit the Taylor & Francis Web site at
http://www.taylorandfrancis.com

and the CRC Press Web site at
http://www.crcpress.com

Foreword

Like Stan Openshaw (1998) in the foreword to the 'Innovations in GIS 5', I have never been asked to write a foreword before, and also like him I am concerned that after you read this one (and who reads forewords anyway?) I may never be invited again. But, as readers of this foreword will be probably be sparse and perhaps limited to the kind of people that read the small print on the backs of cornflakes packets, I can take this opportunity to say more or less anything. So I choose to ruminate on GIS research as seen through the eyes of the very first 1993 GIS Research UK (GISRUK) conference (Worboys, 1994b) and the latest, as represented by contributions in this volume, and discuss one of my pet subjects: the rising star of *time* in GIS research.

GISRUK has become a teenager! In 1993, we set as an objective for GISRUK 'to act as a focus in the UK for GIS research in all its diversity, across subject boundaries and with contributions from a wide range of researchers, from students just beginning their research careers to established experts'. There was at that time a need for a conference that brought together primarily UK researchers and students to discuss the state of GIS research. Indeed, in the original 'Innovations...' 26 out of 31 contributors were from UK institutions. In this latest volume, we count only 10 of the 31 chapter authors as UK-based. So, the conference, or at least the book it has generated has become internationally diverse.

So what are the current research preoccupations, as seen at GISRUK conference and in this volume? The thing that stands out for me, and this partly reflects a personal preoccupation, is the overwhelming importance now given to the temporal dimension in GIS. Time is now a significant partner with space, if not in GI systems, then certainly in the science of GI. Just as space provides the framework for describing the static objects in the world, so the temporal dimension is needed for *occurrent* entities, such as events and processes. Dynamic spatial phenomena require a mix of space and time, leading to so-called spatiotemporal information systems (STIS).

Hägerstrand (1970) had already noted the importance of the temporal dimension in geographical and socio-economic analysis, but it was in the 1990s that STIS really began to take off (Langran, 1993; Worboys, 1994a). Time in this volume has been promoted to the volume theme, 'Dynamic and Mobile GIS', with its focus on the event-oriented aspects of the world. An entire section is devoted to 'Motion, Time and Space', as well as an introductory essay on the technology of space and time (Maguire), discussions on process models (Rietsma and Albrecht) and events (Beard). Almost every chapter, from mobile GIS to disaster management applications, requires an understanding *and efficient implementation* of the temporal dimension in spatial information systems. At last, time is finally being given its true place among those key topics for research in geographic information science.

It is clear that the integrated spatiotemporal dimension is beginning to play the role that 2-dimensional spatial geometry and topology played for GIS at its outset. Applications range from environmental event analysis, disaster management, defense, transportation, and the evolution of a topographic landscape. But whereas with space, the proprietary technology was quickly to hand, for the temporal dimension, even purely temporal databases, let alone spatiotemporal systems, are rare or even non-existent in the marketplace. As Maguire (Chapter 1) states, 'We are just beginning to add support for reading and storing time-series data, but we are still someway off full 4D dynamic modeling within a commercial GIS.' I believe that this is now a matter of timing and economics. The technology is becoming ready, but business cases still need to be made.

What are the current and future issues in STIS research? To my mind, still the really hard question, is what the underlying conceptual model looks like? Or, to use that hackneyed O-word, what is the upper-level ontology of dynamic geographic phenomena? The answer to this question is not just related to the structure of time, but also to the general kinds of dynamic entities that exist in the world: events, processes, actions, trajectories, etc., *and how they are all interrelated*. This question is still wide open.

GIS research, as presented at the GISRUK conference series, and enshrined in the 'Innovations in GIS' book series, is flourishing, and has moved from the relatively narrow national stage to encompass an international participation. Finally, from one of its parents, I wish GISRUK some happy adolescent years, and not too much teenage angst!

Mike Worboys,
University of Maine, USA

References

Hägerstrand, T. (1970) 'What about People in Regional Science?', *Papers of the Regional Science Association*, 24, pp. 7–21.

Langran, G. (1993) 'Issues of Implementing a Spatiotemporal System', *International Journal of Geographical Information Systems, vol.* 7(4): pp. 305–314.

Openshaw, S. (1998) 'Foreword', in Carver, S. (ed.), *Innovations in GIS 5,* London: Taylor and Francis.

Worboys, M. F. (1994a) 'A Unified Model of Spatial and Temporal Information', *Computer Journal,* vol. 37(1), pp. 26–34.

Worboys, M. F. (1994b) (ed.), *Innovations in GIS 1,* London: Taylor and Francis.

Preface

This book's title 'Dynamic and Mobile GIS: Investigating Changes in Space and Time', part of the Innovations in GIS book series, may need some explaining. The technology which will support Mobile GIS is rapidly gaining popularity and effectiveness (PDAs, wireless internet, internet-based GIS, 3G and 4G telecommunications). The application domain of Mobile GIS is wherever important geo-spatial events are taking place – not back at the office. That these events need to be recorded and analysed *in situ* implies that they are rapidly changing (hence dynamic) phenomena. This situation implies technological, databasing, display design and processing constraints requiring investigation and synergistic research and development. To us it seemed appropriate to produce a book linking these dynamic and mobile elements of Geographical Information Science.

Dynamic and mobile GIS is a research area full of good ideas. Some of these emerge from the constraints of current technology; for example, those that seek to solve the problems of limited display (e.g. Anand et al. in Chapter 9) or high volume data transmission (e.g. Li in Chapter 2). Other ideas emerge despite these constraints (e.g. Tsou and Sun in Chapter 12; Laube et al. in Chapter 14). Nevertheless, dynamic and mobile GIS is now an established idea, and, for those researching it, a technology exists that must be acknowledged and understood.

Excluding an Epilogue (Part V), there are four parts to this book. Each is briefly introduced below, although a fuller introduction is provided at the start of each part.

Part I - Technology for Dynamic and Mobile GIS – As Mobile GIS technology already exists we have decided to make this the first part of our book. Chapter 1 'The Changing Technology of Space and Time' by David Maguire, sets the scene by providing an introduction and an overview of both the extant technology and hints of what is to come. This is done within the context of the evolution of: GIS, from 2D to 3D and now, by embracing the time dimension, to 4D; computer systems, from stand-alone systems to distributed, network-centric systems; and miniaturisation wherever more powerful processors are being built into increasingly smart, multi-functional, small and light devices. Chapter 2 'Opportunities in Mobile GIS' by Qingquan Li expands on Maguire's chapter. Li introduces the reader to a very large part of the technology without overwhelming with detail. Thus the reader is left knowing what they ought to know about, and is a most useful guide. Li is very optimistic about the future of mobile GIS and makes this clear through the presentation of successful applications and healthy business projections. Chapter 3 'Location privacy and location-aware computing' by Matt Duckham and Lars Kulik rounds out the book's Part I by raising issues to make us think about some of dynamic and mobile GIS's implications. They suggest that the technology's challenge to our security and privacy needs consideration, and present some solutions. Duckham and Kulik work with researchers active in many applications of spatial information systems for facilities and utilities management, emergency services delivery, and environmental monitoring: all currently exploiting mobile GIS and presenting problems related to privacy.

Part II – Modelling Approaches and Data Models – This part focuses on modelling approaches especially appropriate to depict dynamic processes in GIS. Kate Beard in Chapter 4 proposes an event-based approach in which change itself is the central concept that is modelled. An event-based view provides the foundation for the analysis of dynamic phenomena and is therefore naturally appropriate for dynamic GIS. Femke Reitsma and Jochen Albrecht in Chapter 5 present a new process-based data model called *nen* (after node-edge-node graph representation). While most of the existing theories and models for simulating processes focus on representing the state of the represented system at a moment of time, this approach expresses and represents information about processes themselves. This allows questions to be asked that are not directly answerable with current object-centred formulations. In Chapter 6, Muki Haklay extends the comparison between Map Calculus and Map Algebra in the context of dynamic raster GIS. This chapter focuses on the particular challenges of dynamic modelling in GIS, exploring the ways in which it is implemented in Map Algebra and outlining how such models can be implemented in a Map Calculus-based system. Finally, Peter van Oosterom explores in Chapter 7 issues related to spatial constraints in data models. The chapter argues that constraints should be part of the object class definition, just as with other aspects of that definition, including attributes, methods and relationships. In a dynamic context, with constantly changing geo-information, any changes arising should adhere to specified constraints, otherwise inconsistencies will occur. The chapter demonstrates the need for the integral support of constraints, and proposes a complete description and classification of constraints.

Part III - Display and Visualisation – Although there is a wide range of potential uses for GIS, for many the primary purpose is to display information. Two of the Chapters in Part III examine the need to display an appropriate level of information in a mobile environment, where current displays are of limited size and resolution. In Chapter 8, Malisa Plesa and William Cartwright compare the effectiveness of photorealistic displays with more generalised representations of an urban area. In approaching a conceptually similar issue, Suchith Anand et al. in Chapter 9 proceed by developing procedures for very much simplifying route information so that only a diagrammatic representation of the route or route network is displayed. Britta Hummel on the other hand in Chapter 10 provides insight into solving the not so simple problem of displaying correct vehicle positions in relation to in-car navigation displays when GPS data and map data are not always perfectly matched.

Part IV - Motion, Time and Space - This part focuses on the study of mobility and examples of applications of mobile devices for disaster management and environmental monitoring. Pablo Mateos and Peter Fisher in Chapter 11 start by arguing that mobile phone location might become a new spatial reference system, which the authors call the 'new cellular geography'. However, mobility measurements can be limited by poor accuracy. Chapter 11 therefore presents an evaluation of the accuracy of mobile phone location to determine its appropriate application as an automated method to measure and represent the mobility of people. Ming-Hsiang Tsou and Chih-Hong Sun in Chapter 12 suggest that mobile GIS is one of the most vital technologies for the future development of disaster management systems because it extends the capability of traditional GIS to a higher level of portability, usability and flexibility. The authors argue that an integrated mobile and distributed GIService, combined with an early warning system, is ideal to support disaster management, response, prevention and recovery. Cristina Gouveia et al. in Chapter 13 propose the creation of an Environmental Collaborative Monitoring Network that relies on citizens using either mobile phones or mobile GIS in order to carry out environmental monitoring. The chapter explores the use of mobile computing and mobile communications, together with sensing devices (such as people's own senses like smell and vision), to support citizens in environmental monitoring activities. Patrick Laube et al. in Chapter 14 argue that Geographical Information Science can centrally contribute to discovering knowledge about the patterns made in space-time by individuals and groups within large volumes of motion data. The chapter introduces an innovative approach for analysing the tracks of moving point objects using a methodological approach called Geographic Knowledge Discovery.

The completion of this book leaves us indebted to many people. First of all we wish to thank the 31 contributors, drawn from 11 different countries from all over the world (Australia, Belgium, China, Germany, The Netherlands, New Zealand, Portugal, Switzerland, Taiwan, UK, USA), without whose work this book would not have been possible. Six of the chapters are written by invited experts while eight of the chapters are based on contributions made by authors who participated in the GISRUK 2005 Conference in Glasgow, from 5 – 8th April 2005. Since 1993 these annual conferences have been key events organised by UK universities that have significant interest in Geographical Information Science. The series is considered to represent Europe's premier GIS research conference series.

We are particularly grateful for the excellent editorial work provided by our former colleague David Tait (now of Giffnock Editorial Services: d.a.tait@ntlworld.com) without whom the writing of this book would have been very much more difficult and time consuming. Pierre Hallot (University of Liège) provided special support in the final stages of the preparation of the book and we are very thankful for his help. Generous advice was also provided by our colleague Mike Shand (University of Glasgow). We would like to acknowledge, with tremendous gratitude, the unstinting support of our colleague Anne Dunlop (University of Glasgow), who, although not involved in the editing of this volume, attended to the needs of our students in so many extra ways while we were involved. We are also indebted to those publishers and authors who have granted copyright permission to reproduce extracts from their work for inclusion.

The preparation of this book was, as with GISRUK 2005, the result of collaboration between the Department of Geographical and Earth Sciences (formerly Geography and Geomatics) at the University of Glasgow and the Graduate School of Environmental Studies at the University of Strathclyde, also in Glasgow. To all GIS researchers, academics, practitioners, students and government officials looking to develop dynamic and mobile GIS facilities, we hope you will find this book invaluable in your work and research.

Jane Drummond, Roland Billen, Elsa João and David Forrest

List of Contributors

Jochen Albrecht has been pushing the boundaries of dynamic GIS for the past ten years. His research ranges from philosophical questions such as 'what is change?' to practical implementations in property databases, crime analysis, regional science, and ecological applications. In any of these, the data modelling approaches differ, and Jochen's nirvana lies in finding the underlying commonalities.

Department of Geography, Hunter College, City University of New York, NY 10021, USA; Email: jochen@hunter.cuny.edu

Suchith Anand while writing was a PhD student at the University of Glamorgan working on the application of map generalisation to location based services, but is now a Research Associate in Mobile Location Based Services in The Centre for Geo-spatial Science, University of Nottingham.

Centre for Geo-spatial Science, University of Nottingham, NG7 2RD, UK; Email: Suchith.Anand@nottingham.ac.uk

Kate Beard is a professor in the Department of Spatial Information Science Engineering at the University of Maine. She has been a research faculty member with the National Center for Geographic Information and Analysis (NCGIA) since its beginning in 1989. Her research interests cover multiple representations and cartographic generalisation, investigations of data quality and metadata representation. She also conducts research in digital library issues for geo-spatial information collections which has addressed issues of metadata services, and gazetteer development. Her recent research addresses modelling, analysis and visualisation of space-time events

Department of Spatial Information Science and Engineering, University of Maine, Orono, ME 04469, USA; Email: beard@spatial.maine.edu

Roland Billen is a lecturer of geomatics at the Geography Department of the University of Liège, Belgium. He was previously a lecturer at Glasgow University's Department of Geography and Geomatics (2003-2005). His research interests are in spatial reasoning and analysis, urban GIS (design, implementation, use), 3D modelling, and 2&3D data acquisition (topographic survey, photogrammetry, GPS).

Unité de Géomatique, Département de Géographie, Université de Liège, 6 Allée du 6-Août, 4000 Liège, Belgium; Email : rbillen@ulg.ac.be

António Câmara is a professor at the New University of Lisbon and has been a visiting professor at both Cornell University (1988-89) and MIT (1998-99). He was a senior consultant in the Expo98 project and senior advisor to the National

Geographical Information System (SNIG). He has been YDreams chief executive officer since the company started in June 2000.

Grupo de Análise de Sistemas Ambientais, Faculdade de Ciências e Tecnologia, Universidade Nova de Lisboa, Quinta da Torre, 2795 Monte da Caparica, Portugal; Email: asc@mail.fct.unl.pt

William Cartwright is associate professor of Cartography and Geographical Visualisation in the School of Mathematical and Geo-spatial Sciences at RMIT University. His major research interest is the application of New Media to cartography and the exploration of different metaphorical approaches to the depiction of geographical information.

School of Mathematical and Geo-spatial Sciences, RMIT University, Melbourne, Victoria, Australia; Email: william.cartwright@rmit.edu.au

Beatriz Condessa is a lecturer at the Department of Civil Engineering and Architecture at the Instituto Superior Técnico in Lisbon, having previously worked as a researcher at the National Centre for Geographic Information (CNIG). She has a PhD in Geography from Barcelona University. Her main area of research is urban and regional planning. Other areas of research are GIS, Web mapping and environmental management.

Instituto Superior Técnico, Avenida Rovisco Pais, 1049-001 Lisboa, Portugal; Email: bcondessa@civil.ist.utl.pt

Jane Drummond lectures at Glasgow University in Geomatics topics prior to that being employed at the ITC, Netherlands and NERC's Experimental Cartography Unit, following initial research and practice in photogrammetry. Her present research is in data quality and the integration of primary data into GIS.

Department of Geographical and Earth Sciences, University of Glasgow, Glasgow G12 8QQ, UK; Email: jane.drummond@ges.gla.ac.uk

Matt Duckham is a lecturer in GIS at the Department of Geomatics of the University of Melbourne. Prior to this he worked as a postdoctoral researcher at the NCGIA, Department of Spatial Information Science and Engineering, University of Maine and at the University of Keele, following a PhD at Glasgow University. His research centres on computation with uncertain geographic information, especially within the domain of mobile and location-aware systems. With Mike Worboys, he has co-authored a major GIS text (*GIS: A Computing Perspective*).

Department of Geomatics, Faculty of Engineering, University of Melbourne, Victoria 3010, Australia; Email: matt@duckham.org

Peter Fisher is professor of Geographical Information Science at City University, having previously worked as Professor in Geography at the University of Leicester. He is the editor of the International Journal of Geographical Information Science.

Department of Information Science, City University, Northampton Square, London EC1V 0HB, UK; Email: pff1@city.ac.uk

Alexandra Fonseca is a researcher at the Centre for Exploration and Management of Geographic Information at the Portuguese Geographical Institute (IGP), having previously worked as a researcher at the National Centre for Geographic Information (CNIG). With a PhD in Environmental Systems from the New University of Lisbon, her research areas include using information and communication technologies in environmental management and planning, multimedia GIS, geovisualisation and public participation in decision making.

Instituto Geográfico Português (IGP), Rua Artilharia Um, 107, 1099-052 Lisboa, Portugal; Email: afonseca@igeo.pt

David Forrest is a senior lecturer in the Department of Geographical and Earth Sciences of the University of Glasgow where he researches in the use of mapping and cartographic expert systems, subjects he also teaches, along with GIS, National Spatial Data Infrastructure, geo-spatial visualisation, map production and hydrographic survey. He is a former President of the British Cartographic Society.

Department of Geographical and Earth Sciences, University of Glasgow, Glasgow G12 8QQ, UK; Email: david.forrest@ges.gla.ac.uk

Cristina Gouveia is a researcher at the Centre for Exploration and Management of Geographic Information at the Portuguese Geographical Institute (IGP), having previously worked as researcher at the National Centre for Geographic Information (CNIG). Currently completing a PhD at the New University of Lisbon, her main area of research is the use of information and communication technologies to support environmental management.

Instituto Geográfico Português (IGP), Rua Artilharia Um, 107, 1099-052 Lisboa, Portugal; Email: cgouveia@alum.mit.edu

Muki Haklay is a lecturer in Geographical Information Science in the Department of Geomatic Engineering at University College London (UCL). He holds a PhD in Geography from UCL, an MA in Geography and a BSc in Computer Science and Geography from the Hebrew University of Jerusalem. His research interests include public access to environmental information, public participation GIS, human-computer interaction in GIScience, and spatial databases and data models.

Department of Geomatic Engineering, University College London, Gower St, London WC1E 6BT, UK; Email: m.haklay@ucl.ac.uk

Britta Hummel is a researcher at the Institute of Measurement and Control Theory at the University of Karlsruhe. Currently completing her PhD in vision-based driver assistance systems, she obtained a Diploma in Computer Science in 2003. Her research interests include computer vision, pattern recognition, knowledge representation and reasoning, particularly in the field of autonomous vehicles.

Institute of Measurement and Control, University of Karlsruhe, Engler-Bunte-Ring 21, 76131, Germany; Email: hummel@mrt.uka.de

Stephan Imfeld is a senior researcher at the GIScience Center at the Department of Geography of the University of Zurich and at the Pulmonology Department at the University Hospital in Zurich. He has received a PhD in Natural Sciences from the University of Zurich and a MD at the University of Basel, Switzerland. His main interests are in methodological aspects of GIS applications in biological sciences and in applied medical research.

Department of Geography, University of Zurich, Winterthurerstrasse 190, 8057 Zurich, Switzerland; Email: imfeld@geo.unizh.ch

Elsa João is a lecturer and the director of research of the Graduate School of Environmental Studies (GSES) at the University of Strathclyde in Scotland. At GSES she is responsible for the PhD programme and helps run the MSc in Environmental Studies. Her research interests include the use of GIS for Strategic Environmental Assessment (SEA) and Project Environmental Impact Assessment (EIA). She is particularly interested in spatial data quality and scale issues.

Graduate School of Environmental Studies (GSES), University of Strathclyde, Level 6, Graham Hills Building, 50 Richmond St, Glasgow G1 1XN, UK; Email: elsa.joao@strath.ac.uk

Lars Kulik is a lecturer in the Department of Computer Science and Software Engineering, University of Melbourne. His current research focuses on mobile and location-aware computing. His goal is to develop a spatial foundation for information systems facilitating mobile systems responsive to space and time and offering intelligent location based services. This has led from interests in human-centered algorithms; privacy and imprecision in mobile computing and algorithms simplifying visual representation.

Department of Computer Science and Software Engineering, University of Melbourne, Victoria 3010 Australia; Email: lkulik@cs.mu.oz.au

Patrick Laube researches at the Spatial Analysis Facility, University of Auckland. His Geography PhD is from the University of Zurich. His co-authored chapter presents a synopsis of his thesis *Analysing Point Motion – Spatio-temporal data mining of geo-spatial lifelines*. His main research interests lie in the integration of knowledge discovery techniques in Geographical Information Science.

Spatial Analysis Facility, School of Geography and Environmental Science, The University of Auckland, Private Bag 92019, Auckland, New Zealand; Email: p.laube@auckland.ac.nz

Qingquan Li directs the Research Center for Spatial Information and Network Communication at the State Key Laboratory for Information Engineering in Surveying, Mapping and Remote Sensing (LIESMARS), Wuhan Univesity. His research interests include GIS in higher education, modelling of traffic information and applications of mobile GIS for vehicle navigation and information service.

State Key Laboratory of Information Engineering in Surveying, Mapping and Remote Sensing, Wuhan University, Wuhan 430072, P.R.China; Email: qqli@whu.edu.cn

David Maguire is Director of Products at the Environmental Systems Research Institute (ESRI) in Redlands, California, since 1997. Prior to that he was in ESRI-UK. From 1987 to 1991 he directed the GIS MSc at Leicester University. David is a member of ESRI's senior management team with responsibility for coordinating product development. He has published widely being a founder editor of the 1999 'Big Book' of GIS *GIS: Principles, Techniques, Management & Applications* and co-author of the fast-selling GIS textbook *Geographical Information Systems and Science* (2001, 2004). He has wide interests in all aspects of spatial analysis and GIS, and is particularly interested in new developments of GIS software in terms of system architectures, spatial databases, mobile devices and location based services.

ESRI, 380 New York Street, Redlands, CA 92373, USA; Email: dmaguire@esri.com

Pablo Mateos is a PhD student at the Centre for Advanced Spatial Analysis (CASA), University College London, as well as a joint research associate at the University College London and Camden PCT (National Health Service). His research interests are the applications of GIS and geodemographics in urban and social geography to reduce socio-economic inequalities.

Centre for Advanced Spatial Analysis, University College London, 1-19 Torrington Place, London WC1E 6BT, UK; Email: p.mateos@ucl.ac.uk

Malissa Ana Plesa completed the Bachelor of Applied Science in Geo-spatial Science (Hons) at RMIT University in 2004. Her research focus was on the use of non-realistic representations on mobile devices. She currently works at Lonely Planet Publishing in Melbourne, Australia.

School of Mathematical and Geo-spatial Sciences, RMIT University, Melbourne, Victoria, Australia; Email: Malisa.Plesa@lonelyplanet.com.au

Ross Purves is a lecturer in the GIS division of the Department of Geography at the University of Zurich. His main research interests lie in the fields of environmental modelling and geographic information retrieval.

Department of Geography, University of Zurich, Winterthurerstrasse 190, 8057 Zurich, Switzerland; Email: rsp@geo.unizh.ch

Femke Reitsma has recently become a lecturer at the University of Edinburgh, after completing a PhD that involved trying to squeeze time out of space. Her interests revolve around issues of representation of spatial data: how different approaches to modelling the world have an impact on our ability to understand and explain. In particular, she is interested in representations of space, time and change.

Institute of Geography, School of Geosciences, The University of Edinburgh, Edinburgh, UK; Email: femke.reitsma@ed.ac.uk

Chih-Hong Sun is professor of Geography at the National Taiwan University and is the project leader for the design of Taiwan's National Geographic Information System. His research areas are in geographic information science, decision-support systems, hazard mitigation and sustainable development. His recent research concentrates in developing spatial decision support systems for natural hazard mitigation and sustainable development.

Department of Geography, National Taiwan University, Taipei, Taiwan; Email: chsun@ntu.edu.tw

George Taylor is a professor and the head of GIS Research group in the School of Computing, University of Glamorgan. Current research is focused on the fusion of Geographical Information Systems (GIS) with Global Navigation Satellite Systems (GNSS), such as the Global Positioning System (GPS). Recent projects have focused on the development of point positioning algorithms that integrate raw GPS receiver data with data extracted from digital topographic and height maps.

School of Computing, University of Glamorgan, Wales, CF37 1DL, UK; email: getaylor@glam.ac.uk

Ming-Hsiang (Ming) Tsou is associate professor in the Department of Geography at San Diego State University, USA. As a Cartographer and GIS specialist, his research and teaching interests include Internet mapping, wireless mobile GIS, distributed GIS applications, multi-media cartography, user interface design and software agents. In 2003 he co-authored the book *Internet GIS* with Dr. Zhong-Ren Peng from Wisconsin University at Milwaukee.

Department of Geography, San Diego State University, San Diego, CA 92182-4493, USA; Email: mtsou@mail.sdsu.edu

Peter van Oosterom is professor in GIS technology at the Delft University of Technology (Faculty of Technology, Policy and Management and OTB research institute). His interests are in spatial databases (3D, performance, constraints, temporal), GIS architectures, spatial analysis, generalisation, querying and presentation, Internet/interoperable GIS and cadastral applications.

Delft University of Technology, Section GIS-technology, Jaffalaan 9, 2628 BX Delft, The Netherlands; email: oosterom@otb.tudelft.nl

J. Mark Ware is reader in the School of Computing, University of Glamorgan. Areas of special interest include terrain modelling, automated cartography (map generalisation), automated environmental change detection, spatial data error modelling, spatial indexing, multi-scale data structures, location based services (mobile GIS) and analysis and mapping of crime data.

School of Computing; University of Glamorgan, Glamorgan, CF37 1DL, Wales, UK; email: jmware@glam.ac.uk

Robert Weibel is a professor of Geographical Information Science at the University of Zurich. He has been one of the principal investigators on the recently completed European project WebPark (www.webparkservices.info) in which a mobile information system for protected areas was built and investigated. Besides spatial information use in mobile GIS and spatiotemporal analysis of motion data, he is particularly interested in issues of scale changing and automated map generalisation.

Department of Geography, University of Zurich, Winterthurerstrasse 190, 8057 Zurich, Switzerland; Email: weibel@geo.unizh.ch

Acronyms

2.5D	2.5 dimensional
2D	two-dimensional
3D	three-dimensional
3GPP2	Third Generation Partnership Project 2
AFLT	Advanced Forward Link Trilateration
A-GPS	assisted GPS
ALI	Automatic Location Information
AMPS	Advanced Mobile Phone System
AOA	angle of arrival
API	application programming interface
AVHRR	Advanced Very High Resolution Radiometer
BP	beep pagers
CAD	computer-aided design
Cal(IT)2	The California Institute of Telecommunication and Information Technology
Caltran	The Department of Transportation in California
CCVQ	Cadastral Constraint Violation Queries
CDMA	Code Division Multiple Access
CNIG	National Centre for Geographic Information (Portugal)
COO	cell of origin
CPUs	Central Processing Units
DAMPS	Digital Advanced Mobile Phone System
DBMS	Data Base Management System
DDL	Data Definition Language
DRM	Digital rights management
EC	European Commission
ECMN	Environmental Collaborative Monitoring Networks
E-OTD	Enhanced Observed Time Difference
EPA	Environmental Protection Agency
ESRI	Environmental Systems Research Institute, Inc.
ESTDM	Event-based Spatiotemporal Data Model
ETSI	European Telecommunications Standards Institute
FCC	Federal Communications Commission
FCT-UNL	College of Sciences and Technology from the New University of Lisbon
FDNY	Fire Department of New York.
FEMA USA	Federal Emergency Management Agency
FPE	College of Psychology and Education from the University of Lisbon
GASA	Environmental Systems Analysis Group of the New University of Lisbon (Portugal)

GEM	geo-spatial event model
GIS	Geographic Information System(s)
GIScience	Geographic Information Science
GIServices	Geographic Information Services
GML	Geography Markup Language
GoMOOS	Gulf of Maine Ocean Observing
GPRS	General Packet Radio Service
GPS	Global Positioning System
GSM	Global System for Mobile Communications
HEC-HMS	Hydrologic Engineering Center-Hydrologic Modeling System
HEC-RAS	Hydrologic Engineering Center-River Analysis System
HTTP	HyperText Transport Protocol
ICT	Information and Communication Technologies
ID	identifier
IDIN	Integrated Disaster Information Network
IDW	Inverse Distance Weighted
IEEE	Institute of Electrical & Electronic Engineers
IES	Institute for Environmental Sustainability
IETF	Internet Engineering Task Force
IGP	Portuguese Geographical Institute
IP	Internet Protocol
IPv6	Internet Protocol Version 6
JRC	Joint Research Centre
LAN	Local Area Network
LBS	location based services
LIF	Location Interoperability Forum
MAN	Metropolitan Area Network
MEIS	Mobile Environmental Information Systems
MHz	megahertz
MIS	Management Information Systems
MMP	Municipal Master Plans
MMS	Multi-media Messaging System
MODIS	Moderate Resolution Imaging Spectroradiometer
MTUP	modifiable temporal unit problem
NAPHM	National Science and Technology Program for Hazards Mitigation (Taiwan)
NASA	National Aeronautics and Space Administration
NCS	National Science Council (Taiwan)
nen	node-edge-node
NetCDF	Network Common Data Format
NGO	Non-governmental organisations
NIMBY	not in my backyard
NPR	National Public Radio
NTU	National Taiwan University

OCL	Object Constraint Language
OGC	Open Geo-spatial Consortium
OMG	Object Management Group
PC	personal computer
PDA	Personal Digital Assistant
PDRM	Personal Digital Rights management
PEOPLE	Population Exposure to Air Pollutants in Europe project
PIDF	presence information data format
PLD	Personal Location Device
QA/QC	Quality Assessment/Quality Control
QuikSCAT	Quick Scatterometer
RAM	Random Access Memory
RAN	Radio Access Network
RCEW	Reynolds Creek Experimental Watershed
RePast	Recursive Porous Agent Simulation Toolkit
REPAST	Recursive Porous Agent Simulation Toolkit
RF-ID	Radio Frequency Identification
SDK	Software Development Kit
SeaWIFs	Sea-viewing Wide Field-of-view Sensor
SHOE	Simple HTML Ontology Extensions
SMG	Special Mobile Group
SMS	Short Messaging System
SMS/MMS	Short Messaging Service/ Multi-media Messaging Service
SOA	service-oriented architecture
SOAP	Simple Object Access Protocol
SQL	Structured Query Language
STEFS	Software Tools for Environmental Study
STH	spatiotemporal helix
SVG	Scalable Vector Graphics
TACS	Total Access Communication System
TADMDS	Taiwan Advanced Disaster Management Decision Support System.
TB	terabyte
TBT	Tributyltin
TCP/IP	Trasmission Control Protocol/Internet Protocol
TDOA	Time Difference of Arrival
TD-SCDMA	Time Division-Synchronous CDMA
TIN	Triangulated Irregular Network
TLS	Time Location Stamp
TMCX	Topographic Model Constraints in XML
TOA	time of arrival
TPSP	Third-Party Service Provider
UCGIS	University Consortium for Geographic Information Science
UML	Unified Modeling Language

UMTS	Universal Mobile Telecommunications System
VR	Virtual Reality
VRML	Virtual Reality Markup Language
W3C	World Wide Web Consortium
WAP	Wireless Application Protocol
WCDMA	Wide-band CDMA
WFS	Web Feature Server
WiFi	wireless fidelity
Wi-Fi	wireless fidelity
WiMAX	Worldwide Interoperability for Microwave Access.
WinCE	embedded Windows system
WLAN	Wireless Local Area Networks
WPAN	Wireless Personal Area Network
WWW	World Wide Web
XMI	XML Metadata Interchange
XML	extensible markup language
XSD	XML Schema Document
ZKPs	Zero Knowledge Proofs

Table of Contents

Part I

Technology for Dynamic and Mobile GIS

Part I of this book, 'Dynamic and Mobile GIS: Investigating Changing Space and Time' deals with technology in this active GIS research area. The question may be raised as to whether GIS research is technology or ideas driven. As applied scientists we may find new technology 'sparks' new ideas, but, on the other hand, recognise that many GI ideas pre-date information technology by centuries.

David Maguire sets the scene with a chapter providing a clear overview (1: 'The Changing Technology of Space and Time'). As one of the authors of 'the big book' and its 'spin offs' this is to be expected (Maguire et al., 1991; Longley et al., 2005). The chapter is essential for readers new to Dynamic and Mobile GIS, and by focusing on recent developments in Computer Systems, Computer Networks and Computer Software, it provides sufficient background for anyone wishing to delve into successful applications (Chapters 11–14) as quickly as possible. It is dynamic events in an area ('S') that require rapid recording and analysis. Mobile GIS can allow this to happen at 'S', rather than at 'U' (the location of a conventional user), 'D' (the location of the database) or at 'P' (a separate processing location). Maguire reminds us that sophisticated analyses (for example exploratory spatial analysis) and modelling have often been avoided in static GIS—despite its ability to handle voluminous data and exploit large format screens. Nevertheless, in a fire simulation model provided in his chapter, an approach which involves iterating around parts of the model has been shown to successfully model a highly dynamic event (see Figure 1.9), at 'S'.

Qingquan Li's chapter (2: 'Opportunities in Mobile GIS') presents the opportunities that have already been grasped by researchers in the field, and the technology they have accessed to do so. It also highlights research that must be completed if mobile GIS is to realise, fully, its potential. To a large extent this chapter indicates, in a useful manner, many technological options, and the user is directed towards the Internet and other sources for the deeper explanation, if required. The impression left by this chapter is that many successful mobile GIS projects have already been completed, but its upbeat character leads us to expect more still, and with few problems on the horizon. Li also stresses the importance of understanding Location Based Services (LBS) if one is to understand Mobile GIS. It is worth drawing the reader's attention to this. LBS is a slightly older idea than Mobile GIS, but both require 'location aware' users if they are to be successful. Implied here is a plea for new workers not only to look at Mobile GIS literature, but also that of LBS.

In Chapter 3 ('Location Privacy and Location-aware Computing') Duckham and Kulik question the awareness of mobile technology's enthusiasts of the onslaught on our privacy that Dynamic and Mobile GIS may bring. This may have negative effects on the uptake of Mobile GIS particularly if the vulnerability of users to location based 'spam', personal safety and other intrusions into our private lives is not considered. The authors present potential solutions to location privacy issues, but indicate that the technology is not yet sufficiently proven to support these. This chapter is a thought provoking counter to Chapter 2 in which Li tells how much Mobile GIS has achieved with current technology. Duckham and Kulik remind us how much will *not* be achieved if solutions (including both technological and structural solutions) to privacy issues are not addressed. To quote: 'location privacy lies at the intersection of society and technology', and opens up another, potentially rich, research theme (Duckham and Kulik, 2005).

The three introductory chapters, we hope, will provide the reader with a thought provoking appetizer for this field.

References

Duckham, M. J. and Kulik, L. (2005) 'Simulation of obfuscation and negotiation for location privacy', in Mark, D. M. and Cohn, A. G. (eds.) COSIT 2005, vol. 3693 of *Lecture Notes in Computer Science*, pp. 31–48, Berlin: Springer.

Longley, P., Maguire, D. J., Goodchild, M. F. and Rhind, D. W. (2005) *Geographical Information Systems and Science,* Chichester, W. Sussex: John Wiley and Sons, UK.

Maguire, D. J., Goodchild, M. F. and Rhind, D. W. (1991). *Geographical Information Systems*, Harlow, Essex: Longman Group, UK.

Chapter 1

The Changing Technology of Space and Time

David Maguire

Director of Products ESRI, California, USA

1.1 Introduction

Geography as a discipline, like scientific investigation in general, is changing dramatically. As many commentators have noted, for example Laudan (1996), Hills (1999) and the Committee on Facilitating Interdisciplinary Research (2004), the focus of much leading edge research is now interdisciplinary, applied and distributed. This has important ramifications for the way we think about, conduct research into, and implement Geographic Information Systems (GIS). It comes as no surprise to geographers that scientists now argue that many of science's most challenging questions are multi-faceted and that understanding them requires the knowledge, tools and techniques of many disciplines. Similarly, the idea that the distinction between pure science (so-called positive uses) and applied problem solving (so-called normative uses) has effectively disappeared has been a central tenet of work in the GIS field for many years. For example, conservation of cougar (mountain lion) habitats requires data on cougar movements, prey-predator relationships, cougar reproduction and the impact of the environment (terrain, vegetation, fire, etc.). Such data can be collected and managed within a GISystem. Effective understanding of current patterns and prediction of future activities additionally requires a scientific model of the wildlife biology of cougars (GIScience) that uses the data and tools of the GISystem. The mutual dependence of GI Systems and Science has been argued many times, most notably by Longley et al. (2005), and is central to the discussion in this chapter.

Early GIS projects represented the real world as two-dimensional (2D) map coordinates (X, Y), in no small part because of the technical difficulties of dealing with the third (Z) dimension. Many of the earliest attempts at GIS modelling were little more than 2D map analysis (Longley et al., 2005) and this remains the 'bread and butter' of much of today's work. For example, Figure 1.1 shows a simple site suitability project for San Diego County, USA that is based on 2D map analysis. A simple cartographic model takes roads and vegetation as input (blue ellipses), creates a buffer around the roads, and combines the road buffer with vegetation (yellow boxes) to create a map of the vegetation types affected by the proposed roads (green ellipse). The proposed roads (red lines) and buffers (light blue) are shown on the map. This is obviously a very simple example, but the basic approach

Dynamic and Mobile GIS: Investigating Changes in Space and Time. Edited by Jane Drummond, Roland Billen, Elsa João and David Forrest. © 2006 Taylor & Francis

illustrated here has been used many times to create quite sophisticated process models with hundreds of inputs and transformation tools.

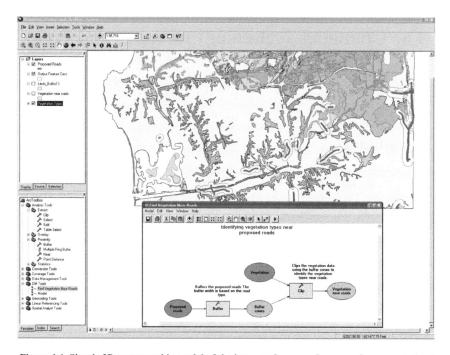

Figure 1.1. Simple 2D cartographic model of the impact of proposed new roads on vegetation in San Diego County, USA. See colour insert following page 132.

In recent years, GISystems and GIScience have started to make the jump first from 2D to 3D (Maguire et al., 2005) and most recently from 3D to 4D, that is, from static to dynamic systems that incorporate a temporal element (Peuquet, 2002; Breman, 2002). GIS are also becoming increasingly distributed and mobile. Goodchild (in Longley et al., 2005) offers an interesting perspective on the implications of GIS becoming more distributed. He suggests that there are four distinct locations of significant to distributed GIS: the location of the GIS user and user interface, denoted by U; the location of the data being accessed by the user denoted by D; the location of data processing, denoted by P; and, finally, the area that is the focus of a GIS project, denoted by S. Traditionally, in GIS projects $U=D=P \neq S$, that is, the user interface, the data and data processing all occur at the same location, and these occur in a laboratory rather than at a field site (S). In the new era of distributed and mobile GIS, it is possible for $U \neq D \neq P = S$, that is, the user interface, the data and data processing can be at different locations, and some or all of them can be in the field.

This chapter begins with discussion of trends in computer systems that affect dynamic and mobile GIS (Section 1.2). Next the implications of the move to distributed, network-centric systems are analysed (Section 1.3). In Section 1.4, all important recent software developments are described. Finally, some outstanding questions are presented in Section 1.5.

1.2 Recent developments in computer systems

For most GIS users, the desktop PC (personal computer) is the primary hardware experience. The last few years have been a period of comparative stability in the desktop hardware community with mainly incremental improvements in speed, storage capacity and reduced power consumption. Processor speeds for desktop and server machines continue to improve at a rate that approximates Moore's Law (the number of components per unit area of chip doubles every 18 months which means speed doubles for the same cost), but in truth most of the speed improvements have been provided by increases in the speed of the buses - the connectors that peripherals use to communicate with the chip. A recent development of some significance is the move to multi-core processors (two or more processors on the same chip), which gives enhanced performance, reduced power consumption, and more efficient simultaneous processing of multiple tasks.

In the last few years many of the peripherals – such as large-format scanners and printers – that were once the domain of cartographers, CAD and graphics users have become commodity items with mass-market appeal. Interestingly, digitizing tables did not make the same transition. Once the mainstay of data capture projects in specialist 'sweat shops', they are now being replaced by software solutions that rely on on-screen, heads-up digitizing and line following algorithms.

A significant hardware trend worth noting is the re-establishment of servers as important platforms for GIS. The centralization of many processing operations that are often connected to public and private networks is a major plank of distributed computing architectures. From an IT (information technology) point of view the basic advantages of servers versus desktops relate to ease and cost of management: since all computer resources are at a single location it is easy to upgrade the system (new OS, new user, new processor, etc.). From a business perspective shared centralized resources typically work out to be more cost effective than distributed personal workstations. Several years ago the swing from desktops to servers began with the development of Internet GIS servers. These were engineered as all new software solutions that were optimized to publish maps on the Internet. A more recent generation of such systems has extended the capabilities from simple mapping to more advanced full-featured GIS services (Maguire, 2003). Today's systems, for example, offer access to a full GIS data model (not just simple vector features and images), as well as capabilities for high-quality production standard cartography, remote data management (e.g. multi-user editing, data delivery and system tuning), and advanced spatial analysis and modelling. There are many applications of enterprise GIS servers including: geoportals that provide access to a catalogue of information sources (Maguire and Longley, 2005), Web services (Tait, 2005), as well as centralized replacements for more traditional mapping and editing systems.

An important trend in server hardware is the wider use of blade servers. A blade server is a collection of blades – self-contained circuit boards with one or more processors, memory and disk – that function as a whole system. The use of standard low-cost components and their modular nature make blade servers easily scalable and very cost effective. Blade servers are often used with clustering software that supports load balancing, failover (ability to switch to a secondary mirror system in the event of failure of a primary system) and virtual allocation of processes for flexible configuration.

Perhaps the most exciting area of computer system development continues to be in hand-held devices. There is a much greater variety in form factor (size, configuration or physical arrangement of a computer hardware), chip type and operating system than on desktop and server systems that have standardized on the Windows, Linux and Unix operating systems and very similar form factors. Seldom do hand-held GIS exist in isolation; rather they represent the user's interaction with a wider system (Li and Maguire, 2003) that in its most complete form comprises the following key elements (Figure 1.2): a hand-held client device with in-built location technology (e.g. GPS); a GIS application server with mapping, geoprocessing and data management capabilities (usually provided by a separate data server); and a wireless / wire-line network for device-server communication.

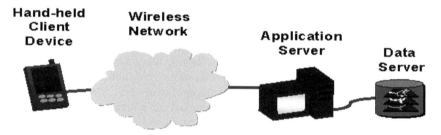

Figure 1.2. Mobile GIS platform.

There is a wide array of hand-held devices that can be classified into three types based on weight, power, cost and functional capabilities: Portable PCs, PDAs and Mobile Phones.

☐ **Portable PCs.** These are powerful devices with advanced processors and local data storage and processing capabilities. Such systems can operate for extended periods disconnected from a network because they have local storage and processing capabilities. They are able to host advanced GIS data models and functions, and are suitable for advanced data collection tasks. Tablet PCs and laptops running full-featured desktop GIS products on the Windows operating system fall into this category. Unfortunately there is a cost to using such systems – they tend to be heavy and have restricted battery life (4–6 hours). As a consequence they are often used in vehicles or for specialist tasks of short duration (e.g. updating work orders with 'as-built' information or dynamic fleet-vehicle routing).

❑ **PDAs (personal digital assistants).** These are medium capacity devices that balance weight/power/cost with functionality. The PDA devices that run the Windows Mobile operating system are archetypal examples of this middle category. With a small form factor, battery life in excess of 8 hours and a sub-$500 price tag, these systems are the mainstay of personal GIS data collection and mapping (Figure 1.3). Specialist hand-held GIS software solutions (e.g. ESRI ArcPad) have been developed that exploit the capabilities and deal with the restrictions (medium speed processors, limited screen size and resolution, and no keyboard) inherent in PDA hardware devices. A major feature of significance in PDA devices is that they have interfaces for peripheral devices. Initially, serial ports were used, but cable unreliability and inconvenience has seen an almost complete shift to the use of wireless connectivity using, for example, Bluetooth. The range of peripherals of interest to geographers includes GPS, digital cameras, barcode readers and laser range finders.

❑ **Mobile Phones.** These are lightweight, personal hand-held devices. This category is dominated by mobile telephones and similar devices (e.g. Blackberry pagers). Such devices assume an always-connected model because they have limited local storage and processing capabilities, and therefore rely on services provided by servers. The availability of mobile phones with embedded GPS and advances in server/network location fixing technologies have opened up a wide range of geographic uses for these devices. The devices in this class of hand-held system are most suitable in situations where mobility (lightweight, long battery life) is of paramount importance, and where there is a wireless connection to a server. Paradoxically, wireless signals are least reliable in urban canyons where most mobile telephone users are based, and in remote areas, where the advantages of lightweight devices and long battery life are most important.

An interesting trend of the last few years is the fusing of PDA and mobile phone technologies to create hybrid devices that have both good connectivity and local processing and storage. The connectivity is usually provided by a wireless telephone service (e.g. GSM - Global System for Mobile communication), as well as local area network access (e.g. 802.11 or WiFi). The standard devices have a ¼ VGA resolution screen and 256 MB RAM storage, with at least a 600 MHz processor. These devices are capable of running quite powerful hand-held GIS mapping and data collection applications (Figure 1.3).

Figure 1.3. Hand-held GIS on a smartphone and Windows Mobile PDA.

A key feature of mobile, hand-held GIS is their ability to determine their location. Several technologies are available for this (Spinney, 2003; Li and Maguire, 2003). Some, such as GPS, are embedded in the hand-held device where location is exposed through mobile software development kits (SDKs), while other methods use the wireless network to query the device—usually accessible through server APIs. Handsets with GPS typically offer the highest accuracy and accelerated time-to-fix through the use of network aiding-GPS servers. Network solutions such as AFLT (Advanced Forward Link Trilateration) vary in speed and accuracy depending on the wireless technology they employ. The Cell-ID of a mobile phone is easy and quick to estimate, but has a comparatively low accuracy (100–10,000 m) depending on cell size. Often multiple handset and networked-based solutions are used together and complement each other depending on the specific application location accuracy requirements.

1.3 Recent developments in computer networks

In the past decade there has been no greater influence on GIS architectures than the enormous improvements in networking. First wired local area networks and then the Internet changed forever the architecture of enterprise computer systems. At the present time we are in the midst of a similar radical shift from wired to wireless networks. Just as the wireless telephone network has replaced wired networks as the standard in telephony, so it will be in digital computing. Networks are not just a way to move data between existing computers; they are central organising principles at the very heart of distributed computing. Major organisations now develop IT strategies around the network not the desktop, and server computers and terms like 'cyberinfrastructure' (Berman and Brady, 2005) and Service-Oriented Architecture (Erl, 2005) have been coined that reflect the centrality of the network in system architectures.

Web services are a central element of SOA and form the foundation of the Internet computing platform. In simple terms a Web service is nothing other than a software application that can be called programmatically over the Web (the programmable equivalent of a url). The real significance lies in the fact that Web services are technology platform neutral and that they can be discovered and called on the fly (that is, without the need to be tightly bound during system compilation). Together these features allow systems to be assembled flexibly in the distributed, loosely-coupled Internet world. There are both geographic Web services technologies for building systems, and pre-built hosted Web services that can be used directly over the Web. Good examples of the latter are:
ESRI ArcWeb Services (http://www.esri.com/software/arcwebservices/index.html), Google Earth (http://earth.google.com/) and
Microsoft MapPoint.Net (http://www.microsoft.com/mappoint/default.mspx).

A world that is networked, especially one in which wireless communication dominates, offers some very interesting possibilities for distributed computing. Several of these have already been discussed, and one other trend of significance is the development of the sensor Web (Reichardt, 2003; Delin et al., 2005). A sensor Web is a collection of typically small, low-cost sensor devices that communicate between each other or to one or more central servers. According to Delin et al., the purpose of a sensor Web is to extract knowledge that can be used to react intelligently and adapt to changing surroundings. Sensor Web capabilities are useful in a diverse set of outdoor applications ranging from critical infrastructure protection, to at-risk disaster management and crowd monitoring. They can form a sophisticated sensing mesh that can be draped over an environment allowing identification of anomalous or unexpected events. In this type of system only the sensor is in the area of study, all other components of a distributed system can be located on a network.

One of the key reasons for the success of the Internet has been its ability to overcome distance: typically you do not know whether the Web site you are using is located in the same town or in another town halfway across the world. Francis Cairncross has written about what she calls the 'death of distance' caused by the Internet (Cairncross, 2001). It is now clear that while the Internet has certainly changed the impact of geography on business, government, education, etc., it has certainly not rendered it irrelevant (see The Economist [2003] for a response to Cairncross' arguments). In fact in recent years there have been several attempts to link the virtual world of the Internet with the real geographic world. Some notable examples include:

❑ Geolocation – mapping the physical infrastructure of the Internet usually based on IP address (quova.com, digitalenvoy.com, netgeo.com). This has applications in advertising, e-commerce and security.
❑ Reverse geolocation – finding Internet infrastructure based on a real-world address, for example, the closet WiFi 'hotspot' (wifinder.com, hotspotlist.com).
❑ Geoparsing – a geographic text search engine for Web documents that is able to find information on the Web based on geographic filters (metacarta.com).
❑ Geocaching – a game that involves searching for objects listed on a Web site using GPS (geocaching.com).
❑ Geoencryption – a technique that only allows decoding of encrypted documents in certain locations determined by GPS.

1.4 Recent developments in computer software

Modelling geographic patterns and processes effectively in both space and time requires an integrated combination of GIS-compatible hardware, network and software. GIS software must be able to read, store, edit, visualise and analyse 4D data in order to deal with dynamic geographic objects (e.g. vehicle deliveries, groundwater resources, and atmospheric pollution). There is a considerable variety of specialist software systems for dealing with each of these tasks, for each type of spatial and temporal data. However, the cost of integrating disparate systems, or moving data between them means that only the most advanced users or large projects are able to make them work together. Maguire (2005) reviews the state of the art in linking commercial GIS and specialist spatial analysis and modelling systems. He concludes that although there has been much recent progress in adding spatial analysis and modelling capabilities to commercial GIS (and vice versa), and linking both types of system together, there is no ideal solution that spans both areas.

Maguire et al. (2005) describe some examples of how ArcGIS and RePast (Recursive Porous Agent Simulation Toolkit) have been linked together to take advantage of ArcGIS's data management, transformation and visualisation capabilities, and RePast's dynamic simulation tools. RePast was originally developed by Sallach, Collier, Howe, North and others at the University of Chicago, and is one of the leading agent-based modelling toolkits (Tobias and Hofmann, 2004). The RePast system, including the source code, is available directly from the Web (repast.sourceforge.net/index.html). There are interfaces for the Java, .Net and Python languages.

Agent Analyst is a free extension to ArcGIS developed by Argonne National Labs in collaboration with ESRI that embeds the RePast agent tool kit inside the framework of ArcGIS. Figure 1.4 shows the results of a land use simulation implemented in the combined system. The base data from Stowe, Vermont and the rules were derived from Dawn Parker's SLUDGE land use change model (Parker et al., 2003). In this simulation each land use polygon is an agent and interaction rules are derived from cost to market (a raster surface), as well as value of land use types (agriculture and urban).

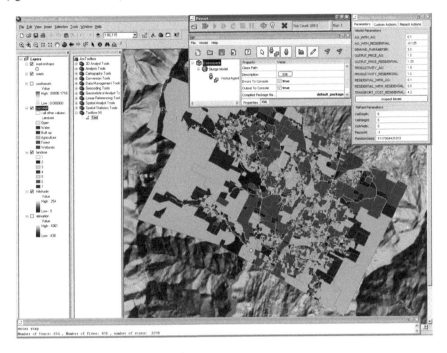

Figure 1.4. ArcGIS-RePast integration showing simulated land use patterns for Stowe, Vermont. The two menus at top right and the feedback window at the bottom are part of the Agent Analyst implementation.

Current commercial GIS software is adept at handling most aspects of 2D and 2.5D (TIN or raster surfaces that have a single Z at each X, Y point) data. Advanced GIS software can deal well with 3D objects on surfaces (Figure 1.5), but to get access to true volumetric analysis specialist domain-specific packages are required. These can be linked to GIS software in order to take advantage of the excellent GIS data management, integration and data dissemination capabilities. There are few examples of robust general purpose integrations, but given the move to component-based architectures with public APIs, and the availability of open transfer formats (e.g. VRML – virtual reality modelling language) integration is less of a problem than it used to be.

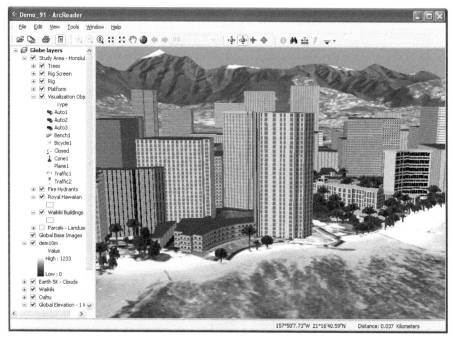

Figure 1.5. Desktop GIS viewer showing 3D building objects draped on a surface (Honolulu, Hawaii).

Recently there has been significant progress in visualising and exploring geographic data in 2.5D, largely as a result of advances in hardware/network performance, OpenGL-based graphics engines and browser-based Web clients. A new generation of geographic exploration systems is being developed that could set new standards for ease of use. The basic idea is that large global data sets can be assembled and hosted as a Web service. Lightweight, low-cost, easy-to-use viewers can access the data over the Web. The emphasis in such situations is on answering geographic questions through data exploration. Geographic exploration systems have two key elements:

❑ Hosted Globe Services that have rich GIS content (global terrain models, imagery and vector overlays). Globe services are built and published using GIS server technologies that can assemble, manage and publish vast quantities (TBs) of global data over the Web

❑ Client software applications that are able to access globe services over the Web. Typically, the clients are free and are easy to use (Figure 1.5)

To be successful replacements for conventional desktop and Web GIS systems, these geographic explorers must meet a series of exacting requirements:

❑ Free, easy and fun to use

❑ Very fast 2D and 3D visual exploration of massive global data sets over the Web

❑ Free rich GIS content, with additional services for a fee

- Ability to load standard GIS data on top of base globes (whole Earth 3D models)
- Direct use of many GIS formats (both open and proprietary)
- Fusion (visual overlay) of multiple globes service to combine data from many distributed systems
- Accessible, embeddable and extendable using industry-standard developer tools (.Net, Java, XML, etc.)

Handling temporal (time series) data in GIS is a similar story. We are just beginning to add support for reading and storing time-series data, but we are still someway off full 4D dynamic modelling within a commercial GIS. Progress has been made in reading and displaying time-series data (e.g. stream discharge for one or more gauging stations, near real-time feeds from aircraft location sensors and earthquake seismic readings). The NetCDF (Network Common Data Format) format (http://my.unidata.ucar.edu/content/software/netcdf/index.html) offers interesting possibilities for storing and transferring multi-dimensional data sets. It has a very flexible, platform neutral, direct access structure that can accommodate X,Y,Z,T and attributes. For example, the author has successfully worked with Tsunami simulation data in a NetCDF file which models wave height (Z), for 750 time slices (T) for a regular grid around Sumatra (X and Y). This data can be displayed in ArcGIS as maps, charts and animations (Figure 1.6).

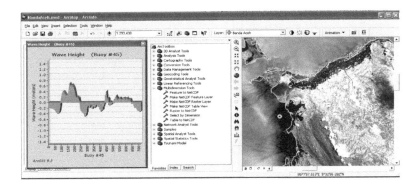

Figure 1.6. ArcGIS map and chart derived from a NetCDF simulation file of Banda Aceh, Indonesia. See colour insert following page 132.

Unfortunately, there are limited capabilities for analysing and modelling multi-dimensional data sets such as this in commercial GIS software products and modellers need to extend commercial GIS or integrate with specialist modelling software. Maidment and his team have been notably successfully at both of these (e.g. Maidment, 2002; Maidment et al., 2005). Hydrologic processes such as conversion of rainfall to runoff or flow routing down rivers can be linked to GIS by calling hydrologic simulation models as tools from the GIS. In their work they have linked hydrologic models HEC-HMS (Hydrologic Engineering Center-Hydrologic Modeling System) and HEC-RAS (Hydrologic Engineering Center-River Analysis

System) with ArcGIS in a case study of flood simulation for Rosillo Creek in San Antonio, Texas.

Much has been written about the value of exploratory spatial data analysis (ESDA) tools in spatial analysis and modelling; see, for example, Anselin (2005) and Dykes et al. (2005). The foundation of such systems is synchronized maps and charts that can be manipulated quickly to provide alternative views of some or all of the data. Figure 1.7 shows how a map and a scatterplot can be linked together. Each census tract is linked to the corresponding point on the scatterplot, which is depicted using the same colour. A selection set made in either the map or chart is highlighted on both.

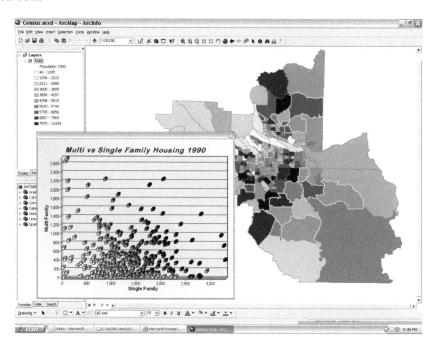

Figure 1.7. ESDA within ArcGIS.

Historically, much of what passed as modelling in commercial GIS was static 2D map analysis. Although very valuable in its own right, it did not support dynamic, probabilistic modelling. These capabilities are now being added to systems like ArcGIS. The dynamic element is provided by the ability to iterate around parts of the model while changing geographic and/or attribute parameters. For example, Johnston et al. (2005) describe a fire growth model that is composed of a series of rules defining how a fire will grow in each model iteration or time step. The input into the first iteration of the model is location where a fire has started. The fire grows during each time step according to the model rules. The result of the first iteration defines the current state of the fire $(t + 1)$, which becomes the input into the next time step of fire growth. The rules are then applied to the larger fire in the second time step, and the fire will either continue to grow or not. The results of time step two are input into time step three. This process continues for the desired number of time steps. Figure 1.8 shows a fire simulation model at step 37. The large red patch represents the main fire. Smaller fire patches to the north-east result from random sparks that 'jump' from the main fire.

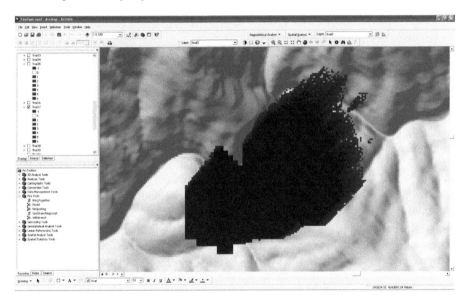

Figure 1.8. Visualisation of one step of a dynamic fire simulation mode. See colour inset following page 132.

1.5 Conclusion and future developments

It is clear that enormous progress has been made in building and applying dynamic and mobile GIS is the last few years. There have been significant advances in computer hardware, networks and especially software. Collectively, these have 'raised the bar' for modelling space and time in GIS. However, much research remains to be done if we are to exploit fully the opportunities that new technologies and computer platforms offer. This section discusses some of the key areas where new technical research is required.

True 3D/4D modelling. Extending GIS from 2.5D to true 3D and 4D (X,Y,Z and T) remains a challenge. This is especially the case for those interested in modelling the lithosphere (e.g. mining geology), hydrosphere (e.g. ocean ecosystems) and atmosphere (e.g. global circulation). There are some promising areas of work that are extending the current frontiers in the areas of representation and visualisation (e.g. Maidment et al. [2005] in the area of time-series hydrology and Breman et al. [2002] in marine data modelling), but much more remains to be done.

Error and uncertainty. The subjects of error and uncertainty are fundamental to dynamic spatial analysis and modelling (Zhang and Goodchild, 2002). There is a clear need to study and understand the mechanics of how uncertainty arises in geographic data and how it is propagated through GIS-based data analyses. We need to develop techniques for reducing, quantifying and visually representing uncertainty in geographic data and for analysing and predicting the propagation of this uncertainty through GIS-based data analyses. The work of Krivoruchko and Gotway (2005) and Anselin (2005), among others, demonstrates that software solutions to this problem are tractable.

Dynamic feedback/simulation modelling. Today's GIS are very much a product of their roots in static map-based analysis and their considerable success at managing natural and physical resources as assets. The real world, however, is fuzzy, uncertain and dynamic, and to be successful at characterising and simulating real-world processes, GIS must be able to incorporate multi-dimensional space–time modelling. The absence of these subjects is all the more surprising given the richness of implementations in non-geographic modelling and simulation software systems, for example, GoldSim (Miller et al., 2005) and STELLA (Maxwell and Voinov, 2005). There are some encouraging signs now that GIS such as ArcGIS, IDRISI and PCRaster support some dynamic simulation capabilities through scripting.

Spatial analysis and spatial statistics. It is now well understood that much of classical statistics is inappropriate for exploring, describing and testing hypotheses on geographic data (Bailey and Gatrell, 1995; O'Sullivan and Unwin, 2003). There is a real need for GIS to support, directly or indirectly through an interface to external systems, advanced spatial analysis and statistical functions. At the most basic level, the requirement is for descriptive and exploratory spatial data analysis tools of the sort described by Anselin (2005). The need also extends to improved geostatistical estimation procedures as discussed by Krivoruchko and Gottway (2005) as well as confirmatory spatial statistical procedures. Significant progress has been made on adding spatial interaction, location–allocation, and operational research optimization techniques to GIS software (e.g. ArcGIS 9.1), but much more remains to be done before commercial GIS can be effective in these domains.

Mobile GIS. Mobile GIS remains in its infancy in so many ways. The computer systems are progressing quite nicely, but it is the area of application that significant research questions are opening up. How can we apply adaptive sampling strategies that are based on the ability to compute sample error and population variance dynamically in the field? In what ways will sensors Webs and other real-time measurement systems alter the way we survey, sample and record information about static and dynamic (over space and time) objects? What are the social, economic and scientific implications of people having real-time locational information? The following chapters in this book begin to address these and other important topics.

Acknowledgements

David Maguire would like to acknowledge the following: Jeff Logan, Pacific Disaster Center for providing Tsunami simulation data, Jon Spinney for advice on mobile location technologies, Kevin Johnson for assistance with Agent Analyst, and Ismael Chivite for help with spatial analysis and modelling examples.

References

Anselin, L. (2005) 'Spatial statistical modeling in a GIS environment', in Maguire, D. J., Batty, M. and Goodchild, M. F. (eds.) *GIS, Spatial Analysis, and Modeling*, pp. 93-112, Redlands, CA: ESRI Press.

Bailey, T. C. and Gatrell, A. C. (1995) *Interactive Spatial Data Analysis*, Harlow: Longman Scientific and Technical.

Berman, D. and Brady, H. (2005) *Final Report: NSF SBE-CISE Workshop on Cyberinfrastructure and the Social Sciences*, [Online], Available: www.sdsc.edu/sbe/.

Breman, J. (ed.) (2002) *Marine Geography: GIS for the Oceans and Seas*, Redlands, CA: ESRI Press.

Breman, J., Wright, D. and Halpin, P. N. (2002) 'The inception of the ArcGIS Marine Data Model', in Breman, J. (ed.) *Marine Geography: GIS for the Oceans and Seas*, pp. 3–9, Redlands, CA: ESRI Press.

Cairncross, F. (2001) *The Death of Distance 2.0: How the Communications Revolution Will Change Our Lives*, Harvard: Harvard Business School Press.

Committee on Facilitating Interdisciplinary Research (2004) *Facilitating Interdisciplinary Research*, National Academy of Sciences, National Academy of Engineering, and Institute of Medicine of the National Academies: The National Academies Press, Washington D.C.

Delin, K. A., Jackson, S. P., Johnson, D. W., Burleigh, S. C., Woodrow, R. R., McAuley, J. M., Dohm, J. M., Ip, F., Ferré, T. P. A., Rucker, D. F. and Baker, V. R. (2005) 'Environmental studies with the sensor web: Principles and practice', *Sensors*, 5, pp. 103–117.

Dykes, J. A., MacEachren, A. M. and Kraak, M-J. (2005) *Exploring Geovisualization*, Amsterdam: Elsevier.

The Economist, (2003) *The revenge of geography*, March.

Erl, T. (2005) *Service-Oriented Architecture: Concepts, Technology, and Design*, Upper Saddle River, NJ: Pearson Education.

Hills, G. (1999) 'The University of the future', in Thornes, M. (ed.) *Foresight: Universities in the Future London*, pp. 213-32, Department of Trade and Industry.

Johnston, K. M., Kopp, S. M. and Tucker, C. (2005) 'Process, Simulation, Error, and Sensitivity Modeling Integrated in a Modeling Environment', *Conference Proceedings of GeoComputation 2005*, Ann Arbor, Michigan.

Krivoruchko, K. and Gotway Crawford, C. A. (2005) 'Assessing the uncertainty resulting from geoprocessing operations', in Maguire, D. J., Batty, M. and Goodchild, M. F. (eds.) *GIS, Spatial Analysis, and Modeling*, pp. 67-92, Redlands, CA: ESRI Press.

Laudan, L. (1996) *Beyond Positivism and Relativism: Theory, Method, and Evidence*, Boulder, CO: Westview Press.

Li, C. and Maguire, D. J. (2003) 'The handheld revolution: towards ubiquitous GIS', in Longley, P. A. and Batty, M. (eds.) *The CASA Book of GIS*, pp. 193–210, Redlands, CA: ESRI Press.

Longley, P. A., Goodchild, M. F., Maguire, D. J. and Rhind, D. W. (2005) *Geographical Information Systems and Science*, 2nd Edition, New York: John Wiley & Sons Inc.

Maguire, D. J. (2003) 'Enterprise geographic information servers', *GIS@development*, 7, 8, pp. 16–18.

Maguire, D. J. (2005) 'Towards a GIS platform for spatial analysis and modeling', in Maguire, D. J., Batty, M. and Goodchild, M. F. (eds.) *GIS, Spatial Analysis, and Modeling*, pp. 19–40, Redlands, CA: ESRI Press.

Maguire, D. J., Batty, M. and Goodchild, M. F. (eds.) (2005) *GIS, Spatial Analysis, and Modeling*, Redlands, CA: ESRI Press.

Maguire, D. J. and Longley, P. A. (2005) 'The emergence of geoportals and their role in spatial data infrastructures', *Computers, Environment and Urban Systems*, 29, pp. 3–14.

Maidment, D. R. (2002) *Arc Hydro: GIS for Water Resources*, Redlands CA: ESRI Press.

Maidment, D. R., Robayo, O. and Merwade, V. (2005) 'Hydrologic modeling', in Maguire, D. J., Batty, M. and Goodchild, M. F. (eds.) *GIS, Spatial Analysis, and Modeling*, pp. 319–332, Redlands, CA: ESRI Press.

Maxwell, T. and Voinov, A. (2005) 'Dynamic, geospatial landscape modeling and simulation', in Maguire, D. J., Batty, M. and Goodchild, M. F. (eds.) *GIS, Spatial Analysis, and Modeling*, pp. 131–146, Redlands, CA: ESRI Press.

Miller, I., Knopf, S. and Kossik, R. (2005) 'Linking General Purpose Dynamic Simulation Models with GIS', in Maguire, D. J., Batty, M. and Goodchild, M. F. (eds.) *GIS, Spatial Analysis, and Modeling*, pp. 113–130, Redlands, CA: ESRI Press.

O'Sullivan, D. and Unwin, D. J. (2003) *Geographic information analysis*, Hoboken, NJ: John Wiley and Sons.

Peuquet, D. (2002) *Representations of Space and Time*, New York: Guilford.

Parker, D. C., Manson, S. M., Janssen, M. A., Hoffmann, M. and Deadman, P. (2003) 'Multi-agent systems for the simulation of land-use and land-cover change: A review', *Annals of the Association of American Geographers*, 93, 2, pp. 314–37.

Reichardt, M. (2003) 'The sensor web's point of beginning', *Geospatial Solutions*, April, p. 40.

Spinney, J. E. (2003) 'Mobile positioning and LBS applications', *Geography*, 88, 4, pp. 256–65.

Tait, M. G. (2005) 'Implementing geoportals: applications of distributed GIS', *Computers, Environment and Urban Systems*, 29, pp. 33–47.

Tobias, R. and Hofmann, C. (2004) 'Evaluation of free Java-libraries for social-scientific agent-based simulation', *Journal of Artificial Societies and Social Simulation*, vol. 7, no. 1, [Online], Available: jasss.soc.surrey.ac.uk/7/1/6.html.

Zhang, J. X. and Goodchild, M. F. (2002) *Uncertainty in Geographical Information*, New York: Taylor and Francis.

Chapter 2

Opportunities in Mobile GIS

Qingquan Li

State Key Laboratory of Information Engineering in Surveying, Mapping and
Remote Sensing, Wuhan University, P. R. China

2.1 Introduction

Mobile information technology has emerged from a combination of the Internet and
wireless communication. This combination provides information services via
mobile terminals—anywhere and at anytime. Mobile GIS, itself, comes into being
whenever mobile information services and GIS are combined.

Mobile GIS has the following characteristics:

1. Mobility. Mobile GIS can operate on many kinds of mobile terminals that offer
    mobile information services to users through the interaction of wireless
    communication and remote servers. This makes GI constantly available for
    those such as field personnel, maintenance operatives, etc. who are always on
    the move.
2. Dynamic and operating in real time. As a service system, a mobile GIS
    responds to users' requirements and provides dynamic and current information
    about, for example, accidents, travel status and traffic jams.
3. Supports applications. In daily life, more than 80% of available information
    relates to spatial position; the geographic information resource is abundant and
    dispersed. Analysing this information and serving dispersed applications is the
    core task of mobile GIS.
4. Depends on location information. To provide mobile services, mobile GIS
    requires knowledge of the location of users, in real time.
5. Diverse mobile terminal technologies. There are various classes of terminals,
    including mobile computers, personal digital assistants (PDAs), mobile
    telephones, beep pagers and vehicle terminal devices. Furthermore, because of
    different manufacturers, different technologies and the fact that both spatial and
    non-spatial information is transmitted to and from terminals, technological
    diversity is further augmented.

Therefore, the arrival of mobile GIS can be considered to be GIS's 'new age'.
Because of the above characteristics, mobile GIS builds up the professional,
commercial and public service GIS sectors by integrating modern mobile
communication technologies and GIS technology. It has changed the GIS
application scene, bringing into being new application fields and adding value to
otherwise routine services.

Dynamic and Mobile GIS: Investigating Changes in Space and Time. Edited by Jane Drummond, Roland
Billen, Elsa João and David Forrest. © 2006 Taylor & Francis

This chapter starts by assessing, in Section 2.2, the development of relevant technologies (wireless communication, mobile positioning, mobile terminal technology and the emergence of mobile GIS). The applications of dynamic and mobile GIS are then presented in Section 2.3 and Section 2.4 introduces market opportunities. Finally, Section 2.5 addresses future research and conclusions are drawn in Section 2.6.

2.2 The development of related technologies

Several technologies now contribute to the development of dynamic and mobile GIS. These are briefly introduced in the following sections, prior to discussing the emergence of mobile GIS, from Web GIS, in Section 2.2.4.

2.2.1 Wireless communication technology

At present, spatial information transmission is a key technological requirement for mobile GIS. By using wireless communication, the connection between mobile terminals and spatial servers on the Internet, is enabled. Current mobile communication networks include: the *first generation* mobile communication system (TACS and AMPS are similar mobile cellular telecommunication systems; typical terminals are trunked telephones and cordless telephones); the *second generation* mobile communication network (the digital cellular system characterised by narrow-band digital technology, such as GSM, IS54 DAMPS, and IS95 CDMA); and **2.5G** systems (including GPRS and CDMA). *Third generation* mobile communication systems (CDMA2000, WCDMA and TD-SCDMA) are now developing fast (Hasan and Lu, 2003; Haung and Ho, 2005)

Digital cellular systems, including 2G and 2.5G mobile communication systems, which serve as the main communication platform for mobile GIS, cannot support large-volume spatial information services, so it is necessary to reduce the quantity of spatial data displayed on hand-held or pocket devices. The prominent design characteristics of **3G** mobile communication is that information communication and transmission be possible 'anyway, anywhere, anytime'. The rate of mobile multi-media communications via satellite is 96 KB per second, which is twice the rate of current 2 and 2.5G mobile communication systems. Within 3G systems, the rate of data delivery is 144 KB per second when mobile terminals move at the speed of a normal vehicle, 384 KB per second when sitting or walking outdoors and 2 MB/second when based indoors (Choi et al., 2000; 3GPP2, 2002; Li et al., 2002; Casademont et al., 2004). Because the service quality of 3G can be comparable to that of fixed systems, it can fit well with cellular, cordless, satellite, PSTN, Internet or IP/data networks; that is 3G provides globally seamless coverage with roaming terminals. Thus 3G mobile communication systems will work effectively for even large-volume spatial information transmission.

Mobile IPv6 technology in a **4G** mobile network is able to support advanced position and location based services using Internet Protocol (IP) to combine different radio access networks. Radio Access Network (RAN) consists of physical entities that manage radio resources and provide users with a mechanism to access both core and packet-switched network services; furthermore, RAN can be adapted to maintain real-time network services via the mobile Internet (Bravo, 2004). The development of wireless Internet technology offers new ways for the transmissions required by mobile GIS. The continuing improvement of wireless access technology, such as WAP, i-Mode, SMS/MMS and so on, provides an excellent communication platform for the development of mobile GIS.

2.2.2 Mobile positioning technology

Generally speaking, mobile positioning technology can be presented as three classes: network-based, terminal-based, and integrated technology. The first includes COO (cell of origin) positioning, TOA (time of arrival), AOA (angle of arrival), TDOA (time difference of arrival), and E-OTD (enhanced-observed time difference) positioning technologies. The second includes the Global Positioning System (GPS). The third includes wireless Assisted-GPS (A-GPS) and combines the positioning function of mobile terminals with functions of the network. In A-GPS and GPS, GPS receiving modules (receivers) must be added to mobile terminals, and thus the receiving antenna will be altered. However, terminals do not, themselves fully determine the position information, they only transfer information received from the GPS to the wireless communication network. The network's positioning servers then calculate the receiver's position and return it to the mobile terminal.

2.2.3 Mobile terminal technology

Mobile terminals are responsible for communication with users and for retrieval of spatial information. User terminals include hand-held computers, personal digital assistants (PDAs), mobile phones, intelligent watches, vehicle computers and so on. Terminal devices may deploy many kinds of embedded operating systems (the operating system running in a terminal), such as the embedded Windows system (WinCE), VxWorks, Palm OS, EPOC, uC/OS-II, the embedded Linux system, the QNX system and so on. Microsoft's WinCE, which is designed for a platform with limited resources, provides multi-thread, full priority and multi-task services. VxWorks is characterised by extensive intertask communications and synchronization facilities, a high-performing multi-tasking kernel and a user friendly development environment; it can be deployed in different terminals fulfilling tasks varying from anti-lock brake systems to space exploration. Palm OS is another embedded operating system commonly deployed in PDAs, which is compatible with a variety of hardware, its program execution is efficient, it supports many third-party manufacturer and software applications and has low resource consumption. The hardware devices that the Palm OS supports include smart-phones, hand-helds, multi-media devices, game players, industrial, scientific and educational tools. The Palm OS system offers features such as compatibility with Microsoft Windows and other major enterprise standards; multi-tasking, multi-threading; memory protection; support for more memory and larger screens; industry standards-based security; extensible communication and multi-media frameworks capable of handling multiple connections simultaneously. EPOC (reputedly an acronym for 'Electronic Piece of Cheese') is the PDA operating system produced by Symbian, a joint venture between Psion, Nokia, Ericsson, Motorola and Panasonic. EPOC is a three-tier system consisting of a base, middleware and an EIKON GUI. EPOC's third-party support is certainly as extensive as that for Windows CE. EPOC has enormous potential and its PDAs remain a very popular choice in the UK and Europe. WinEpoc is a powerful Windows-like desktop for Symbian/EPOC-equipped pocket computers. It provides a familiar workplace allowing intuitive control without losing any advantages of the powerful EPOC operating system. µC/OS-II (pronounced: 'micro C O S version 2') is a portable, ROMable, pre-emptive, real-time and multi-tasking kernel. The execution time for almost every service provided by µC/OS-II is both deterministic and constant. µC/OS-II allows the user to: create and manage up to 63 tasks, delete tasks, change the priority of tasks; suspend and resume tasks; create and manage binary or counting semaphores; delay tasks for an integral number of time periods ('ticks'), or for a user-specified number of hours, minutes, seconds and milliseconds; lock/unlock the scheduler; create and manage fixed-sized memory blocks and send messages from an ISR or a task to other tasks. The embedded Linux system is an open system available from many different suppliers, and it supports POSIX, an industry-standard program application interface, as well as standard, open interfaces for networking and graphics. Developers are protected from dependency on a single vendor's future directions and successes because their applications can easily be moved to Linux systems from multiple suppliers, as well as to other UNIX and compatible systems.

All these embedded operating systems provide not only system and hardware support for mobile services, but also facilitate an applications development environment for mobile terminals.

2.2.4 From WebGIS to mobileGIS

'Dynamic and multi-dimensional GIS is a technology in demand for the 21st century, and therefore it will be a key research direction,' Bergougnoux has claimed (Bergougnoux, 2000). Using WebGIS technology, no matter at which Internet node users find themselves, they can browse spatial data at their WebGIS site, make thematic maps, and perform many operations such as spatial query and spatial analysis. However, there are problems with WebGIS. For example, it relies on the network environment, needs to transmit large data volumes and its structure does not fit the wireless communication network.

With the development of mobile Internet technology, applications of WebGIS in the mobile environment, can, potentially, boom. Compared to traditional GIS, Mobile GIS seems closer to many users' work situation and will attract, potentially, more user groups. But, due to the very mobility of its terminals, mobile internet provides challenging problems for WebGIS systems with regard to bandwidth, transfer of larger data volumes, increasing expense, response speeds and so on. As a result, Mobile GIS has had to be developed step-by-step, combining the real-time dynamic environment with carriers' characteristics in a manner to satisfy users' needs. Nevertheless, Mobile GIS is fast becoming operational with the development of wireless communication, mobile positioning technology, mobile terminals, and the distributed management of spatial data.

2.3 The applications of mobile GIS

The basic functions of GIS system are: (i) storage; (ii) processing; (iii) management; (iv) analysis; and (v) displaying spatial information by involving computer-assisted cartography and a spatial database. Considering the requirements in application fields such as urban planning and management, transportation management and environment monitoring, GIS provides the powerful functionality of spatial analysis and decision support. Early applications of GIS in these fields were limited and simple, however, mobile GIS changes the application pattern of GIS so that users can free themselves from desktop computers via mobile terminals. It therefore shortens the distance between GIS applications and users, and, based on the above list of functions, mobile GIS can provide very many more services than static GIS, even when considering the limitations of data volume and unstable communication.

Summarising, the characteristics of mobile GIS applications:

1. Reduced hardware configuration requirements of terminal devices. Usually an embedded processor has a small volume memory and low CPU frequency, and supports 'mini' peripheral equipment at the mobile terminal device. Compared with desktop computers, the performance of the hardware configuration is much lower, but mobile GIS can still execute the basic GIS functions.
2. Wireless networks as carriers. At present, though the wireless network is used to carry spatial information under conditions of unstable communication and considerable expense, it can be improved with the developing wireless communication systems.
3. It is easy for traditional GIS to manage distributed spatial data via the Internet/Intranet while, for mobile GIS, the management of massive data sets is difficult. As mobile GIS needs real-time information about location, spatial data management should be improved in distributed and dynamic computing environments.
4. Mobile GIS relies on real-time position information. High-quality mobile GIS services can be offered only when the terminals are supported with location information, because most spatial information provided by Mobile GIS relates to users' current locations.
5. The User Interface must be very user friendly in mobile GIS. Traditional GIS software is designed for professionals, and its operation and interface can be complex. But mobile GIS is oriented to the public, so the operations should be necessarily simpler and with simpler interfaces than traditional GIS because of small display screen of mobile terminals.
6. Location based services (LBS) emerge as pivotal in converting GIS from a professional application to a public service industry. LBS mean easy information provision on the basis of location defined by different kinds of indexing and navigation systems. For example, location based services can be

provided using the location data of a mobile phone as the search criterion (Jensen et al., 2003).

According to Agrawal and Agrawal (2003), the applications of LBS are:
1.   Destination guidance with maps and directions;
2.   Location-based traffic and weather alerts;
3.   Wireless advertising and electronic promotions;
4.   Movie, theatre and restaurant location and booking;
5.   Locating stores offering the cheapest prices for brand-name items;
6.   Child or car finders;
7.   Telematics-based roadside assistance, utilising location information that is currently provided by the users but, in future, by GPS or similar;
8.   Personal messaging (live chats);
9.   Mobile yellow pages;
10. Information services (news, stocks, sports); and
11. Personalized content, e.g. wireless portals may have personal information about the preferences of a subscriber and may serve (push) relevant content to that subscriber. In yet another implementation, the subscriber may obtain (pull) content that is of interest to them.

Other LBS applications that can be added, are:
1.   Location-specific health information (local diseases rates and guidelines, health risks and hazards, pollen and air condition alerts and maps);
2.   Real-time in-car navigation systems within intelligent navigation systems;
3.   Location tracking services;
4.   On-line decision making, especially for emergency applications such as forest fire and floods;
5.   In-door positioning for portable wireless device (Di Flora et al., 2005);
6.   Mobile charts; and
7.   Mobile games.

Some typical applications of mobile GIS are described in the following sections.

On-line services and navigation for traffic information. Mobile GIS can be applied as an on-line service providing real-time traffic information. When a traffic accident takes place or a driver becomes acutely unwell, information about the location, the vehicle and the rescue activity can be provided by pressing a button. Thus, any alarm response will be much more rapid than otherwise, with traffic jams resolved and the death toll reduced. By sharing real-time traffic information between, for example, traffic services and fleet management departments, information about current and future traffic conditions, weather and environmental conditions can be delivered to mobile users by means of mobile GIS. Decision support is thus provided to travellers. 'Support for navigation is an indispensable part of any Intelligent Transportation Systems (ITS), where it assists drivers in designing routes and in modifying these routes in response to new information'

(Goodchild, 2000). Some functions can already be provided by common navigation systems, including route planning, audio guidance, 2-D and 3-D display, enlarged intersection maps, points of interest (POI) query, traffic and location information, navigation, data updating, etc.

Public information services. In the Service sector, mobile GIS can provide mobile map services and map-based value-added services, for example:

❑ Location-tracking services for public security, banking, logistics, vehicle fleet management, prisoners, children, senior citizens and the disabled; and

❑ Points of interest (POIs), such as hospitals, restaurants, hotels, cinemas, leisure centres, government centres, gas stations, convenience stores, dry-cleaners, recreation-parks, information on prices, etc.

as well as (1), (3), (8), (9), (10) and (15) listed above. All this information can be provided with user-friendly graphics (see Figure 2.1).

(a) map operations   (b) POI information   (c) mobile map service in phone

Figure 2.1. Mobile map services and POI searching. (a) The system gives the user six map operations, namely, load map, zoom in, zoom out, scale, roam and extents—to show the whole map. (b) Shows the POI information: name, address, post code and telephone, etc. (c) Schematic map showing in mobile phone display (Note: all original dialogue is in Chinese).

Urban disaster management. When urban disasters strike, the disaster events can, by means of mobile GIS, inform decision makers in a timely manner. With location information marked on the map of a hand-held device alarm messages can be sent to the emergency service and the shortest route can be computed and displayed, enhancing public security, by improving the efficiency of the emergency services. When an incident takes place, the police, for example, can be quickly informed and reinforcements appropriately deployed. Other disasters such as the spread of infectious diseases, tsunamis, tornados, typhoons and floods (see Figure 2.2) can be well managed via emergency response systems.

(a) Flood publishing software in PDA   (b) Flood submerging area

Figure 2.2. Flood management. (a) The flood publishing software offers five functions, namely, map viewing, flood analysis and simulation, flood information querying, navigation to different monitoring sites and system calibration. (b) Inundation map. See colour insert following page 132.

Field data collection. Field data collection is carried out by means of pen or wearable computer tools providing digital topographic and thematic maps as well as input masks for attribute information (Lam and Chen, 2001). Mobile GIS, running on hand-held personal computers or personal digital assistants (PDAs), can be deployed as a low-cost data collection strategy (Figure 2.3). If a GPS receiver is mounted on a mobile terminal, real-time positioning will be possible. Information can be added and deleted conveniently and databases accessed via wireless terminals. 'GISPAD is an exemplary application for field data acquisition that provides such functionalities' (Pundt, 2002). Mobile GIS can be used to collect and update field data in real-time on traffic, travel, planning, real estate and resources, the environment, ocean, surveying, electric power, etc.

2.4 Market opportunities

'Many computing and spatial information business analysts believe that LBS represent the ideal means by which spatial information can be provided to a wide range of public users. For mobile workers, all information needed to undertake the fieldwork may be accessed from their mobile device' (Dao et al., 2002). 'While carrier deployments are escalating globally, the real money is in the [LBS] services,' comments Edward Rerisi of ABI Research, and 'LBS enables a carrier to

Figure 2.3. Mobile GIS for field data collection. (a) PDA hardware configuration for field data collection. (b) Land use maps and database from collected field data.

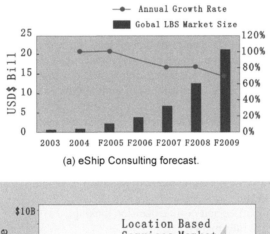

(a) eShip Consulting forecast.

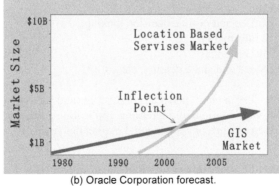

(b) Oracle Corporation forecast.

Figure 2.4. Market forecasts for location based services (eShip, 2005) and http://www.oracle.com.

raise their ARPU (average revenue per user) by offering value-added location services that will also fuel demand for higher-priced data services. If deployed successfully, it can give a significant boost to the top line' and that 'Global GPS market will exceed $22 billion by 2008. Handset and people tracking markets will experience the largest growth rates, significantly outpacing the overall market growth' (ABI Research, 2004). Beverly Volz, president of Volz & Associates, has said, 'strategy analytics [have] forecast a $16 billion European and North American market for wireless location based services by 2005' (Volz, 2005). More conservatively Van der Meer (2001), of Airbiquity Inc., a developer of wireless data communication solutions, '... predicted that by 2005 the location based service (LBS) market [would] exceed $11 billion in revenue ...'. Figure 2.4 presents some market forecasts of location based services.

The market opportunities are further considered in the following section.

2.4.1 Booming market from increasing personal, industrial and commercial applications

An increasing number of personal applications of mobile GIS have emerged with the availability of PDAs, such as personal navigation in the field, querying points of interest in cities, designing rational tour routes between different destinations, on-line bookings, reserving theatre or dinner seats, secure guidance for the blind, the aged, the young and so on. These personal applications provide services without the resource implications of accessing transportation systems. Turning to current industrial and commercial applications of mobile GIS, examples include:

❑   Field data collection and update for land and building surveying;
❑   Criminal identification for police department;
❑   Pollution monitoring and response by environmental protection departments;
❑   Traffic-flow based navigation systems for vehicles, traffic flow collection and update;
❑   Mobile office and on-line decision making;
❑   Tourism information;
❑   Public bus routing;
❑   Mobile stock and share dealing, etc.

These represent just a few personal, industrial and commercial mobile GIS applications, but with continually increasing applications, this service industry will boom, being driven by current and potential market demands.

2.4.2 Development opportunities for 'high-tech' enterprises

The emergence of mobile GIS will encourage growth in a series of 'high-tech' industries, including the telecommunication, mobile hardware and the service sectors, as well as the GI industry itself. Telecommunication provides communication support and data transmission, mobile hardware offers GIS terminals and the service sector grows by disseminating its services more widely. The GI industry provides the basic spatial information for mobile GIS applications. In general, 'high-tech' enterprises are presented with good development opportunities to add on value to their products. For example, vehicle DVD player manufacturers can profit by assembling intelligent navigation systems within their product. On the one hand, high-tech enterprises can enlarge their market by perfecting their product while on the other hand they can do this by integrating different technologies around mobile GIS applications.

2.4.3 New GIS employment opportunities

With the development of mobile GIS applications new jobs emerge. Technology, adapted for mobile GIS services by developers who know their individual technologies but need to collaborate with other technological groups, will create new employment. And with changing user demands, also more and more developers and sellers will be needed to satisfy these. A new service industry, The Mobile GIS Service Industry, different from the standard IT industry, will appear. The managers in this industry will know the basic characteristic of mobile GIS applications and distinguish their sub-sector from the rest of the IT sector, employing from those appropriately qualified. New employment opportunities for managers, developers, service engineers and sales personnel, with a specific mobile GIS background, can be expected.

2.5 Future research directions

Considering the diversity of mobile terminals and their hardware and software configurations, several potential research directions emerge for the GIS community, if such devices are to solve problems using mobile GIS and improve its performance.

2.5.1 Standards

The following standards are associated with mobile GIS:
1. GSM, widely used and important standards, presented by the European Telecommunications Standards Institute (ETSI) Special Mobile Group (SMG) in 1990;
2. CDMA/W-CDMA specifications (the third-generation partnership project, 3GPP) (Adams et al., 2004);
3. UMTS (Universal Mobile Telecommunications System) technical specifications (which include standards for the 3G mobile Internet);
4. OpenGIS specifications of OGC;
5. GML(Geography Markup Language, developed by Open GIS Consortium) incorporated with SVG(Scalable Vector Graphics);
6. Simple Object Access Protocol (SOAP), one of a handful of standards behind the industry move toward building Web services software, published by World Wide Web Consortium (W3C);

7.   Mobile Location Protocol Specification (Location Interoperability Forum (LIF) within the OMA Location Working Group);
8.   Mobile Web Initiative (making Web access from a mobile device as simple, easy and convenient as Web access from a desktop device);
9.   Wireless application protocol (WAP), etc.

All these specifications can be modified and developed to meet GIS service requirements in the mobile environment in terms of spatial information abstraction, spatial data compression, mobile positioning and data transformation, etc.

2.5.2 Key technologies

Several key technologies relevant to mobile GIS, such as data fusion, spatial data transmission and efficient server systems, should be researched and developed for the increasing demands of mobile GIS applications. The data fusion function is needed to supply mobile GIS terminals with semantic information and decision-making support for both professional GIS workers and the public. Spatial data transmission technology should exploit new wireless communication technology, such as wireless Internet, wireless LAN and mobile communication systems. The development of wireless communication protocols and technologies offers good opportunities for mobile GIS to promote itself. These include: geo-coding/decoding in the mobile environment, compressing and transferring multi-source spatial information in 3G/4G communication systems; dynamic update of spatial information based on 3G/4G wireless communication systems; and the supporting theories and techniques for terminal positioning in the 3G/4G mobile communication environment. Concerning efficient server systems, the server should support both dynamic and static geo-databases. Owing to the vastness of spatial information sets generated by large user groups, there are some wide open issues in current GIS theory and methods which need to be investigated further, including: optimisation algorithms; the rapid processing of spatial information accommodating small capacity memory; fast extraction and compression of spatial information in the context of large user groups; and concurrent data manipulation.

2.5.3 Professional application systems

Mobile technology in general and mobile GIS in particular has proven itself to be an invaluable tool for utilities. From simple attribute updates to complex geometry editing, users are eliminating many steps that are potential error sources while simplifying and expediting fieldwork and decreasing its cost. The functions and advantages of mobile GIS can be further extended to many applications in various fields to form professional mobile GIS application systems, such as: Intelligent navigation systems and advanced driver assistance systems (ADAS) in ITS (intelligent transportation systems), disaster management (infectious disease, tsunami, tornado, typhoon, flood, coast, and so on) and its emergency response systems, logistics management and scheduler systems for chain stores, and marketplace enterprise groups to control and reduce cost. For example the RDS-TMC service in Spain provides the ALERT C service (Arbaiza and García, 2004);

ACTMAP's developers have introduced a renovation mechanism for current navigation map databases, which has improved the robustness of that system (Bastiaensen, 2004), mobile oil spill response system and portable utilities map application (PUMA).

Also the mobile GIS function of field data access, remote GIS functionality and real-time data capture can be used in the fields of tourism, urban planning, real estate, resource inventory, environment protection, oceanography, field surveying and information update, utilities management, enterprise publicity and criminal control. On one hand, these applications play an important role in mobile GIS theory and technology, and on the other hand, the development of mobile GIS theory and technology facilitate applications needing powerful functions, and convenient and simple operations.

2.5.4 Ethical industrialisation policy

The development of mobile GIS and LBS has an impact on national information safety and privacy. To harmonize the development of the industry in accordance with national and international law, policies for the management of navigation/map data need to be studied. In order to be used legally and reasonably, spatial information should be provided to user units, which can be vehicles, offices or individuals. The type and value of information and the nature of its provision should be considered in terms of ethics (these issues are further addressed in this book's next chapter). Furthermore, a study on spatial information safety policy should be carried out and spatial information concerning LBS in the context of geographic information resources, the exchange and sharing of services and the infrastructure of information safety should be considered and standardized appropriately.

2.6 Conclusion and future developments

Prompted by a variety of well-established and more recent data transmission technologies, such as radio (modern mobile communications and satellite communication) and the Internet, mobile GIS applications have developed from simple GIS applications to LBS. With the powerful stimulus of modern information technology and a huge potential market, great changes are occurring in terms of the environment, the pattern of and the application fields for mobile GIS and LBS. In order for mobile terminals to perform well, some issues must be considered, such as kernel mechanism, data structure, storage and algorithms. As a result, the functions and advantages of mobile GIS can be further extended to almost all fields of human endeavour, but particularly including government agencies, intelligent transportation systems, emergency response, tourism, planning, real estate, resource inventory, environment protection, oceanography, field surveying and information update, utilities management, enterprise publicity and vehicle navigation.

More research should be carried out on standards for LBS, key technologies of mobile GIS, developing professional application systems in different fields and ethical industrialization policies for mobile GIS. All of this research can lead to the development of theories supporting mobile GIS methods, which form the industrial chain of location based services (LBS), and will thereby contribute to both societal and economic development.

References

3GPP2 (2002) *3rd Generation Partnership Project 2: A report on Issues Identified with IOS V4.1*, [Online], Available: http://www.3gpp2.org/.

ABI Research (2004) *Location Based Services Making A Humble Comeback, Declares ABI Research*, [Online], Available: http://www.directionsmag.com/press.releases/index.php?duty=Show&id=9222&trv=1&PRSID=e9 43ab44d.

Adams, P. M., Ashwell, G. W. B. and Baxter, R. (2004) 'Location-based services — an overview of the standards', *BT Technology Journal*, vol. 21, no. 1, pp. 671–682.

Agrawal, S. C. and Agrawal, S. (2003) *Location Based Services*, Tata Consultancy Services, [Online], Available: http://www.tcs.com/0_whitepapers/htdocs/atc/location_based_services_sep03.pdf.

Arbaiza, A. and García, J. F. (2004) 'The New Spanish TMC Location Table (Tmc04): Methodology and Auditing', *11th World Congress on ITS*, Nagoya, Aichi on 18–24 October 2004.

Bastiaensen, E. (2004) 'ACTMAP – Incremental Map Updates for Advanced in-vehicle Applications', *11th World Congress on ITS*, Nagoya, Aichi on 18–24 October 2004.

Bergougnoux, P. (2000) 'Editorial: A perspective on dynamic and multi-dimensional GIS in the 21st century', *GeoInformatica*, vol. 4, no. 4, pp. 343–348.

Bravo, A. M., Moreno J. I. and Soto, I. (2004) 'Advanced positioning and location based services in 4G mobile-IP radio access networks', *Personal, Indoor and Mobile Radio Communications, 2004 (PIMRC 2004)*, vol. 2, pp. 1085–1089.

Casademont, J., Lopez-Aguilera, E., Paradells, J., Rojas, A., Calveras, A., Barcelo, F. and Cotrina, J. (2004) 'Wireless technology applied to GIS', *Computers & Geosciences*, vol. 30, no. 6, pp. 671–682.

Choi, W., Kang, B. S., Lee, J. C. and Lee, K. T. (2000) 'Forward Link Erlang Capacity of 3G CDMA System', *3G Mobile Communication Technologies, Conference Publication No.471.IEE*, pp. 213–217.

Dao, D., Rizos, C. and Wang, J. (2002) 'Location-based services: technical and business issues', *GPS Solutions*, vol. 6, no. 3, pp. 169–178.

Di Flora, C., Ficco, M., Russo, S. and Vecchio, V. (2005) 'Indoor and outdoor location based services for portable wireless devices', *First International Workshop on Services and Infrastructure for the Ubiquitous and Mobile Internet (SIUMI) (ICDCSW'05)*, pp. 244–250.

eShip (2005) [Online], Available: http://www.eship.cn/sec-yj.asp?yjid=43 (in Chinese).

Goodchild, M. F. (2000) 'GIS and transportation: Status and challenges', *GeoInformatica*, vol. 42, pp. 127–139.

Hasan, C. and Lu, W. W. (2003) 'Guest editorial: 3G wireless and beyond', *Computer Communications*, vol. 26, pp. 1905–1906.

Haung, Y-R. and Ho, J-M. (2005) 'Overload control for short message transfer in GPRS/UMTS networks', *Information Sciences*, vol. 170, pp. 235–249.

Jensen, C. S., Kligys, A., Pedersen, T. B. and Timko, I. (2003) 'Multidimensional data modeling for location-based services', in *Proceedings of the Tenth ACM International Symposium on Advances in Geographic Information Systems*, McLean, VA, USA, November 8-9, pp. 55–61.

Lam, S. Y. and Chen, Y. Q. (2001) 'Ground-based positioning techniques', *Geographical Data Acquisition*, pp. 85-97, Wien, New York: Springer.

Li Deren, Li Qingquan, Xie Zhiying and Zhu Xinyan. (2002) 'The technique integration of spatial information and mobile communication', *Geomatics and Information Science of Wuhan University*, vol. 27, No.1, 1–6.

Pundt, H. (2002) 'Field data collection with mobile GIS: Dependencies between semantics and data quality', *GeoInformatica*, 6(4), pp. 363–380.

Ralston, B. A. (2000) 'GIS and ITS traffic assignment: Issues in dynamic user-optimal assignments', *GeoInformatica*, 4(2), pp. 231–243.

Van der Meer, J. (2001) 'Will wireless location-based services pay off?', *Business Graphics*, [Online], Available: http://www.geoplace.com/bg/2001/0201/0201pay.asp.

Volz, B. (2005) 'First Understand Your Market', *Sun developer network— product and technologies*, [Online], Available: http://developers.sun.com/foryourbusiness/mobility/market.html.

Chapter 3

Location Privacy and Location-Aware Computing

Matt Duckham and Lars Kulik

University of Melbourne, Australia

3.1 Introduction

Combined technological advances in location sensing, mobile computing and wireless communication are opening up new and exciting opportunities in the domain of location-aware computing. Many of these opportunities are explored elsewhere in this book (e.g. Chapters 2, 11–13); others are already being developed into practical applications that will provide benefit to a wide cross section of society, such as elder care (Stanford, 2002), emergency response and E911 systems (Werbach, 2000), and navigation systems for the visually impaired (Helal et al., 2001).

Despite the undoubted future potential of location-aware computing, location awareness also presents inherent future threats, perhaps the most important of which is location privacy. Most people would not feel comfortable if regularly updated information about their current location were made public, any more than we would feel comfortable if information about our home address, telephone number, age or medical history were public. Our precise location uniquely identifies us, more so than our names or even our genetic profile.

This chapter examines the foundations of location privacy: the factors that affect location privacy and the strategies for managing location privacy. The development of location-aware computing technology and mobile GIS is changing forever the way we interact with information, our physical environment and one another. How we deal with location privacy issues will be a determining factor in the ultimate direction of those changes.

This chapter begins by exploring the different concepts of privacy and their relevance to location-aware computing and mobile GIS (Section 3.2). Section 3.3 reviews the important privacy characteristics of one of the key enabling technologies for location-aware computing: positioning systems. The four classes of privacy protection strategy, which form the basis of any location privacy protection system, are introduced and described in Section 3.4. Section 3.5 concludes the chapter with an examination of some future challenges for location privacy research.

Dynamic and Mobile GIS: Investigating Changes in Space and Time. Edited by Jane Drummond, Roland Billen, Elsa João and David Forrest. © 2006 Taylor & Francis

3.2 Background and definitions

The term 'privacy' covers a wide range of concepts, and many different definitions of privacy have been proposed. An initial distinction is often made between *bodily privacy* (concerned with protection from physically invasive procedures, such as genetic testing), *communication privacy* (concerned with security of communications, like mail and email), *territorial privacy* (concerned with intrusions into physical space, like homes and workplaces) and *information privacy* (concerned with the collection and handling of personal data) (Rotenberg and Laurant, 2004). Under the heading of 'information privacy', one of the most influential and commonly quoted definitions was developed by the privacy pioneer Alan Westin:

> Privacy is the claim of individuals, groups, or institutions to determine for themselves when, how, and to what extent information about them is communicated to others (Westin, 1967, p 7).

Correspondingly, *location privacy* can be defined as a special type of information privacy which concerns the claim of individuals to determine for themselves when, how and to what extent location information about them is communicated to others. In short, *control* of location information is the central issue in location privacy.

Location privacy is especially important (to this book, specifically, and at this time, generally) as a result of the development of location-aware computing. *Location awareness* concerns the use of information about an individual's current location to provide more relevant information and services to that individual (Worboys and Duckham, 2004). Location awareness is a special type of *context-awareness*. The term 'context' is used to encompass the entire characteristics of an individual's physical, social, physiological or emotional circumstances (Schmidt et al., 1999). Location information is one of the most important aspects of an individual's (physical) context (see, for example, Ljungstrand's discussion of context awareness and mobile phones, Ljungstrand, 2001). Thus, location-aware computing environments offer the capability for automatic, regular and real-time sensing of a person's location with a high degree of spatial and temporal precision and accuracy. Together with technological advances in mobile computing and wireless communication, which enable rapid processing and communication of location information, these developments allow the location of mobile individuals to be tracked in a way never before possible.

3.2.1 The right to location privacy

Privacy is regarded as a fundamental human right, internationally recognised in Article 12 of the UN Universal Declaration of Human Rights (General Assembly of the United Nations, 1948). The history and development of privacy rights have been examined from many different perspectives in the literature (e.g. see Langheinrich [2001] for a concise overview of the history of privacy from the perspective of ubiquitous and location-aware computing).

Not all authors agree that privacy should be regarded as an inalienable right. Some authors, for example Brinn (1999) and Etzioni (1999), have argued for greater transparency in place of privacy. Proponents of greater transparency cite the practical difficulties of protecting privacy in the face of changing technological capabilities—encapsulated in the now infamous remark by Sun CEO Scott McNealy: 'You have zero privacy anyway, get over it!' (Sprenger, 1999)—and the public benefits that may be accrued through the relaxation of some privacy protections, for example, saving infant lives through the disclosure of positive HIV test results of pregnant mothers (Etzioni, 1999b).

Studies of users' attitudes to location privacy issues often provide some support for these views. Evidence presented in Beckwith (2003) and Kaasinen (2003) indicates a lack of awareness or even moderate indifference to location privacy issues amongst the general public. Other studies have painted a more complex picture. For example, Barkuus and Dey (2003) found that concern about location privacy can be dependent on the type of application, with applications that track users' movements over a period of time causing more concern than simple positioning applications.

Attitudes to privacy have changed in the past and will continue to change over time. As an example of how attitudes have changed in the past, J.B. Rule quotes the 1753 bill to establish a census in Britain (Rule, 1973): the bill was defeated as being 'totally subversive of the last remains of English liberty'. In the same 1973 book, Rule himself discards as 'unhelpfully rash speculations' Westin's vision of a future credit system, in which all transactions are digital and individuals can be tracked through their spending habits. By today's standards, this 'future' credit system seems rather conventional and unremarkable.

Although the need for a right to privacy will continue to be debated, in the shorter term at least there would seem to be a pressing need for privacy protection measures able to cope with a rapidly changing technological landscape. Concerns about protecting the individual's right to privacy have previously appeared in connection with numerous other new technologies, including GIS (Onsrud et al., 1994), the Internet (Ackerman et al., 1999), and collaborative user interfaces (Hudson and Smith, 1996). The need for location privacy is recognised in some of the earliest literature on information privacy (e.g. Westin, 1967) and location-aware computing (e.g. Harper, 1992; Harper et al., 1992; and Schilit and Theimer, 1994). Looking at more recent literature, it is possible to identify at least three key negative effects associated with failures to protect location privacy within a location-aware computing environment (e.g. Gruteser and Grunwald, 2004; Schilit et al., 2003; and Kaasinen, 2003).

1. **Location based 'spam':** Location could be used by unscrupulous businesses to bombard an individual with unsolicited marketing for products or services
2. **Personal well-being and safety:** Location is inextricably linked to personal safety. Unrestricted access to information about an individual's

location could potentially lead to harmful encounters, for example stalking or physical attacks.

3. **Intrusive inferences:** Location constrains our access to spatiotemporal resources, like meetings, medical facilities, our homes, or even crime scenes. Therefore, location can be used to infer other personal information about an individual, such as that individual's political views, state of health or personal preferences.

High-profile media coverage of accusations of location privacy infringements is indicative of increasing public awareness of location-privacy issues. For example, rental companies who use GPS to track their cars and then charge renters for infringements of their rental agreement have resulted in a flush of media articles and legal cases, e.g. James Turner versus Acme car rental (Canny, 2002; *Chicago Tribune*, 2001). Similarly, Samsung in Korea attracted media attention when it allegedly used a 'Friend finder' service to track its own employees with the aim of blocking the establishment of a labour union (Lee, 2004). In the future, greater familiarity with cheaper, more reliable location-aware technology is likely to amplify location-privacy concerns. These issues have already created a perception that inadequate privacy protection is retarding the uptake of location based services, and has led location privacy to be elevated to one of the key research challenges in pervasive computing (Muntz et al., 2003). In short, there is strong evidence that location privacy will be a key issue for the future of location-aware computing systems, including dynamic and mobile GIS.

3.3 Positioning systems and location privacy

In addition to the social constraints on location privacy, discussed in the previous section, location-aware computing environments place certain technical constraints on location privacy. The primary technical constraints arise from the positioning systems themselves. Hightower and Boriello (2001) provide a survey of the wide variety of positioning systems currently in use. In addition to the familiar GPS, positioning systems in the literature and in common usage include triangulation of RF wireless LAN signals (e.g. Bahl and Padmanabhan, 2000), proximity to infrared beacons (e.g. Want et al., 1992), scene analysis and computer vision (e.g. Krumm et al., 2000), and inertial tracking (e.g. Scott-Young and Kealy, 2002). New positioning systems, such as audio-based positioning (Beresford and Stajano, 2003b; Scott and Dragovic, 2005) and radio signal profiles (LaMarca et al., 2005), are continually being developed.

Positioning systems vary widely in their accuracy and precision characteristics. Accuracy and precision of location have implications for location privacy. For example, a positioning system that locates an individual to a precision of 200 m is generating less information about location (and so can potentially be less invasive of location privacy) than a positioning system that locates an individual to a precision of 2 m. Other characteristics of the positioning system may also present constraints to location privacy, such as the extent of the coverage of the positioning system (e.g. global or local) or the accuracy and precision of the positioning system relative to the density of geographic features (e.g. a location precision of 100 m in a dense downtown area of a city may be considered more private than a location precision of 100 m in a desert).

There exist several classifications of positioning systems. For example, a top-level distinction is often made between *active* positioning systems, which rely on the establishment of beacons to operate (such as WiFi signal triangulation, GPS, infrared proximity sensors), and *passive* positioning systems, which require no beacons (such as inertial navigation, scene analysis and audio-based positioning, see Worboys and Duckham (2004) for more information). However, from a privacy perspective, positioning systems are more usefully classified into *client-based*, *network-based* and *network-assisted* systems (Schilit and Theimer, 1994).

❑ In client-based positioning systems, mobile clients autonomously compute their own location (for example, GPS and inertial navigation). It is technically possible in a client-based positioning system for a client to compute its location, without ever revealing that location to any other entity.

❑ In network-based positioning systems, the network infrastructure is responsible for computing a mobile client's location. Cell phone positioning using CGI (cell global identity) is an example of network-based positioning. In network-based positioning systems, the network infrastructure administrator must hold information about the location of mobile clients.

❑ In network-assisted positioning systems, a combination of client-based and network-based computation is required to derive a client's location. For example, A-GPS (assisted GPS) combines network-based CGI positioning to increase the speed of GPS positioning. In network-assisted positioning systems, some information about a mobile client's location must reside in the network infrastructure, although this information may be less precise than the information held by the mobile client itself.

Client-based positioning systems inherently allow for greater location privacy than network-assisted or network-based positioning systems. In a client-based positioning system it is technically possible for the client to have complete control over information about its location, possibly to the extent that the client becomes the only entity with information about its own position.

One potential solution to location privacy issues, therefore, is to use only client-based positioning, perform all processing of location information locally on the mobile device, and never share any personal location information with other entities, whether centralized servers of peer-to-peer clients (cf. Marmasse and Schmandt, 2000). However, adopting this completely client-oriented, centralized model of mobile computing presents several drawbacks:

❑ Mobile devices typically possess limited processing and storage capacity, making it inefficient to perform complex calculations on voluminous spatial data directly on the mobile device.

❑ Spatial data sets remain expensive to collect and collate, despite continuing advances in positioning systems. The companies who collect this data would usually be reluctant to make their valuable data sets available in their entirety to mobile users.

❑ Downloading spatial data sets from a remote service provider will be subject to wireless network bandwidth limitations and may provide an indication of the user's location (either by inferring location from knowledge of the data sets of interest to the user or by positioning using a client's mobile IP address, as in Dingledine et al. [2004]). Alternatively, storing all potentially useful spatial data in a user's mobile device leads to the data integrity and currency issues that are inevitably associated with maintaining copies of the same data sets across multiple clients.

In summary, the different types of positioning system place some inherent constraints on the privacy characteristics of location-aware computing environments. Irrespective of these constraints, as mobile computing environments move toward increasingly distributed models of computation, the need to share personal information about location with a variety of remote location based service providers increases correspondingly.

3.4 Location privacy protection strategies

Having identified location privacy as a key issue for location-aware computing and outlined some of the technical aspects of location privacy, the next step is to ask what mechanisms exist for location privacy protection. The different strategies that exist for protecting a mobile individual's location privacy can be classified into four categories: *regulatory*, *privacy policies*, *anonymity* and *obfuscation* strategies. In this section each type of strategy is reviewed in turn.

3.4.1 Regulatory strategies

Regulatory approaches to privacy involve the development of rules to govern fair use of personal information. Most privacy regulation can be summarised by the five principles of *fair information practices*, originally developed as the basis of the U. S. privacy legislation (U.K. Department of Health, 1973; U.S. Department of Justice, 2004):

1. **Notice and transparency:** Individuals must be aware of who is collecting personal information about them and for what purpose.
2. **Consent and use limitation:** Individuals must consent to personal information being collected for particular purposes, and the use of personal information is limited to those purposes.

3. **Access and participation:** Individuals must be able to access stored personal data that refers to them, and may require that any errors be corrected.
4. **Integrity and security:** Collectors must ensure personal data is accurate and up-to-date and protect against unauthorized access, disclosure, or use.
5. **Enforcement and accountability:** Collectors must be accountable for any failures to comply with the other principles.

Although these principles of fair information practice are at the core of most privacy regulation (e.g. Organisation for Economic Co-operation and Development, 1980; U.K. Government, 1998), there are a variety of ways in which these rules have been implemented. In general, regulatory frameworks aim to adequately guarantee privacy protection for individuals without stifling enterprise and technology. The concept of *co-regulation*, which aims to encourage flexible self-regulation on top of legal enforcement of minimum privacy standards, is one example of a mechanism for achieving such a balance (Clarke, 1999).

The concept of fair information practices is usually applied to 'personal information' in general, not specifically to location information. Personal information can be defined as 'information ... about an individual whose identity is apparent, or can reasonably be ascertained, from the information ...' (Australian Government, 1988). In this respect, location information is usually treated as one type of personal information, like age, gender or address. A small number of privacy regulations have been developed to address location privacy issues explicitly, for example, proposed location tracking legislation in Korea (Park, 2004) and the discontinued AT&T 'Find Friends' location based service (Strassman and Collier, 2004).

Although regulation lies at the foundations of any privacy protection system, there are at least four reasons for believing that, on their own, regulations do not represent a complete solution to location-privacy concerns. First, regulation itself does not prevent invasions of privacy, it simply ensures that there exist mechanisms for 'enforcement and accountability' when unfair information practices are detected. Second, the development of regulation may lag behind innovation and new technology. Third, regulation applies 'across the board', making a satisfactory balance between guaranteed levels of privacy protection and freedom to innovate and develop new technology difficult to achieve, even using models such as co-regulation. As a consequence, other privacy protection mechanisms are needed in addition to regulation. Finally, abiding by fair information practice principles can give rise to practical problems with respect to location awareness. For example, Ackerman et al. (2001) examine the difficulties created by the requirements for notice and consent for user interfaces and HCI in context-aware computing environments (e.g. overwhelming users with frequent, disruptive and complex consent forms or notice information).

3.4.2 Privacy policies

Privacy policies are trust-based mechanisms for proscribing certain uses of location information. Whereas regulation aims to provide global or group-based guarantees of privacy, privacy policies aim to provide privacy protection that is flexible enough to be adapted to the requirements of individual users and even individual situations and transactions. Overviews of a range of different privacy policy systems can be found in Görlach et al. (2004). In this section we summarise three of the major privacy policy initiatives currently underway that illustrate the range of approaches that privacy policies can take.

IETF GeoPriv The Internet Engineering Task Force (IETF) is an international consortium concerned with future Internet architectures. The IETF's GeoPriv working group is adapting PIDF (presence information data format) as a privacy policy system for location privacy. PIDF is an IETF XML dialect for instant messaging, which includes a mechanism for exchanging information about the presence of a person (or place or thing) (Peterson, 2004). The GeoPriv specification additionally includes information about the location of that person, effectively annotating location data with metadata about the fair uses of that location data. In order to protect location privacy, the GeoPriv specification defines a *location object* that encapsulates both an individual's location and their privacy policy. At the centre of the privacy policy are *usage rules* that describe acceptable usage of the information, such as whether retransmission of the data is allowed or at what date the information expires, and must be discarded. Further, location objects can be digitally signed, making the privacy policy resistant to separation from the location information (Myles et al., 2003).

W3C P3P The World Wide Web Consortium (W3C) has developed the platform for privacy preferences project (P3P) as a simple mechanism for communicating information about Web-based privacy policies (WorldWideWeb Consortium, 2005). In contrast to the IETF approach, where users attach privacy policies to their data, the focus of P3P is to enable service providers to publish their data practices. The data practices may include for what uses personal data is collected, for how long it is held, and with what other organisations and entities it may be shared. Users of a particular service can then decide whether these data practices fit with their own requirements (Cranor, 2001). Typically, this process is achieved automatically using software agents with access to users' profiles. P3P does not provide any mechanisms for encrypting privacy protection within location data (like those found in IETF GeoPriv specification) and does not explicitly address location issues. However, because P3P is XML-based it can be easily extended for location-aware computing environments. For example, Langheinrich (2002) describes an architecture (the privacy awareness system, pawS) that uses P3P to enable location aware system users to keep track of the storage and usage of their personal location information. IBM's enterprise privacy authorization language (EPAL) is a different XML-based dialect with similar goals to P3P (IBM, 2004).

PDRM Digital rights management (DRM) concerns the technical efforts by some intellectual property vendors and other organisations to enforce intellectual property protection (for example, protection from piracy). PDRM (personal DRM) adopts a similar approach for personal data. When applied to location privacy, the PDRM approach is closer to the 'user-oriented' IETF GeoPriv model than the P3P 'provider-oriented' model. For location-aware systems, location data is treated as the property of the person to whom that data refers. PDRM then aims to enable that person to 'license' the personal data for use by a location based service provider (Gunter et al., 2004). So, for example, an entity wishing to use an individual's location data may first need to demonstrate their willingness to agree to the licensing, which may set limits on that entity's ability to share or process the data.

Policy-based initiatives for privacy protection, like PDRM, P3P and GeoPriv, are continuing to develop. However, there are again reasons for believing that policy-based initiatives provide only a partial answer to the question of location privacy protection. First, privacy policies are often highly complex and their practicality for use in location-aware environments with frequently updated highly dynamic information remains, as yet, unproven. Second, privacy policies systems generally cannot enforce privacy, instead relying on economic, social and regulatory pressures to ensure privacy policies are adhered to. Consequently, privacy policies are ultimately vulnerable to inadvertent or malicious disclosure of personal information (Gruteser and Grunwald, 2004; Wu and Friday, 2002).

3.4.3 Anonymity

Anonymity concerns the dissociation of information about an individual, such as location, from that individual's actual identity. A special type of anonymity is pseudonymity, where an individual is anonymous, but maintains a persistent identity (a pseudonym) (Pfitzmann and Köhntopp, 2001). For example, Espinoza et al. (2001) describe a location-aware system for allowing users to leave and read digital notes at specific locations ('geonotes'). One of the ways users can protect their privacy is to associate an alias (pseudonym) with a note in place of their real name.

An explicitly spatial approach to providing anonymity in location-aware computing environments is presented in Gruteser and Grunwald (2003). Gruteser and Grunwald used a quadtree-based data structure to examine the effects of adapting the spatial precision of information about an individual's location according to the number of other individuals within the same quadrant, termed 'spatial cloaking'. Individuals are defined as *k-anonymous* if their location information is sufficiently imprecise in order to make them indistinguishable from at least k-1 other individuals. The authors also explore the orthogonal process of reducing the frequency of temporal information, termed 'temporal cloaking'.

There are several disadvantages to using anonymity-based approaches. First, anonymity-based approaches often rely on the use of a trusted anonymity 'broker', which retains information about the true identity of a mobile individual, but does not reveal that identity to third-party service providers (e.g. Gruteser and Grunwald, 2004). Second, anonymity often presents a barrier to authentication and personalization, which are required for a range of applications (Langheinrich, 2001; Hong and Landay, 2004). Pseudonymity does allow some personalization and is therefore sometimes preferred to general anonymity in order to combat this problem. For example, Rodden et al. (2002) use a randomly generated pseudonym that is held by a trusted information broker and persists only for the duration of the provision of a particular service (like a location-aware taxi collection system). A promising new research direction that may help overcome these limitations is *zero-knowledge interactive proof systems* (see Goldwasser et al., 1985, described in more detail below).

Zero knowledge proofs The idea of a zero-knowledge proof is to prove the knowledge of a certain fact without actually revealing this fact. Zero-knowledge proofs (ZKPs) involve a *prover*, who attempts to prove a fact, and a *verifier*, who validates the prover's proof. The verifier may determine the correctness of the proof, but not does learn *how* to prove the fact or anything about the fact itself. Fiat and Shamir (1986) developed the first practical zero-knowledge proof system in 1987.

ZKPs often appear somewhat counter-intuitive at first, so consider the following simple example. Person A claims to know the secret combination to a safe. Person B deposits a valuable item in the safe, locks the safe, and leaves the room without the safe. Person B does not know the combination to the safe. If person A is able to present the item locked in the safe to B, then A has proven to B that A knows the combination to the safe without revealing the actual combination. In ZKP terminology, the proof is interactive because the verifier (person B) *challenged* the prover (person A) and the prover must *respond* to the verifier.

In a ZKP, a prover may provide the correct response to a challenge purely by chance. To combat this possibility, there are usually several rounds of challenges and responses in a ZKP. As the number of rounds increases, the probability that the prover will give the correct answer in every round decreases. Typical ZKPs will verify a proof with a probability of $1-1/2^n$, where n is proportional to the number of rounds used.

There are two distinct application scenarios for ZKPs:

1. **Authentication:** Prover P is able to prove to verifier V that P is authorized to access information without requiring any knowledge about P's identity.
2. **Identification:** Prover P can prove to verifier V that P is P, but no party Q is able to prove to V that Q is P.

The first application scenario that uses ZKPs without revealing an individual's identity is *anonymous digital cash* (Brands, 1994). To date, ZKPs have not been widely researched within the domain of location-aware computing. However, clearly ZKP-based authentication and identification might also be used with location based services, and initial work in this area is beginning to appear (e.g. Canny, 2002).

There is one further, explicitly spatial problem facing any anonymity-based system for location privacy: a person's identity can often be inferred from his or her location. Consequently, anonymity strategies (even those employing pseudonymity or ZKPs) are vulnerable to data mining (Duri et al., 2002). Beresford and Stajano (2003) have used simulated historical data about anonymized individual's movements to investigate ways of subverting anonymity-based privacy protection. Their results show how simple heuristics can be used to de-anonymize pseudonyms, providing users with much lower levels of location privacy than they might naively expect. Thus, anonymity alone cannot hope to provide total location privacy protection.

3.4.4 Obfuscation

Obfuscation is the process of degrading the quality of information about a person's location, with the aim of protecting that person's location privacy. The term 'obfuscation' is introduced in Duckham and Kulik (2005a) and Duckham and Kulik (2005b), but several closely related concepts have been proposed in previous work. The 'need-to-know principle' aims to ensure that individuals release only enough information that a service provider needs to know in order to provide the required service (Hutter et al., 2004). The idea of a need-to-know principle is closely related both to obfuscation and the fundamental fair information practice principle of consent and use limitation (Section 3.4.1). Snekkenes (2001) investigates a privacy policy-based approach to enforcing the need-to-know principle in location-aware computing by adjusting precision of location information. In the domain of anonymity-based approaches, the work of Gruteser and Grunwald (discussed in Section 3.4.3) aims to enforce the 'principle of minimal collection' (Grutesar and Grunwald, 2003), again akin to obfuscation. On a slightly different theme, Jiang et al. (2002) discuss the 'principle of minimal asymmetry', which aims to ensure that the flow of personal information away from an individual is more closely matched by the information flow back to that individual about who is using that information for what purposes.

It is possible to identify three distinct mechanisms (types of imperfection) in the literature for degrading the quality of location information: *inaccuracy*, *imprecision* and *vagueness* (see Worboys and Clementini, 2001; Duckham et al., 2001; Worboys and Duckham, 2004). Inaccuracy concerns a lack of correspondence between information and reality; imprecision concerns a lack of specificity in information; vagueness concerns the existence of boundary cases in information. Any combination of inaccuracy, imprecision and vagueness may be used as the basis for an obfuscation system. An inaccurate description of an agent's location means that the agent's actual location differs from the conveyed location: the agent is 'lying' about its current location. An imprecise description of location might be a region including the actual location (instead of the location itself). A vague description would involve linguistic terms, for example that the agent is 'far' from a certain location. Most research to date has looked at the use of imprecision to degrade the quality of location information (e.g. Snekkenes, 2001; Gruteser and Grunwald, 2003; Hong and Landay, 2004; Duckham and Kulik, 2005a). However, the use of inaccuracy has also been investigated and compared with imprecision in Duckham and Kulik (2005b).

The work in Duckham and Kulik (2005a) develops and tests an algorithmic approach to obfuscating *proximity queries* (e.g. 'where is the closest...?') based on imprecision. A simplified version of the algorithm introduced in Duckham and Kulik (2005a) is summarised in Figure 3.1.

The algorithm assumes a graph-based representation of a geographic environment (for example, a road network). An individual protects his or her location privacy by only reporting a set O of locations (*an obfuscation set*), one of which is that individual's actual location (Figure 3.1a). For an obfuscation set O, the location based service provider must compute the relation δ (Figure 3.1b), where $o\delta p$ means $o, p \in O$ are most proximal to the same point of interest (POI). The algorithm then proceeds according to three possibilities. First, all the locations in the obfuscation set may be most proximal to a single POI ($O \in O/\delta$), in which case that POI can be returned to the user (Figure 3.1c). Second, the individual may agree to reveal a more precise representation of his or her location, in which case the algorithm can reiterate (Figure 3.1d). Otherwise, the best estimate of the most proximal POI is returned (Figure 3.1e). The analysis in Duckham and Kulik (2005a) shows that efficient mechanisms for computing the relation δ can ensure that the entire algorithm has the same computational (time) complexity as a conventional algorithm for proximity queries, and that the algorithm must terminate in a finite number of iterations.

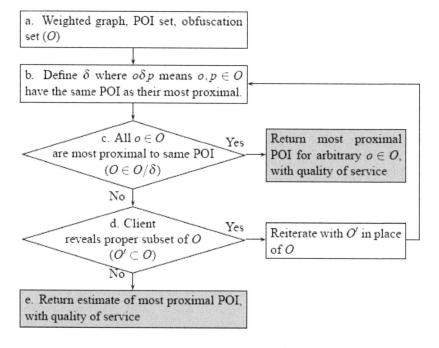

Figure 3.1. Summary of simplified obfuscation algorithm, after Duckham and Kulik (2005a).

Obfuscation has several important advantages that complement the other privacy protection strategies. Obfuscation and anonymity are similar, in that both strategies attempt to hide data in order to protect privacy. The crucial difference between obfuscation and anonymity is that while anonymity aims to hide a person's identity, obfuscation is an explicitly spatial approach to location privacy that aims to allow a person's identity to be revealed. Potentially, this combats one of the key limitations of anonymity approaches: the need to authenticate users. At the same time, degrading the quality of location information makes inferring identity from location more difficult. Obfuscation is flexible enough to be tailored to specific user requirements and contexts, unlike regulatory strategies; does not require high levels of complex infrastructure and is less vulnerable to inadvertent disclosure of personal information, unlike privacy policies; and is lightweight enough to be used without the need for trusted privacy brokers, unlike many anonymity approaches.

Obfuscation aims to achieve a balance between the level of privacy of personal information and the quality of service of a location based service. Current research has indicated that there exist many situations where it is possible to expect high quality location based services based on low-quality positional information (see Duckham and Kulik, 2005b). Consequently, in situations where the user requires a higher quality of service than can be achieved at a user's minimum acceptable level of privacy, then other privacy protection strategies must be relied upon instead. Further, obfuscation assumes that the individual is able to choose what information about his or her location to reveal to a service provider. While this may be realistic when using client-based or network-assisted positioning systems and when sharing location information with a third-party location based service provider, dealing with the entities that administer network-based positioning systems still requires privacy protection based on regulatory or privacy policy approaches.

3.5 Conclusion and future developments

Location privacy lies at the intersection of society and technology. This chapter has reviewed the reasons why location privacy is becoming such an important topic in society, and the technological constraints to location privacy. When considering the strategies that can be used to protect an individual's location privacy, it becomes clear that no single strategy currently available is capable of providing a complete solution to location privacy protection. Each approach has distinct advantages and disadvantages. Therefore, it seems likely that the future of location privacy protection involves combinations of the approaches: regulation, privacy policies, anonymity and obfuscation.

There remain many challenges for privacy researchers. For example, for information to be worth protecting, it must also be worth attacking. Current research tends to be biased toward privacy protection. By contrast, it is also important to understand the techniques a hostile agent might employ in order to invade a person's privacy (circumventing location privacy protection and attempting to discover an individual's exact location). In this respect, privacy research is analogous to cryptology, which comprises both cryptography (code making) and cryptanalysis (code breaking).

As this chapter has shown, location information differs from many other types of personal information. Consequently, future research aimed specifically at location privacy will need to focus on specialized privacy protection techniques for several reasons. First, unlike many other types of personal information, identity may be inferred from location. Such inferences are especially likely where a history of locations can be derived (for example, my patterns of movement over the course of a week). These types of inferences make anonymity and pseudonymity much harder to maintain than in other privacy applications, such as Internet use.

Second, information about personal location is highly dynamic. By contrast, current research approaches to location privacy are usually fundamentally static in nature, modelling the movement of an individual as a sequence of static snapshot locations. Many aspects of location privacy demand models that provide a more faithful representation of the temporal aspects of LBS. For example, counter-strategies for invading an individual's privacy can be devised by making assumptions about an individual's maximum or minimum speeds of movement. Understanding such counter strategies requires the development of truly spatiotemporal models of location privacy. Further, the potential uses and privacy implications of dynamic location information change over time. Current privacy protection strategies, such as regulation and privacy policies, tend to make no distinction between static information (such as an individual's date of birth) and dynamic information (such as an individual's location). Thus, these approaches may ignore the dynamic aspects of location information, making it difficult to definite privacy policies that have a temporal component, for example, where acceptable uses change over time.

Finally, the potential uses of spatial information are highly varied. Correspondingly, the potential benefits of invading an individual's location privacy may be higher than for some other types of information. Without proper protection, the location information generated by location-aware systems could conceivably be abused or unfairly used in almost any domain of human, social or economic activity, including marketing, insurance, surveillance, harassment, social security, politics, law enforcement, health or employment. Indeed, it is this very feature of location information that makes location information so vital to our future information systems.

Acknowledgments

Dr. Duckham is partially supported by funding from the Australian Academy of Science, an Edward Clarence Dyason Fellowship and an International Collaborative Research Grant from the University of Melbourne. Dr. Duckham and Dr. Kulik are further supported by Early Career Researcher Grants from the University of Melbourne and by an ARC Linkage Grant, entitled 'CEWAY: Cognitively ergonomic wayfinding directions for location based services'.

References

Ackerman, M.S., Cranor, L. F. and Reagle, J. (1999) 'Privacy in e-commerce: Examining user scenarios and privacy preferences', *Proc. 1st ACM Conference on Electronic Commerce*, pp. 1–8, ACM Press.

Ackerman, M.S., Darrell, T. and Weitzner, D. J. (2001) 'Privacy in context', *Human Computer Interaction*, vol. 16, (2, 3, & 4), pp. 167–176.

Australian Government. *Privacy Act.* (1988). [Online], Available: http://www.privacy.gov.au/act. [25 July 2005].

Bahl, P. and Padmanabhan, V. N. (2000) 'Radar: An in-building RF-based user location and tracking system', *Proceedings IEEE INFOCOM 2000*, vol. 2, pp. 775–784.

Barkuus, L. and Dey, A. (2003) 'Location-based services for mobile telephony: A study of users' privacy concerns', in *Proc. INTERACT 2003*, 9th IFIP TC13 International Conference on Human-Computer Interaction.

Beckwith, R. (2003) 'Designing for ubiquity: The perception of privacy', *IEEE Pervasive Computing*, 2(2), pp. 40–46.

Beresford, A. R. and Stajano, F. (2003) 'Location privacy in pervasive computing', *IEEE Pervasive Computing*, 2(1), pp. 46–55.

Beresford, A. R. and Stajano, F. (2003b) 'Using sound source localization in a home environment', in Gellersen, H. W., Want, R. and Schmidt, A. (eds.) Pervasive 2005, vol. 3468 of *Lecture Notes in Computer Science*, pp. 19–36, Berlin: Springer.

Brands, S. (1994) 'Untraceable off-line cash in wallet with observers', *CRYPTO '93: Proc. 13th Annual International Cryptology Conference on Advances in Cryptology*, pp. 302–8, Berlin: Springer.

Brin, D. (1999) *The Transparent Society*, Reading, MA: Perseus Books.

Canny, J. (2002) *Some techniques for privacy in ubicomp and context-aware applications*, [Online], Available: http://guir.berkeley.edu/pubs/ubicomp2002/privacyworkshop/papers/ubicomp.pdf [4 Oct 2005].

Chicago Tribune (2001) 'Rental firm uses GPS in speeding fine', (2 July 2001), p. 9.

Clarke, R. (1999) 'Internet privacy concerns confirm the case for intervention', *Communications of the ACM*, vol. 42 (2), pp. 60–67.

Cranor, L. F. (2001) 'P3P: The platform for privacy preferences project', in Garfinkel, S. and Spafford, G. (eds.) *Web Security, Privacy, and Commerce*, 2nd edition, pp. 699–707, Sebastopol, CA: O'Reilly.

Dingledine, R., Mathewson, N. and Syverson, P. (2004) *Tor: The second-generation Onion router*, [Online], Available: http://tor.eff.org/cvs/tor/doc/design-paper/tor-design.pdf [4 October 2005].

Duckham, M. and Kulik, L. (2005a) 'A formal model of obfuscation and negotiation for location privacy', in Gellersen, H. W., Want, R. and Schmidt, A. (eds.), Pervasive 2005, vol. 3468 of *Lecture Notes in Computer Science*, pp. 152–170, Berlin: Springer.

Duckham, M. and Kulik, L. (2005b) 'Simulation of obfuscation and negotiation for location privacy', in Mark, D.M. and Cohn, A.G. (eds.), COSIT 2005, vol. 3693 of *Lecture Notes in Computer Science*, pp. 31–48, Berlin: Springer.

Duckham, M., Mason, K., Stell, J. and Worboys, M. (2001) 'A formal approach to imperfection in geographic information', *Computers, Environment and Urban Systems*, vol. 25, pp. 89–103.

Duri, S., Gruteser, M., Liu, X., Moskowitz, P., Perez, R., Singh, M. and Tang, J-M. (2002) 'Framework for security and privacy in automotive telematics', *Proc. 2nd International Workshop on Mobile Commerce*, pp. 25–32, ACM Press.

Espinoza, F., Persson, P., Sandin, A., Nyström, H., Cacciatore, E. and Bylund. M. (2001) 'GeoNotes: Social and navigational aspects of location-based information systems' in Abowd, G. D., Brumitt, B. and Shafer, S. (eds.), Ubicomp 2001: Ubiquitous Computing, vol. 2201 of *Lecture Notes in Computer Science*, pp. 2–17, Berlin: Springer.

Etzioni, A. (1999) 'A contemporary conception of privacy', *Telecommunications and Space Journal*, vol. 6, pp. 81–114.

Etzioni, A. (1999b) 'Less privacy is good for us (and you)', [Online], Available: http://speakout.com/activism/opinions/3729–1.html [5 October 2005].

Fiat, A. and Shamir. A. (1986) 'How to prove yourself: Practical solutions to identification and signature problems', *Proc. on Advances in Cryptology—CRYPTO '86*, pp. 186–194, Berlin: Springer.

General Assembly of the United Nations (1948) 'Universal declaration of human rights', *United Nations Resolution 217 A (III)*, December, 1948.

Goldwasser, S., Micali, S. and Rackoff, C. (1985) 'The knowledge complexity of interactive proof-systems', STOC '85: *Proceedings of the Seventeenth Annual Acm Symposium on Theory of Computing*, pp. 291–304, New York, NY: ACM Press.

Görlach, W. W., Terpstra, A. and Heinemann, A. (2004) 'Survey on location privacy in pervasive computing', *Proc. First Workshop on Security and Privacy, Conference on Pervasive Computing (SPPC)*, 2004, [Online], Available: http://www.ito.tu-darmstadt.de/publs/index_en_html [5 October, 2005].

Gruteser, M. and Grunwald, D. (2003) 'Anonymous usage of location-based services through spatial and temporal cloaking', *Proc. MobiSys '03*, pp. 31–42.

Gruteser, M. and Grunwald, D. (2004) 'A methodological assessment of location privacy risks in wireless hotspot networks' in Hutter, D., Müller, G. and Stephan, W. (eds.) Security in Pervasive Computing, vol. 2802 of *Lecture Notes in Computer Science*, pp. 10–24, Berlin: Springer.

Gunter, A., May, M.J., and Stubblebine, S.G. (2004) 'A formal privacy system and its application to location-based services', Proc. 4th International Workshop, Privacy Enhancing Technologies, Toronto, Canada, May 26–28, vol. 3424 of *Lecture Notes in Computer Science*, pp. 256–282, Berlin: Springer.

Harper, R. H. R. (1992) 'Looking at ourselves: An examination of the social organisation of two research laboratories', *Proc. 1992 ACM Conference on Computer Supported Cooperative Work*, pp. 330–337, New York: ACM Press.

Harper, R. H. R., Lamming, M. G. and Newman, W. M. (1992) 'Locating systems at work: Implications for the development of active badge applications', *Interacting with Computers*, vol. 4 (3), pp. 343–363.

Helal, A., Moore, S. and Ramachandran, B. (2001) 'Drishti: An integrated navigation system for visually impaired and disabled', *Proceedings of Fifth International Symposium on Wearable Computers*, Zurich, Switzerland, 2001 [Online], Available:
http://www.harris.cise.ufl.edu/projects/publications/wearableConf.pdf [5 October 2005].

Hightower, J. and Boriello, G. (2001) 'Location systems for ubiquitous computing', *IEEE Computer*, vol. 34 (8), pp. 57–66.

Hong, J. I. and Landay, J. A. (2004) 'An architecture for privacy-sensitive ubiquitous computing', *Proc. 2nd International Conference on Mobile Systems, Applications, and Services*, pp. 177–189, ACM Press.

Hudson, S. E. and Smith, I. (1996) 'Techniques for addressing fundamental privacy and disruption tradeoffs in awareness support systems', *Proc. ACM Conference on Computer Supported Cooperative Work*, pp. 248–257, ACM Press.

Hutter, D., Stephan, W. and Ullmann, M. (2004) 'Security and privacy in pervasive computing: State of the art and future directions', in Hutter, D., Müller, G. and Stephan, W. (eds.) Security in Pervasive Computing, vol. 2802 of *Lecture Notes in Computer Science*, pp. 284–289, Berlin: Springer.

IBM (2004) *The enterprise privacy authorization language (epal 1.1)*, [Online], Available: http://www. zurich.ibm.com/security/enterprise-privacy/ [2 August 2005].

Jiang, X., Hong, J. I. and Landay, J. A. (2002) 'Approximate information flows: socially-based modeling of privacy in ubiquitous computing', in Borriello, G. and Holmquist, L. E. (eds.), Proc. 4th international conference on Ubiquitous Computing, vol. 2498 of *Lecture Notes in Computer Science*, pp. 176–193, Springer: Berlin.

Kaasinen, E. (2003) 'User needs for location-aware mobile services', *Personal and Ubiquitous Computing*, vol. 7(1), pp. 70–79.

Krumm, J., Harris, J., Meyers, S., Brumitt, B., Hale, M. and Shafer, S. (2000) 'Multicamera multi-person tracking for EasyLiving', in *Proceedings Third IEEE Workshop on Visual Surveillance VS2000*, pp. 3–10.

LaMarca, Y., Chawathe, S., Consolvo, J., Hightower, I., Smith, J., Scott, T., Sohn, H., Howard, J., Hughes, F., Potter, J., Tabert, P., Powledge, G., Borriello, G. and B. N. Schilit. (2005) 'Place lab: Device positioning using radio beacons in the wild' in Gellersen, H. W., Want, R. and Schmidt, A. (ed.) *Pervasive 2005*, vol. 3468, pp. 116–133, Berlin: Springer.

Langheinrich, M. (2001) 'Privacy by design—principles of privacy-aware ubiquitous systems' in Abowd, G. D., Brumitt, B. and Shafer, S. (eds.) Ubicomp 2001: Ubiquitous Computing, vol. 2201 of *Lecture Notes in Computer Science*, pp. 273–291, Berlin: Springer.

Langheinrich, M. (2002) 'A privacy awareness system for ubiquitous computing environments' in Borriello, G. and Holmquist, L. E. (eds.) UbiComp 2002: Ubiquitous Computing, vol. 2498 of *Lecture Notes in Computer Science*, pp. 237–245, Berlin: Springer.

Lee, J-W. (2004) 'Location-tracing sparks privacy concerns', *Korea Times*, 16 November 2004 [Online], Available: http://times.hankooki.com. [26 July 2005].

Ljungstrand, P. (2001) 'Context awareness and mobile phones', *Personal and Ubiquitous Computing*, vol. 5 (1), pp. 58–61.

Marmasse, N. and Schmandt, C. (2000) 'Location-aware information delivery with comMotion' in *Proceedings 2nd International Symposium on Handheld and Ubiquitous Computing (HUC)*, Bristol, UK, pp. 157–171.

Muntz, R. R., Barclay, T., Dozier, J., Faloutsos, C., Maceachren, A. M., Martin, J. L., Pancake, C. M. and Satyanarayanan, M. (2003) *IT Roadmap to a Geospatial Future*, Washington, DC: The National Academies Press.

Myles, G., Friday, A. and Davies, N. (2003) 'Preserving privacy in environments with location-based applications', *Pervasive Computing*, vol. 2 (1), pp. 56–64.

Onsrud, H. J., Johnson, J. and Lopez, X. (1994) 'Protecting personal privacy in using geographic information systems', *Photogrammetric Engineering and Remote Sensing*, 60(9), pp. 1083–1095.

Organisation for Economic Co-operation and Development (OECD) (1980) *Guidelines on the protection of privacy and transborder flows of personal data*, [Online], Available: http://www.oecd.org [25 July 2005].

Park, C. (2004) 'Location-based information service due next year.' *Korea Times* (2 July 2004), [Online], Available: http: //times.hankooki.com [26 July 2005].

Peterson, J. (2004) *A presence-based GEOPRIV location object format*, [Online], Available: http://www. ietf.org/internet-drafts/draft-ietf-geopriv-pidf-lo-03.txt [5 October 2004].

Pfitzmann, A. and Köhntopp, M. (2001) 'Anonymity, unobservability, and pseudonymity—a proposal for terminology' in Federrath, H. (ed.) Designing Privacy Enhancing Technologies, vol. 2009 of *Lecture Notes in Computer Science*, pp. 1–9, Berlin: Springer.

Rodden, T., Friday, A., Muller, H. and Dix, A. (2002) 'A lightweight approach to managing privacy in location-based services', *Technical Report Equator-02–058*, University of Nottingham, Lancaster University, University of Bristol.

Rotenberg, M. and Laurant, C. (2004) *Privacy and human rights 2004: An international survey of privacy laws and developments*, [Online], Available: http://www. privacyinternational.org/survey/ [26 July 2005].

Rule, J. B. (1973) *Private Lives and Public Surveillance*, London: Allen Lane.

Schilit, N., Hong, J. I. and Gruteser, M. (2003) 'Wireless location privacy protection', *IEEE Computer*, 36(12), pp. 135–137.

Schilit, B. N. and Theimer, M. M. (1994) 'Disseminating active map information to mobile hosts', *IEEE Network*, 8(5), pp. 22–32.

Schmidt, A., Beigl, M. M. and Gellerson, H-W. (1999) 'There is more to context than location', *Computer and Graphics Journal*, 23(6), pp. 893–902.

Scott, J. and Dragovic, V. (2005) 'Audio location: Accurate low-cost location sensing' in Gellersen, B., Want, R. and Schmidt, A. (eds.) *Pervasive 2005*, vol. 3468, pp. 1–18, Berlin: Springer.

Scott-Young, S. and Kealy, A. (2002) 'An intelligent navigation solution for land mobile location based services', *Journal of Navigation*, 55, pp. 225–240.

Snekkenes, E. (2001) 'Concepts for personal location privacy policies' in *Proc. 3rd ACM Conference on Electronic Commerce*, pp. 48–57, ACM Press.

Sprenger, P. (1999) 'Sun on privacy: "Get over it"', *Wired*, January 26, 1999.

Stanford, V. (2002) 'Using pervasive computing to deliver elder care', *IEEE Pervasive Computing*, 1(1), pp. 10–13.

Strassman, M. and Collier, C. (2004) 'Case study: Development of the Find Friend application' in Schiller, J. and Voisard, A. (eds.) *Location-based Services*, Chapter 2, pp. 27–39, San Francisco, CA: Morgan Kaufmann.

U.K. Government. (1998) *Data Protection Act*, London: HMSO.

U.S. Department of Justice, Office of Information and Privacy. (2004) *Overview of the Privacy Act of 1974*.

U.S. Department of Health, Education and Welfare, Secretary's Advisory Committee on Automated Personal Data Systems. (1973). *Records, Computers, and the Rights of Citizens*, Cambridge, MA: MIT Press.

Want, R., Hopper, A., Falcao, V. and Gibbons, J. (1992) 'The Active Badge location system', *ACM Transactions on Information Systems*, 10(1), pp. 91–102.

Werbach, K. (2000) 'Location-based computing: Wherever you go, there you are', *Release 1.0*, 18(6), pp. 1–26.

Westin, V. (1967) *Privacy and Freedom*, New York: Atheneum.

Worboys, M. F. and Clementini, E. (2001) 'Integration of imperfect spatial information', *Journal of Visual Languages and Computing*, 12, pp. 61–80.

Worboys, M. F. and Duckham, M. (2004) *GIS: A Computing Perspective*, 2nd edition, Boca Raton, FL: CRC Press.

WorldWideWeb Consortium (W3C) (2005) *Platform for privacy preferences project (p3p)*, [Online], Available: http://www.w3.org/P3P/ [2 August 2005].

Wu, M. and Friday, A. (2002) 'Integrating privacy enhancing services in ubiquitous computing environments' in *Proc. Workshop on Security in Ubiquitous Computing*, 4th Intl. UbiComp Conference.

Part II

Modelling Approaches and Data Models

Part II of the book 'Dynamic and Mobile GIS' focuses on modelling approaches, especially those appropriate to depicting dynamic processes in GIS. The four contributions present a high level of novelty and suggest a dramatic evolution of data models, spatial analyses and spatial queries.

As opposed to initial GIS approaches considering geographic features and relations among them as independent of time, Kate Beard in Chapter 4 proposes an event-based approach in which change itself is the central concept that is modelled. Events as explicit representations of change with associated attributes of change such as rate of change or rate consistency provide the key units for exploration and analysis of the mechanism of change. In this approach, the time dimension dominates the spatial dimensions as the ordering of events in time is critical. An event-based view provides the foundation for the analysis of dynamic phenomena and is therefore naturally appropriate for dynamic GIS. Chapter 4 includes a very complete state-of-the-art presentation of dynamic process modelling and event-based models. A formal categorization of changes is then proposed, sources of events are described and a method for event visualisation and exploration is outlined.

Femke Reitsma and Jochen Albrecht present in Chapter 5 a new process-based data model called *nen* (after node-edge-node graph representation). While most of the existing theories and models for simulating processes focus on representing the state of the represented system at a moment of time, this approach expresses and represents information about processes themselves. This data model provides advantages in querying, analysis and exploration of process descriptions under computer simulation conditions – or *in silico*. Chapter 5 includes a description of the model and presents its application in a small watershed modelling test case. Part of the originality of *nen* is to provide a new epistemological window on the modelled results, allowing for new process-oriented queries and analyses. It allows questions to be asked (where is a process operating at a particular instant of time? how has the process changed over time? etc.) not directly answerable with current object-centred formulations that focus on the states of a system resulting from the operation of a process.

In Chapter 6, Muki Haklay extends the comparison between Map Calculus and Map Algebra in the context of dynamic raster GIS. Map Calculus is an alternative to current representation in GIS, and is based on the use of function-based layers in GIS. Its main strength is its ability to treat analytical layers in their symbolic form in a similar way to the manipulation of mathematical functions in software packages. This chapter focuses on the particular challenges of dynamic modelling in GIS, exploring the ways in which it is implemented in Map Algebra and outlining how such models can be implemented in a Map Calculus-based system. It appears that Map Calculus allows easier linkage to rapidly changing inputs and easier implementation of dynamic models based on differential equations. The use of Map Calculus poses certain challenges such as the reformulation of common GIS operators and the consideration of optimal visualisation methods. Ultimately, it should make GIS more accessible to domain experts, as they can focus on the construction of the model and not on finding ways to fit a conceptual model within the constraints of GIS.

Finally, Peter van Oosterom explores, in Chapter 7, issues related to spatial constraints in data models. In GIS, constraints are conditions that must always be valid for the model of interest. In a dynamic context, with constantly changing geo-information, any changes arising should adhere to specified constraints; otherwise inconsistencies (data quality errors) will occur. Constraints should be part of the object class definition, just as with other aspects of that definition, such as attributes, methods and relationships. Currently, the implementation of constraints (whether at the front-end, database level or communication level) cannot be driven completely automatically by constraints' specifications within the model. This chapter demonstrates the need for the integral support of constraints through four quite different cases but all dealing with dynamic situations: a VR system for landscape design, cadastral data maintenance, topographic data maintenance and a Web feature service. It proposes a complete description and classification of constraints and describes some solutions for the formalisation and the implementation of constraints in the different presented cases.

Chapter 4

Modelling Change in Space and Time: An Event-Based Approach

Kate Beard

Department of Spatial Information Science and Engineering, University of Maine, Orono, USA

4.1 Introduction

Initially geographic information systems modelled geographic features and the relations among them under the assumption that such features were independent of time. A rationale for this perspective is that many geographic features retain their identity and location for long periods of time. Given the persistence of these fundamental properties, representation of change was not an initial consideration for geographic information systems. Additionally, early spatial data collection methods (primarily photogrammetry) focused on capturing these fundamental properties: identity and location, but were generally too expensive to repeat with a frequency that could support interesting change analysis. More recent GIS research has begun to address models for representing the dynamic and mobile components of geographic features. What has remained constant, however, is that geographic layers or features are the central units of analysis and GIS operate on these units. In an event-based approach geographic features or locations are not the primary focus. This chapter describes an event-based approach in which change itself is the central concept that is modelled and change units are the principal objects and units of analysis. In this event-based approach the time dimension dominates the spatial dimensions as the ordering of events in time is critical. Because the principal unit of analysis and the organising dimension are fundamentally different, new approaches are needed for change objects. Section 4.2 of this chapter discusses previous event-based models and approaches that have appeared in the literature. Section 4.3 presents a categorisation of change, Section 4.4 describes the proposed event model, Section 4.5 describes sources of events, Section 4.6 outlines a method for event visualisation and exploration and Section 4.7 concludes the chapter with a summary and future research challenges.

4.2 Previous approaches to time and dynamic models in GIS

Beginning in the late 1980s and early 90s GIS research began to address the dynamic aspects of geographic features and to include time in GIS (Armstrong,

Dynamic and Mobile GIS: Investigating Changes in Space and Time. Edited by Jane Drummond, Roland Billen, Elsa João and David Forrest. © 2006 Taylor & Francis

1988; Langran, 1992). The addition of time-supported tracking the history of spatial objects and their attributes and the prediction of potential future behaviour and change. The primary focus however remained the geographic feature with the temporal dimension added in the form of a time stamp to monitor and analyse successive states of these features (Abraham and Roddick, 1999). In much of this work the central unit of analysis was viewed as the spatial layer or theme and change was conceived of as modifying the fabric of a layer (Langran, 1992). This view is generally associated with the concept of the spatial snapshot.

The influence of the object-oriented view shifted the association of change to individual objects. Worboys (2005) terms this the object change view. Under this view an object has a unique identifier that it maintains while changes may occur to its spatial and non-spatial properties. A related pre-GIS view was introduced by Hagerstrand (1970) who was interested in analysing the dynamic behaviour of individuals in geographic space. He was particularly interested in the movements of individuals through space and time, a perspective which lead naturally to an object-based view focused on the individual. Hagerstrand's work has been recently promoted by others (Mark, 1998; Miller, 2003) under the concept of geo-spatial lifelines. Hornsby and Egenhofer (2000) address the object change view generally under the concept of identity-based change. A particular challenge for the object change view lies in establishing and maintaining the identity of an individual object over time. Questions naturally arise as to when and what change becomes so substantial that an object is no longer the same object.

Noting limitations in models that simply add time stamps to support the management of versions and state changes to geographic features or locations, several researchers proposed event-based models (Peuquet and Duan, 1995; Claramunt and Thériault, 1995; Worboys and Hornsby, 2004; and Worboys, 2005). Event-based models move from geographic feature identification and location characterisation to an explicit focus on change. Within the event-based view subtly different definitions and perspectives on events exist. Events are variously described as happenings or expressions of change. Claramunt and Thériault (1995) describe events as things that happen. More specifically they note processes lead to changes in entity states and these changes show the result of the process and constitute events. Peuquet (1994) describes an event as denoting some change in some location(s) or some object(s) and Peuquet and Duan (1995) describe an event as representing the spatiotemporal manifestation of some process.

These variations in the notion of an event compare with the variability in event definitions generally. The SHOE General Ontology (SHOE) defines an event as something that happens at a given place in time. In the specifications of the Dublin Core metadata standard, an event is defined as 'a non-persistent, time-based occurrence'. Quine (1985) described events as objects where objects are regions bounded in space and time. He further notes that events can be broken into sub-parts and arranged in a taxonomic hierarchy.

The common factor among the event-based views is the requirement that change be explicitly modelled and they share the objective of facilitating the analysis of change, patterns of change, or happenings through time. Each professes the importance of change or happening as the central unit of analysis.

The Event-Based Spatiotemporal Data Model (ESTDM) proposed by Peuquet and Duan (1995) temporally orders changes to locations within a pre-specified geographical area. It is interesting to note that while change (an event) is explicit in this model it is subtly subservient to a time location and a spatial location (i.e. bound to pixels). Specific changes are associated with a stored temporal location t_i called a time stamp in an event list. Furthermore each event list is associated with a single thematic domain (layer). The ESTDM (Peuquet and Duan, 1995) identifies itself as an event-based data model yet it does not fully support change as the primary unit. The difference is that the focus remains on a value change associated with a location (a pixel) and that change is not the characterised unit. In a pure event view the change unit itself is characterised not the geographic feature or location.

Worboys (2005) and Worboys and Hornsby (2004) describe an event as a happening or *occurrent* to be distinguished from a thing or *continuant*. They note the weakness of snapshot models as lacking explicit representation of events and explicit representation of change. They argue that events are needed to capture the mechanism of change but describe events more as happenings, activities and processes, rather than explicitly characterised change units.

In an object change view we would typically record changes in the properties of an object. Assume that a house is repainted from white to yellow and hence undergoes a non-spatial change to its colour property. The primary object is the house and we would record its new colour, possibly keeping a record of its previous colour. In the event view, 'painting the house' is the recorded unit along with its specific properties such as its time of onset, its duration and perhaps the method by which it was accomplished. In an event view the focus moves from the change to a particular house to an analysis of the change objects themselves, e.g. an analysis of painting events through for example a comparison of their durations, seasonal patterns, differences in methodology. The difference in these perspectives can perhaps be illustrated by considering an example query. Using a GIS, a user might ask for all the houses that were painted over the last year and receive a map showing the geographic distribution of these houses. The focus is on a geographic object and a particular type of change to that object type. In an event view the request would be for the 'house painting events' over the last year. In this case the set of house painting events would have the same geographic distribution but clearly different sets of attributes that could be queried and analysed.

There were early calls for maintaining records of events and processes as the basis for understanding dynamic behaviours (Chrisman, 1998), but the realisation of the event view owes much to new technologies that are now able to deliver a wide range and volume of spatiotemporal information. Environmental monitoring and sensor data streams are creating repositories of information with high temporal resolution that support the analysis of change. Fine temporal resolution data streams begin to provide a picture of how processes operate and a foundation for investigating cause and effect. Because the data are now becoming available to truly realise this analytical challenge there is renewed impetus to develop models and analytical tools for events and processes.

The motivation for an event-based approach has been well articulated in the literature. It supports representation of the dynamic behaviour of geographic phenomena, hypothesis generation, scientific investigation of complex relationships, an ability to investigate causal linkages and associate entities with influences and underlying processes. These outcomes are reason enough for an event-based view yet there is another pragmatic motivation. Fine temporal data streams from monitoring and sensor networks have tremendous scientific value but are not fully exploitable due to difficulties in integrating across the heterogeneous spatial and temporal sampling regimes and assimilating across a large multi-variate space. Geographic information systems have been effective at integrating multiple spatial data sets as long as they are within similar spatial scales. Similarly time-series analysis tools are available to analyse multiple time series but encounter problems in handling multiple time scales and cyclical patterns. In the context of multi-variate, multi-media, multi-resolution space-time series there is a challenge in integrating across this diversity, and currently no system exists to effectively merge and explore collections of heterogeneous multi-variate space-time data. One approach to the integration problem is the transformation of heterogeneous space-time series to a higher level of abstraction – an event. The abstraction derives from meaningful partitioning of space-time series into segments sharing a common property or label that makes them identical at a more abstract level of representation. The abstraction process generates an event data type with a spatial-temporal footprint that effectively normalises across the space-time heterogeneity of diverse sensor data streams creating common space-time units for analysis.

4.3 Categorizations of change

Development of an event view as an explicit change-based view can benefit from a categorization or ontology of change. Several researchers have presented categorizations of change that vary with the perspective on what is undergoing change. Armstrong (1988) presented one of the first categorizations of temporal change to appear in the GIS literature. He identified eight change categories to a simple spatial distribution as including: no change, attribute change, morphology change, topology change and combinations of these three. This perspective aligns with the early view of change associated with a spatial layer or theme. Topology is included which makes sense in the context of layer change and we notice movement is not included which is also consistent with layer-based change (i.e. there is no expectation that layers will move).

The identity-based change model has been articulated by Hornsby and Egenhofer (2000), and also by Roshannejad and Kainz (1995). It assumes an independent and uniquely identifiable entity. The Hornsby and Egenhofer (2000) model focuses on the most essential change behaviour for an entity: its transformation from a non-existent to existent state and vice versa. An entity comes into existence; it has a presence and properties, chief among these being its identity. In its existent state the entity can change any number of its properties except its identity since to do so means it is no longer the same entity.

Yuan (1997) identifies two types of temporal change: mutation and movement and considers Armstrong's categories as subtypes. Frank (2001) in the context of socio-economic units identifies change in two forms: one that he refers to as life of the object and the second as motion or change in the position or geometric form. Claramunt and Thériault (1995) combine most of these notions into a three-level categorization of change. The three levels include:

❑   evolution concerning a single entity,
❑   evolution in the functional relations between several entities and
❑   evolution in spatial structure involving several entities.

Their change classes for a single entity are shown in Table 4.1.

Table 4.1. Categories of change for a single entity based on Claramunt and Thériault (1995).

Basic Changes	Transformations	Movement
appearance	expansion	displacement
disappearance	contraction	rotation
stability (no change)	deformation	

Their second-level change category involves change in the relations among entities and covers group dynamics. They use succession as an example for this change category, a change that may be construed as a decline in one species and replacement by another. This is a broad category of change type but one that may be viewed as compositions of the first-level changes by individual entities. The distinction in their third level is the attachment of specific spatial constraints on a change type. Their examples for this category are splits, unions and real locations (e.g. of parcels). These again can be considered compositions of the primary level change types with the addition of spatial constraints. For example a split most often involves the termination of the original entity and the creation of two new entities with the spatial constraint that they share a common boundary. A union typically involves the termination of two entities and the creation of one new entity within the spatial confines of the original two.

The change categorization presented in this chapter borrows from these previous categorizations and identifies four principal change types: basic change, attribute change, shape change and movement (position or location change). Basic change covers creation and termination. Attribute change includes any changes in non-spatial properties and the last two types are forms of spatial change. Because there can be multiple levels of granularity to change for any interval of time, a change type may not be pure. For example, a bird moults and loses its feathers. This transformation is primarily a non-spatial change; however the loss of feathers could result in subtle change in the bird's shape.

Subtypes of attribute change depend on the measurement level of the attribute. Change in nominal attributes is a discrete change from one class to another. Some classes may be considered more closely related than others and thus for some class changes qualitative differences in degree of change among classes may be recognised. Class changes could be ordered by their likelihood of occurrence but this issue is not addressed further here. Change in interval and ratio level attributes can be characterised as: increasing, decreasing, switching from increasing to decreasing values (peaks) or switching from decreasing to increasing values (troughs).

Shape change can be characterised as a scaling transformation. The subtypes of a scaling transformation are: uniform or non-uniform and expanding or contracting for a combination of four subtypes. Rates of expansion or contraction can be associated with each subtype. Deformation, a subtype noted by Claramunt and Thériault (1995) is a shape change under the constraint that area remains constant.

The subtypes of movement are translation and rotation transformations. The attributes of movements include their azimuth or direction, speed and rate of change in speed (acceleration or deceleration). These are the same parameters ascribed to movement by Laube and Imfeld (2002).

All of the previously cited categorizations included the category of no change and no change is maintained in the current categorization as essential for explicitly noting periods of no change in association with the three categories of attribute change, shape change and movement.

Table 4.2. Illustrates three of the four primary change categories. The basic change category of creation and termination is excluded.

Attribute Change	Shape Change	Movement
Nominal Class Change	**Scaling**	**Translation**
Interval, Ratio	**Regular expansion**	**Rotation**
Increase	**Irregular expansion**	
Decrease	**Regular contraction**	
Peak	**Irregular contraction**	
Trough	**Deformation**	

Table 4.2 summarises the primitive change types excluding the basic change category. A primitive event is a unit of change in a single variable or property (e.g. a pressure decrease, a wind speed increase, a 90 degree rotation). An important point discussed further in Section 4.4 is that these primitive change types are extractable from time series and space-time series. Complex events are compositions of these primitive events. Cognitively people recognise, speak about, and label events of interest such as storms, droughts, species declines and expansions, wildfires, disease outbreaks, economic depressions, crimes and traffic accidents to name a few. An examination of such labelled events can reveal their composition from sets of the primitive change events just described. A storm surge on the Maine coast on March 7, 2001, for example, was described as starting with a low-pressure front, followed by increasing north-easterly winds with wind gusts exceeding 40 knots at several weather stations and high seas with tide gauge readings in Portland Harbor exceeding 12 feet.

4.4 The event model

The event model as presented here treats events as first-class units. In the object-oriented sense each has a unique identity, has attributes, can participate in class hierarchies, may have parts or be aggregated to composite units, and can engage in relationships with other events and entities. In these respects the event model shares the foundations of the GEM model presented by Worboys and Hornsby (2004). Figure 4.1(a,b) illustrates the basic components of the event model. To avoid ambiguity with the object-oriented concept of object, the term entity substitutes for object in the GEM model. The primitive event is noted here as a special subclass of event.

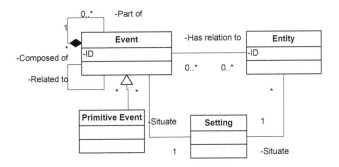

Figure 4.1.a. Event model (see Figure 4.1.b for Setting subclasses).

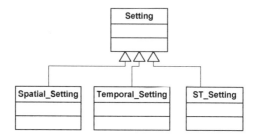

Figure 4.1.b. Subclasses of a setting.

An event has a localisation in space and time that Worboys and Hornsby (2004) describe as its space-time setting. A setting has the subclasses spatial, temporal and spatial-temporal. As Worboys (2005) quoting Kim (1976), notes, many events can occupy the same space-time setting. For example a barometric pressure decrease, wind speed increase and precipitation event can all occupy the same space-time setting. Thus an event instance is associated with one setting but a particular setting can be associated with many events and entities.

The attributes of a primitive event depend upon the change type. As shown in Figure 4.2 change type is modelled as an object with three subclasses: attribute change, shape change and movement as summarised in Table 4.2. A primitive event by definition is associated with one and only one change type.

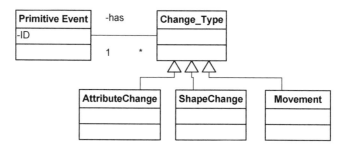

Figure 4.2. Primitive event object showing change_type subclasses.

The next section describes the sources for events and some of the challenges in populating this model.

4.5 Sources of events

We may ask what is the data collection process for events. Knowledge of the world is gained through observations of values of measureable properties. Some observation protocols are designed to capture events explicitly such as flood gauges that record high water marks. More typically observations are part of routine monitoring programs in which events appear only implicitly. Data collected from a range of sensing systems contain a range of possible events. Technology improvements have led to increases in the sampling increments in time and space for many of these sensing systems resulting in the possibility of detecting fine-scale events. The primitive event types identified above can be extracted from the data streams of these sensing systems. Such systems include the climate stations managed by the National Weather Service, stream flow monitoring stations managed by the U.S. Geological Survey, toxic release monitoring stations managed by the Environmental Protection Agency (EPA), and the host of satellite sensor programs managed by NASA (e.g. AVHRR, MODIS) to name a few. Collections of movement data can be obtained from telemetry, GPS and increasingly through cell phones and other devices.

Existing sensing systems can be characterised as generating space-time series of three types. One type is a spatial field of time series, another is a time series of spatial fields and a third is a moving point time series. The first form of space-time series consists of multiple fixed spatial locations at which one or more attributes are measured at regular time intervals creating one or more time series. While each time series is associated with a fixed location, the set of collection points can be considered to create a spatial field and hence a spatial field of time series. A spatial field of time series can be denoted as sequences of the type $(a_i, t_i)_s$ $i=1, \dots n$ where a_i is a measured attribute value at time t_i and indexed by spatial location s. The data from a fixed set of climate stations illustrate this type of space-time series. A time series of spatial fields corresponds to the type (b_i, t_i) where b_i is a spatial field (an array of values for an n x m space) generated at time t_i. The set of spatial fields for t_i $i=1, \dots n$ constitutes the time series (a time series of spatial fields). As an example the time-ordered set of satellite images for a month illustrates this type of space-time series.

For a moving point space-time series the sensor moves and measures its location at regular time intervals. Under this scenario the attribute is fixed and is typically the label for the unit carrying the sensor (e.g. person, animal, car) and the outcome is a set of observed positions for the moving object. A moving point time series can be denoted as a sequence of the type $(s_i, t_i)_m$ $i=1, \dots n$ where s_i is a measured location value at time t_i here indexed by m, a moving object label.

The Gulf of Maine Ocean Observing System (GoMOOS) provides a case study for closer examination of event detection and collection. GoMOOS includes between 10 to 12 moored buoys reporting hourly on up to 15 variables at 3 depths near shore and 5 depths in deeper basins. Measurements at the surface include wind, waves, temperature and visibility. Below the surface measurements include currents, temperature, salinity, ocean colour, turbidity, light attenuation and dissolved oxygen. The set of time series generated at the moored buoy locations represent a spatial field of time series. GoMOOS also includes AVHRR sea surface temperature data collected at approximately 4-6 hour intervals at a spatial resolution of 1.1 km; NASA's SeaWIFs sensor measuring near-surface ocean phytoplankton biomass and composited as 8-day images; and wind data from NASA's SeaWinds instrument on board the QuikSCAT satellite measuring the speed and direction of near-surface winds at a spatial resolution of 0.25 degrees twice daily. These last three sensor streams represent time series of spatial fields.

For each moored buoy, primitive event detection is initially applied to individual time series for each variable at each depth. Since each time series is a sequence of measured attribute values the attribute change primitives are those of interest. The problem involves piecewise segmentation of the time series into the primitives: generally increasing sequences, decreasing sequences, local maxima, local minima, or intervals of nearly constant value. Buoy measurements are collected and reported hourly but there can be periods of instrument or sensor malfunction when faulty or no values are reported. No-data and bad-data 'events' are thus also extracted.

Segmentation of time series can be accomplished by statistical model fitting (Guralnik and Srivastava, 1999) or by template matching (Agrawal et al., 1995). Change point detection methods model time series as mathematical models (Hawkins 1976) and determine the temporal locations where the model or model parameters change abruptly. The models are typically regression models of different orders or auto-regressive models. The change units are characterised by specification of model type: linear, exponential, logarithmic or complex linear; by model coefficients, and may include measures of fit and characterisation of residuals.

If the observed time series were smooth curves, the segmentation into the above primitives would not be difficult. For real-time series the segmentation is challenging. Identification of the event primitives at multiple temporal scales is one challenge. Within a time series, scales of events can vary from diurnal to weekly to monthly to seasonal and beyond depending on the length and temporal resolution of the time series. The diurnal signal shown in Figure 4.3 is embedded in the seasonal spring temperature increase shown in Figure 4.4 along with additional approximately 3-5 day peaks and troughs. A multi-scale segmentation must be applied to extract the different time scale changes (Höppner, 2002).

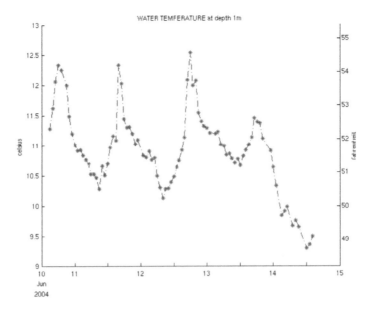

Figure 4.3. Diurnal scale changes in water temperature.

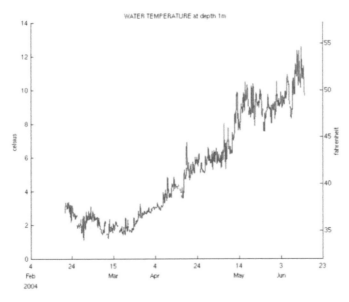

Figure 4.4. From the period of mid March to June the temperature is generally increasing constituting an increase event at the seasonal scale. Finer scale peaks and troughs are apparent in this generally increasing segment.

An additional challenge arises in the space-time localisation. This problem is exemplified by the temperature time series shown in Figure 4.5. Assume we are seeking a segmentation into fall (decreasing), winter (trough), spring (increasing) and summer (peak) sequences. The exact time points at which one of these events starts and ends is difficult to isolate. The uncertainty in the time localisation is managed by reporting the start and end time at an appropriate resolution. The buoy measurements are collected hourly. If an event start cannot be precisely determined hourly it may be reported to the nearest day(s), week(s), or month(s).

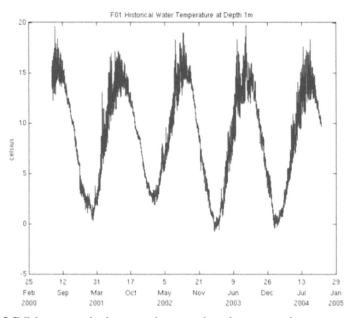

Figure 4.5. Fall decrease, spring increase, winter troughs and summer peaks are apparent in this time series but precise localization of the start and end points for these change segments is difficult to determine due to the variation in the time series.

Primitive events for each time series are extracted and stored. These primitives may include sets of diurnal, weekly, monthly, seasonal or annual change units. Their temporal setting consists of the start and end times and the resolution with which they are reported. Their spatial setting is given by the mooring position and the depth of the sensor on the mooring.

These primitive events provide an initial set of change units for analysis and a set of units that can be investigated for composition into more complex change units. Individual time series yield only attribute change events. By examining the spatial field of point-based time series more information can be gained on the possible spatial extent of an event. An event extracted from one time series may be replicated at depth or in nearby time series if its spatial extent or influence covers more than one time series location or depth.

Additionally if an event is replicated in nearby time series locations with systematic time lags we can potentially make inference of event movement as the event trajectory may cause it to pass one or more buoy locations. A challenge arises in confirming evidence for the 'same' event across locations. If a common event identity can be established across several time series, the common primitive events can be assembled into a composite event.

Shape change and movement events can be more precisely extracted from time series of spatial fields. The approach to event detection in spatial fields equates to object extraction from digital imagery, namely to extract the outline of a uniform region. Image processing techniques such as snakes can be used to detect and extract primitive edges and blobs (Agouris et al., 2001a; 2001b). The outputs are spatial primitives in contrast to the attribute change (temporal) primitives described above. Once the spatial primitives (possible event outlines) are captured at discrete time instances they can be composited using a spatiotemporal helix model (Partsinevelos et al., 2001; Agouris and Stefanidis, 2003; Stefanidis et al., 2003). The helix model assembles possible event outlines in a trace through 3D (x,y,t) space. The spatiotemporal helix *(STH)* captures shape change and movement through a central spine and annotated prongs. The central *spine* models the trajectory of the centre of a possible event as it moves over a temporal interval. It is expressed as a sequence of nodes $S(n^l, \ldots n^n)$ that correspond to breakpoints along the trajectory. Each node n^i is modelled as $n^i(x,y,t,q)$, where *(x,y,t)* are the spatiotemporal coordinates of a node and *q* qualifies the node as an acceleration (q^a), deceleration (q^d), or rotation (q^r) node. Prongs express expansion or contraction of an event outline at a specific temporal instance. Prongs are modelled as $p^i(t,r,a_1,a_2)$ where *t* is the time, *r* is the magnitude of expansion/contraction, and a_1, a_2 is the range of azimuths over which the change occurs.

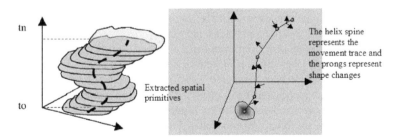

Figure 4.6. Spatial change primitives illustrating shape change and movement extracted from time series of spatial fields

The spatial primitive events extracted from time series of spatial fields need to be assembled across time in contrast to the temporal primitives that are assembled across space for spatial fields of time series. A similar challenge arises in establishing a common event from the collection of spatial primitives.

4.6 Event visualisation and exploration

The event detection methods yield a set of change primitives and valuable information can be gleaned from exploration and analysis of these explicit change units. They provide the material to ask specific questions about change. We may want to investigate whether events exhibit any spatial constraints, whether they exhibit any repeating temporal patterns, and if so further examine the repeat period and its regularity. We may wish to explore spatial relations among events (i.e. are they clustered in space?), temporal relations among them (i.e. are they clustered in time? does one event class typically precede another?) or combined spatialtemporal relations (are they clustered in space and time?)

An event exploration environment (Beard, 2004) based on a set of event bands is being developed to explore change patterns. The approach shares concepts with the TRELLIS display developed by Becker and Cleveland (1996). Events warrant an exploratory environment within which to investigate spatiotemporal patterns and the challenge becomes one of finding representations for displaying combined dimensions of space, theme and time such that possible relationships can be effectively uncovered and analysed. Additionally such an exploratory environment needs to be able to address varying resolutions and scales in time and space.

The basic display problem for the three dimensions parallels the measurement problem articulated by Sinton (1978). He noted that in the presence of dimensions of space, theme, and time not all three dimensions can be simultaneously measured. The solution was to hold one dimension constant, allow a second to vary in a controlled manner and measure the third for variation within the controlled attribute. Mirroring this strategy the Event Explorer supports one measured dimension: the time dimension in this case. One or more dimensions are fixed meaning they assume a single value. The exploratory power comes from the flexible assignment of space, theme and time control (a partitioning into a few discrete values or categories similar to conditioning variables in the Trellis display) to a set of graphic elements: event bands, event band stacks and panels.

An event band is a display container for a temporal sequence of event instances belonging to an event class for a requested time interval. The event band is made up of the band and represents measured time for the requested time interval. Bars within the band represent instances of events. The bars are rendered within the band according to the associated time line to represent the time of occurrence of the event. The width of the bars represents the duration of an event occurrence and a colour coding of the bars could be used to encode various event attributes.

An event band stack is a set of event bands stacked one on top of another. Stacking bands initiates the first level of control that is assigned to the band. The band itself represents control in one of the dimensions of theme (i.e. event class), space or time. Stacks can be composed of any number of bands but examples in Figures 4.8 and 4.9 show stack depths of two bands.

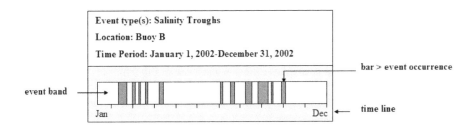

Figure 4.7. Components of an event band.

Assume the user has a comparative task in mind and wishes to compare the patterns of events for two locations. In this example, location becomes the control dimension and it is applied to the bands. For a specific GoMOOS data example, assume the user requests salinity trough events for the year 2002 at buoys A and B. In this example two dimensions are fixed: theme (event type, salinity trough) and time (the year 2002). Location is controlled through the specification of buoys A and B and a location-controlled stack is generated as shown in Figure 4.8. With location-controlled bands a user can examine how similar locations are in terms of event incidence for a specified time interval.

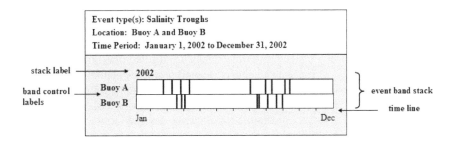

Figure 4.8. Example event band stack illustrating location control.

Instead of comparing event pattern differences with respect to location, a user may be interested in comparing distributions with respect to different times. For this comparative task, time becomes the controlled dimension. Suppose a user wants to examine salinity trough events at Buoy A for years 2002 and 2003. This request generates a time-controlled stack as shown in Figure 4.9 in which location and theme are fixed. From this display the user can evaluate the similarity in the event patterns at one fixed location between years. For example are there more or fewer events per year, are there similar yearly patterns and (if several years of data are available) is there an anomalous year?

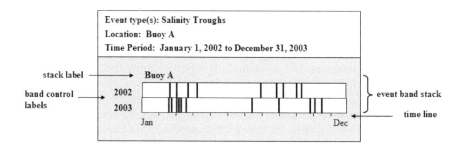

Figure 4.9. Example event band stack illustrating time control.

In a more complex comparison task a user may wish to compare event patterns across two dimensions. Consider a query for salinity events for Buoys A, B and C for years 2002, 2003, 2004. Two dimensions are controlled in this example: buoy location and year. The additional level of control is obtained through the addition of stacks. Instead of a single stack, multiple stacks are formed that reflect the number of control categories. These stacks behave as small multiples (Tufte, 1983) meaning they must share a common graphic framework. For the example query, two options are possible. Time control can be associated with the bands and location control with the stacks or conversely location control can be assigned to the bands and time control to the stacks. A display that assigns time control to bands and location control to stacks is shown in Figure 4.10. This configuration allows direct comparison of yearly differences in salinity events at Buoys A, B and C and comparison of these patterns across buoys.

Converting the display to one in which location control is assigned to bands and time control is assigned to stacks supports the direct comparison of location differences in events from year to year (see Figure 4.11) and comparison of these patterns across years.

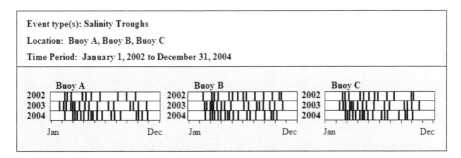

Figure 4.10. Example event band stacks illustrating time control assigned to bands and location control assigned to stacks.

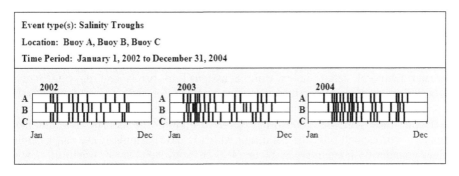

Figure 4.11. Example event band stacks illustrating location control assigned to bands and time control assigned to stacks.

The third graphic object, the panel (a collection of event band stacks) adds another level of control and allows for exploration of patterns of events within and across stacks. Figure 4.12 shows an example in which location control is assigned to the bands, time control is assigned to the stacks and theme (event type) is assigned to the panels.

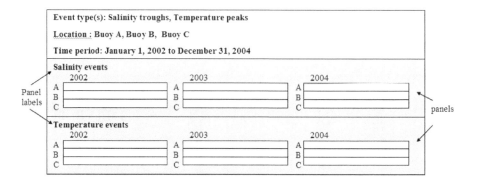

Figure 4.12. Bands, stacks and panels provide three levels of control to investigate event patterns.

A particular advantage of the flexible assignment of control to the three graphic levels is that the same control type (e.g. time) can be applied to all levels to explore changes at different granularities. For example we can assign spring months to bands and years to stacks to examine spring season events from year to year, or for greater temporal detail, hours can be assigned to bands and days to stacks. Constraints are that the control levels must nest hierarchically and when time is the controlled dimension, the interval of the time line on the band must be consistent with the finest temporal control unit (note in Figure 4.14 the time line units are hours to correspond with the band unit which is days). Similarly we can examine finer granularity in event types. Figure 4.13 shows assignment of primitive change types to bands and the measured variable to stacks.

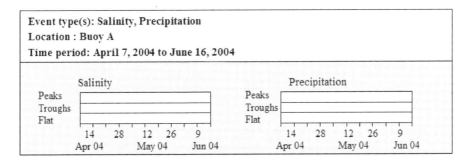

Figure 4.13. An example of assigning theme control to bands and stacks. Event change type is assigned to bands and event variable is assigned to stacks.

The band, stack, panel structure also supports investigation of repeat and cyclical events. Repeat patterns can be explored by controlling time and varying the temporal resolution on the bands. If the user suspects a diurnal repeat cycle, band control can be set to days (where the stack corresponds to a month) and the time line units to hours as shown in Figure 4.14.

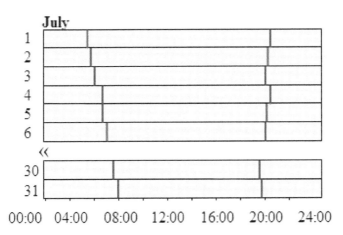

Figure 4.14. An example event band stack configured to check for diurnal repeat cycles using days on the bands and hours as the temporal units on the time line.

A user might know the repeat cycle but want to investigate how regularly it repeats. For example we know spring thaw ice out events on lakes repeat seasonally but we can use the event band/stack/panel structures to examine how regularly they repeat each year as shown in Figure 4.15.

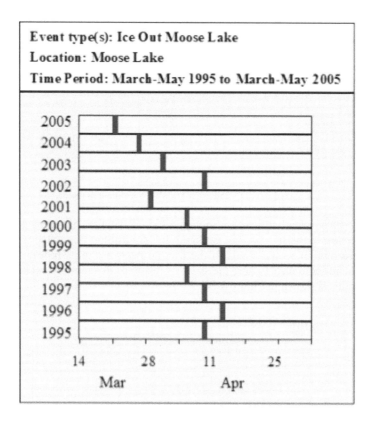

Figure 4.15. An event band stack configured to examine the repeat cycle of ice out events on Moose Lake.

4.7 Conclusion and future developments

Geographic information systems have been useful tools for investigation of spatial patterns but have suffered from a lack of ability to explore the dynamic aspects of geographic phenomena. An event-based view provides the foundation for the analysis of dynamic phenomena. Events as explicit representations of change with associated attributes of change such as rate of change, rate constancy, etc., provide the key units for exploration and analysis of the mechanisms of change. Events can be explored and analysed for various spatial and temporal patterns and for hypothesized causal relationships. The additional pragmatic benefit of an event view is that events provide a basis for the integration of information from heterogeneous spatiotemporal data streams that are currently challenging to integrate due to diverse spatial and temporal sampling protocols. An event-based model provides a starting point but much further work is needed to develop information system support and suites of new analytical tools to handle event objects. Event-based systems require new indexing strategies, new query languages and new visualisation methods and analytical tools. The event viewer presents one possible approach but cannot fully address the spatial detail of movement and shape change events. Additionally new analysis and statistical methods are needed to test the significance and similarities among event patterns.

Acknowledgements

This material is based upon work supported by the National Science Foundation under Grant No. 0429644.

References

Abraham, T. and Roddick, J. (1999) 'Survey of spatio-temporal databases', *Geoinformatica*, vol. 3, pp. 61–69.

Agouris, P. and Stefanidis, A. (2003) 'Efficient summarization of spatio-temporal events', *Communications of the ACM*, vol. 46, no. 1, pp. 65–66.

Agouris, P., Gyftakis, S. and Stefanidis, A. (2001a) 'Quality-aware deformable models for change detection', *IEEE-ICIP01 (International Conference on Image Processing)*, Thessaloniki, vol. 2, pp. 805–808.

Agouris, P., Stefanidis, A. and Gyftakis, S. (2001b) 'Differential snakes for change detection in road segments', *Photogrammetric Engineering and Remote Sensing*, vol. 67, no. 12, pp. 1391–1399.

Agrawal, R., Psaila, G., Wimmers, E. L. and Zait, M. (1995) 'Querying shapes of histories', *Proceedings 21st VLDB Conference*, Zurich, Switzerland, 11–15 Sept, pp. 502–514.

Armstrong, M. P. (1988). 'Temporality in spatial databases', *GIS/LIS 88 Proceedings: Accessing the World*, vol. II, Falls Church, VA, USA: American Society for Photogrammetry and Remote Sensing, pp. 880–889.

Beard, K. (2004) 'A spatio-temporal exploratory framework for events', *Proceedings of the Third International Conference on GIScience*, Springer Lecture Notes.

Becker, R. and Cleveland, W. (1996) 'The design and control of Trellis display', *Journal of Computational and Statistical Graphics*, vol. 5, pp. 123–155.

Chrisman, N. (1998) 'Beyond the snapshot: Changing the approach to change, error, and process', in Egenhofer, M. J. and Golledge, R. G. (eds.), *Spatial and Temporal Reasoning in Geographic Information Systems*, pp. 85–93, New York, NY: Oxford University Press.

Claramunt, C. and Thériault, M. (1995) 'Managing Time in GIS: an event-oriented approach', in Cliffford, J. and Tizhilin, A. (eds.) *Recent Advances in Temporal Databases*, pp. 23–42, Berlin: Springer Verlag.

Frank, A. (2001) 'Socio economic units. Their life and motion', in Frank, A., Raper, J. and Cheylan, J-P. (eds.) *Life and Motion of Socio-Economic Units*, vol. 8, GISDATA, ch. 2, pp. 21–34, London, UK: Taylor and Francis.

Guralnik, V. and Srivastava, J. (1999) 'Event detection from time series data', *Proceedings of the 5th ACM SIGKDD International Conference on Knowledge Discovery and Data Mining*, San Diego, USA, pp. 33–42.

Hagerstrand, T. (1970) 'What about people in regional science', *Papers of the Regional Science Association*, vol. 24, pp. 7–21.

Hawkins, D. (1976) 'Point estimation of the parameters of piecewise regression models', *Applied Statistics*, vol. 25, no. 1, pp. 51–57.

Höppner, F. (2002) 'Time Series Abstraction Methods -- A Survey', in *Proceedings GI Jahrestagung Informatik, Workshop on Knowledge Discovery in Databases*, pp. 777–786, Dortmund: LNI.

Hornsby, K. and Egenhofer, M. (2000) 'Identity based change: a foundation for spatio-temporal knowledge representation', *International Journal of Geographical Information Science*, vol. 14, no. 3, pp. 207–224.

Kim, J. (1976) 'Events as property exemplification', in Brand, M. and Walton, D. (eds.) *Action Theory*, pp. 159–177, Dordrecht: Redal.

Langran, G. (1992) *Time in Geographic Information Systems*, London:Taylor and Francis.

Laube, P. and Imfeld, S. (2002) 'Analyzing relative motion within groups of trackable moving point objects', in Egenhofer, M. and Mark, D. (eds.), *Proceedings GIScience 2002*, pp. 132–144, Berlin: Springer Verlag,.

Mark, D. (1998) 'Geospatial lifelines', in *Integrating Spatial and Temporal Databases*, vol. 98471 Dagstuhl Seminars.

Miller, H. (2003) 'What about people in geographic information science', *Computers, Environment and Urban Systems*, vol. 27, no. 5, pp. 447–453.

Partsinevelos, P., Stefanidis, A. and Agouris, P. (2001) 'Automated spatiotemporal scaling for video generalization', *IEEE-ICIP '01 (Int. Conf. on Image Processing)*, Thessaloniki, Greece, vol. 1, pp. 177–180.

Peuquet, D. (1994) 'It's about time - a conceptual framework for the representation of temporal dynamics in geographic information systems', *Annals of the Association of American Geographers*, vol. 84, pp. 441–461.

Peuquet, D. and Duan, N. (1995) 'An event-based spatiotemporal data model (ESTDM) for temporal analysis of geographical data', *International Journal of Geographic Information Systems*, vol. 9, pp. 7–24.

Quine, W. (1985) 'Events and reification', in LePore, E. and McLauglin, B. (eds.) *Actions and Events. Perspectives in the Philosophy of Donald Davidson*, pp. 162–171, Oxford: Blackwell.

Roshannejad, A. and Kainz, W. (1995) 'Handling Identities in Spatio-Temporal Databases', *ACSM/ASPRS Annual Convention & Exposition Technical Papers*, Bethesda:ACSM/ASPRS, vol. 4, pp. 119–126.

SHOE (n. d.), *General Ontology*, [Online], Available: www.cs.umd.edu/projects/plus/SHOE/onts/general1.0.html [11 Oct. 2005].

Sinton, D. (1978) 'The inherent structure of information as a constraint to analysis: Mapped thematic data as a Case Study', in Dutton, G. (ed.) *Harvard Papers on Geographic Information Systems*, Reading, MA: Addison-Wesley.

Stefanidis, A., Eickhorst, K., Agouris, P. and Partsinevelos, P. (2003) 'Modeling and comparing change using spatiotemporal helixes', in Hoel, E. and Rigaux, P. (eds.), *ACM-GIS '03*, pp. 86–93, New Orleans: ACM Press.

Tufte, E. (1983) *The Visual Display of Quantitative Information*, Cheshire, CT: Graphics Press.

Worboys, M. F. (2005). 'Event-oriented approaches to geographic phenomena', *International Journal of Geographical Information Science*, vol. 19, no. 1, pp. 1–28.

Worboys, M. and Hornsby, K. (2004) 'From objects to events: GEM, the geospatial event model', *Third International Conference on GIScience*, pp. 327–344, Springer Lecture Notes.

Yuan, M. (1997) 'Modeling Semantic, Spatial and Temporal Information in a GIS', in Craglia, M. and Couclelis, H. (eds.) *Progress in Transatlantic Geographic Information Research*, pp. 334–347.

Chapter 5

nen, A Process-oriented Data Model

Femke Reitsma[1] and Jochen Albrecht[2]

[1]Institute of Geography, School of Geosciences, The University of Edinburgh, Scotland

[2]Department of Geography, Hunter College, City University of New York, USA

5.1 Introduction

Thus far, GIScience has lacked an appropriate data model to represent processes; processes such as erosion, migration and pollution dispersal. The need for extending geographic representations for processes has been recognised in GIScience literature (Peuquet, 2001; Raper, 2000; Worboys, 2001) and acknowledged as a key goal in the University Consortium of GIS's (UCGIS) research agenda (McMaster and Usery, 2005). Yuan et al. (2005, p. 132) posit that 'As the conceptual core of a geographic information system, geographic representations determine what information is available for communication, exploration and analysis. Hence, research in extensions to geographic representations is critical to advancing geographic information science'. In order to investigate change in space and time, the theme of this book, we need to be able to explicitly represent change as it occurs.

Existing theories and data models for simulating processes focus on representing the state of the represented system at a moment in time. The future pattern of global temperature from a global climate change model or the distribution of humans in an agent-based simulation of disease spread, for example, only provides information about the status of the attributes of the system at each step of the simulation, attributes such as temperature or agent health at a particular location. Information about the processes defined in the model is typically not expressed or represented in any form. In utilising a process-oriented data model, we gain the advantage of being able to query, analyse and visualise processes.

This chapter presents a new process-oriented data model called *nen*, which can be used to represent process information. The application of the *nen* data model to process modelling offers a set of modelling results that is complementary to those of traditional models. Its novelty is the provision of a new epistemological window on the modelled results, allowing for new process-oriented queries and analysis. The data model is applied to a small watershed modelling test case, which provides initial scope for simulating geographic processes with the new data model. In what

Dynamic and Mobile GIS: Investigating Changes in Space and Time. Edited by Jane Drummond, Roland Billen, Elsa João and David Forrest. © 2006 Taylor & Francis

follows, Section 5.2 describes current approaches to theorising and representing processes in GIScience, forming a framework for discussion of the new data model. Section 5.3 presents an alternative approach, describing the new data model, which is then applied with a prototype implementation of a watershed runoff model in Section 5.4. The results of the *nen*-based approach are then discussed in Section 5.5, followed by consideration of validation of models and results of this method in Section 5.6. Section 5.7 concludes the chapter.

5.2 Process theories and models

Current research into dynamic phenomena in GIScience has focused on the representation of object states at each moment of time and over time. This is built upon long-standing theories defining the entities that populate or compose space and time. What is meant by object, are those things that we typically identify and categorize as existing at an instant of time, such as trees, mountains, barrier islands and political boundaries. These are the things dominating metaphysics (Hartshorne, 1998; Rosenthal, 1999), as well as GIScience ontologies (for example, Casati et al., 1998; Fonseca and Egenhofer, 1999; Smith and Mark, 1998; Thomasson, 2001). Spatiotemporal research in GIScience has consequently focused on the dynamics of these entities, i.e. connecting the states of these entities over time (e.g. Tryfona and Pfoser, 2001), or exploring the relationships between objects and the processes that modify them (Bittner and Smith, 2003; Tomai and Kavouras, 2004).

As a consequence of the focus on static objects, data models for dynamic phenomena centre on state changes of objects. For example, a global climate change model, while containing process information in the model structure, does not represent or store this information for analysis; rather, the states of the climate system are stored at each instant of time. There is no data object that represents a geographic process that changes over space and time (Yuan et al., 2005). This results in a loss of information about the modelled process, which cannot accurately be regained by interpolating between time slices. For example, in global climate modelling virtually the same future state of increased temperature can be modelled as a result of two very different changes to the model, an increase in solar luminosity or an increase in CO_2. It is not immediately obvious which process or processes, such as heat transport or a change in cloud optical depth, caused these results.

The static roots of GIS are found in its cartographic origins, which have formed the intellectual framework for much of GIScience research (Kuhn, 2001; Yuan et al., 2005). Kuhn (2001) notes a number of other reasons for such object orientation in geographic and other information systems, including:

❑   an emphasis on attributes and relationships rather than process and change,

❑   the weakness of logic-based formal languages in dealing with operations and their semantics,

❑   and a presumed priority of objects in human (spatial) cognition.

5.3 An alternative process data model

An alternative data model for the representation of processes is presented in this chapter, which provides advantages in querying, analysis and exploration of process descriptions under computer simulation conditions - or *in silico*. The data model is referred to as a *nen* because its simplest and most abstract graph representation is a node-edge-node triple (Figure 5.1). This simple point process representation, which was used for the watershed prototype described in Section 5.4, can be extended to larger spatial entities, as might be represented by a polygon (Figure 5.2).

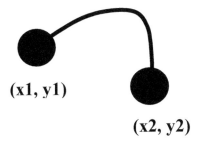

(x1, y1)

(x2, y2)

Figure 5.1. Process representation for point feature.

A more comprehensive representation is in form of a tuple: (x1, y1, x2, y2, t, st, {a1, a2,...}, {r1, r2, ...}). The spatial location of the process is identified by x1, y1, x2, y2, which expresses the spatial extent of the process. The temporal location of the process is defined by t, where a process is represented on a single layer of spatial information rather than lost between time slices. The st represents the spatiotemporal granularity of the process, which may be a function of the amount of energy that initiates the process. For example, given some threshold breaking push, the spatiotemporal granularity expresses how far and over what time period the process will operate in response to that push. The set {a1, a2, ...} defines the attributes of the process. The set {r1, r2, ...} defines the rules of the process that govern its dynamics and interaction with other processes. For example, a set of rules for modelling the process of sediment transport in the longshore may define the spatiotemporal extent of an instance of that process as 5m/hour, depending on various relationships it holds between other processes operating in the nearshore.

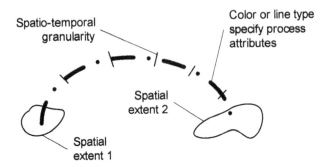

Figure 5.2. Process representation for area feature.

This data model provides a new epistemological window on geographic processes. Simulating processes with a process data model allows us to ask questions that are not directly answerable with current object-centred formulations, which focus on the states of a system that result from the operation of processes. Our new data model allows us to ask questions such as:

❑   Where is a process operating at a particular instant of time?
❑   How has the process changed over time?
❑   What process(es) caused another process to occur?

The answers are not inferred (or interpolated) but are explicitly stored as part of running the process model. How the rules of the process affect the spatial dynamics of the process may therefore also be better explored.

5.4 Watershed modelling application

The theory of taking process as a representational primitive has been prototyped with a watershed model within a simulation environment called Flux.

5.4.1 Simulator

Flux is written in Java and inherits and extends a number of basic operating classes from the RePast (Recursive Porous Agent Simulation Toolkit) library, which is an open source agent-based modelling environment created by Social Science Research Computing at the University of Chicago[1]. RePast is primarily used for its display and scheduling classes, and also has the advantage of containing Java classes for importing GIS raster data (ESRI ASCII raster files). Flux contains a set of interfaces and default classes that define the basic structure of the process model, including methods that must be implemented by an inheriting domain model. The

---

[1] http://repast.sourceforge.net/

objective was to maximise generic functionality within the Flux classes, thereby minimising the code to be developed within the domain model. The output of a simulated model is stored in text files, which can then be queried with a query tool that was developed as part of the initial steps towards process analysis. For a full description of the simulator, see Reitsma and Albrecht (forthcoming).

Figure 5.3 presents a sample simulation using the Flux simulator. Each *nen*, represented by a node-edge-node tuple (as depicted in Figure 5.1), indicates an instance of groundwater flow. The raster backdrop is a digital elevation model of a small sub-watershed, where lighter shades represent higher elevation. At each time step, groundwater flows towards the North-Western corner of the sub-watershed.

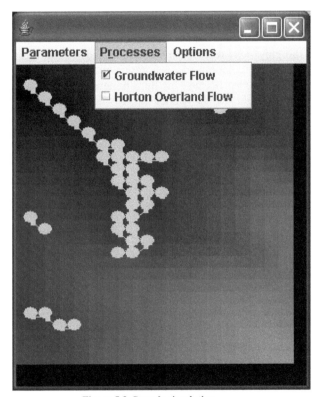

Figure 5.3. Sample simulation.

5.4.2 Model and simulation

For the purposes of testing the methodology a simple watershed model was simulated. The model included the following restricted set of processes: Hortonian overland flow, groundwater flow, infiltration, percolation, saturation excess runoff and surface ponding. The data used to define the parameters for the simulation are taken from the Reynolds Creek Experimental Watershed (RCEW), which is a high-quality long-term dataset created by the U.S. Department of Agriculture Agricultural Research Service's Northwest Watershed Research Center in Boise, Idaho, United States. For a full description of the RCEW, see the special issue of Water Resources Research introduced by Marks (2001).

At each hourly time step the precipitation input is updated, which initiates one of three processes, Hortonian overland flow, infiltration or surface ponding. Each process type has a set of rules defining its behaviour. For example, if the precipitation exceeds the infiltration capacity of the soil and depending on the slope characteristics, an instance of Hortonian overland flow will be generated. Although hydrologically limited, the example explores the advantages of the methodological approach of considering process as a data modelling primitive.

Two time slices of the simulation are presented in Figure 5.4. The black *nens* represent the process of Hortonian overland flow, the dark grey nens represent infiltration, the grey *nens* represent percolation, and the light grey *nens* represent groundwater flow. Percolation and infiltration processes are represented by two nodes on top of each other because the third dimension is not represented. With the *nen* data model, insight can be gained as to where and when certain processes dominate, which may lead to a better understanding of the modelled system and give guidance to better ways of interacting with that system. For example, in Figure 5.4 it is evident that the process of Hortonian overland flow dominates in certain upland parts of Upper Sheep Creek. This is in contrast to typical approaches to modelling that generate results expressing where some energy or mass is at an instant of time within the system, such as water in our watershed, with no information of the processes that caused that state.

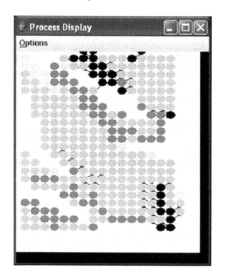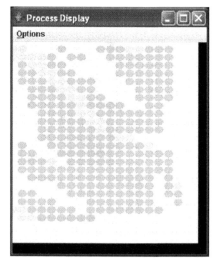

Figure 5.4. Simulation at two time steps, in progressive order from left to right.

5.5 Analysis of results

Without an appropriate data model to represent processes, we cannot easily analyse or visualise the dynamics and interactions of processes for the purpose of understanding the modelled system. Because the *nen* data model represents a process as a spatially extended entity at any moment in time, not only can its state be analysed but also its dynamics. In addition, due to the structure of the data model, namely two nodes connected by an edge, network analysis may also result in new insights into the model results. This may be of particular interest in recording the interaction of processes and provide new patterns of process relationships to be explored and classified, as will be discussed below.

5.5.1 Process state and change

As will be discussed further below, the state information of a process includes all of the components of the data structure, namely:

- the spatial location (x1, y1, x2, y2),
- the temporal location (t),
- the spatiotemporal granularity (st),
- the attributes ({a1, a2,...}),
- and the rules ({r1, r2, ...}).

Furthermore, from the data structure the direction and velocity of the process may be derived. Each of these aspects of the state of the process can be temporally extended such that processes can be queried for change. For example, the change in direction of groundwater flow or change in the mass of water involved in this process can be queried.

The location of individual or interacting processes can be analysed spatially, spatiotemporally or temporally. Discovering spatial, spatiotemporal or temporal clusters of processes may provide new insights into thresholds and critical combinations of processes. Spatial clusters of processes may indicate the dominance of processes in certain locations over time, such as erosion on a certain part of a hill slope. Spatiotemporal clusters of processes are the spatial clustering of processes at certain times, where we may use different notions of time, such as linear or cyclic; for example, analysing the results of our model may result in findings of new large-scale recurrent weather patterns such as El Nino. Modelled processes might be widely distributed with no evident spatial pattern, yet we might find temporal clusters that indicate that these processes are temporally correlated in some way; for example, ocean thermohaline circulation has a significant effect on global climate change (Knutti et al., 2004). In these three cases, we may find interesting new patterns among process instants of the same type or among different processes.

The attributes of the modelled processes can be analysed for variations in magnitude, or specific values of interest. Certain magnitudes may dominate in certain types of processes or be correlated with other processes. The dynamics of the magnitude of groundwater in the process of groundwater flow, for example, may be of interest in understanding the impact of soil structure on groundwater flow.

The rules of the process may also be of interest for analysis. Although typically the rules or mathematical functions defining the behaviour of the process are static, they may also be evolutionary. Genetic algorithms, for example, allow us to evolve rules. We may find that certain types of rules dominate, or particular patterns of rules or cycles of rules may develop.

Because the data model is spatiotemporally extended, the difference between one location and the other can be used to provide information on direction and velocity of processes. Determining the average direction and average velocity of a certain type of process may be of particular import to analysing and understanding climate processes. The direction and velocity of climate processes, for example, may be correlated with certain types of erosion or vegetation growth processes at a certain location. They also assist in the identification of when model rules need to change as small-extent *nens* move into a new geographic regime; the effect of tropical hurricanes on previously unaffected deciduous forests as a result of large-extent global warming would be an example for that.

Each of these dimensions, location (spatial, temporal and spatiotemporal), direction, velocity, attributes and rules can be combined, as is reflected in Figure 5.5. Some of these variables may be held constant, others may vary. The example provided in the figure illustrates a case where analysis is undertaken on the relationship between direction and attributes of a process. A yet unresolved challenge is how we visualise all of these dimensions of analysis, either individually or combined.

	Location	Direction	Velocity	Attributes	Rules
Location	■				
Direction		■		X	
Velocity			■		
Attributes				■	
Rules					■

Figure 5.5. Matrix of dimensions of process analysis.

5.5.2 Process interaction and causality

In order to analyse the interaction of processes, the data model has another advantage of supporting network analysis. Network analysis describes the structure of a network based on the number of nodes, links and the attributes associated with the nodes and/or links. It includes a large range of measures that are applied in fields as disparate as sociology (e.g. Wasserman and Faust, 1994) and physics (e.g. Dorogovtsev and Mendes, 2002). The network described by *nens* may be of a single type of process, such as Hortonian overland flow, or of a collection of different processes, such as those operating within a watershed. Analysing the network of *nens* allows us to explore the relationships among processes. The application of network analysis to networks of interacting processes may provide new measures of process patterns, and perhaps, as with recent discoveries of patterns in animate and inanimate networks (Barabasi, 2002), new insights into the systems that we model.

Tracing the complex interactions among processes of different types in our model also allows us to monitor causality. In Figure 5.6, for example, five interacting processes are schematically displayed, with the x-axis defining the temporal extent and the y-axis a set of discrete rules. The interaction of processes is indicated by spatial coincidence of some part of the *nen* data model representing the process. In this figure: *nen* 1 interacts with *nen* 2 according to rules 4 and 5; the process represented by *nen* 1 is followed by *nen* 3, which is followed by *nen* 4, this is evident by the (x2, y2) of *nen* 1 being equivalent to the (x1, y1) of *nen* 3, and the (x2, y2) of *nen* 3 being the same as the (x1, y1) of *nen* 4; *nen* 2, *nen* 3, and *nen* 4 interact with the long-term process *nen* 5; *nen* 3 starts as a point process and ends as an area process.

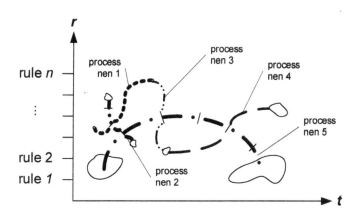

Figure 5.6. Five interacting processes.

5.6 Validation of model and results

As with analysis, without an appropriate data model we cannot easily validate the spatial behaviour of our modelled processes. For example, while a lumped hydrological model may produce a hydrograph that concurs with the measured discharge of the watershed, all of that modelled discharge may have resulted from Hortonian overland flow, whereas in reality it may have been a mixture of processes such as groundwater flow and saturated excess flow. Without a data model to represent these processes, we cannot easily tell which processes caused the final modelled state. This problem is well recognised by watershed modellers as that of equifinality, which describes the situation where the same system state can result from many different sets of processes (Bevan, 2000).

The *nen* data model allows us to visualise and analyse the dynamics of the processes in the model, facilitating the validation of the definitions, in rules of mathematical formulas, of the processes in the model. Furthermore, the *nen* provides the basis for testing and comparing different definitions of processes. By visualising and measuring how descriptions of processes within the model compare to other definitions and known spatial dynamics of processes, modellers can test whether their mathematical or rule-based formalisms act in expected and realistic ways.

A process data model also enhances the ability to compare models, lending itself to model inter-comparison studies. The *nen* allows us to compare distribution, quantity and dynamics of processes among models. This contrasts with traditional approaches to model inter-comparison, which analyse the state of the modelled system at the end of the simulation or over specified time steps (for example Dutay et al., 2002).

In validating the results of a *nen*-based model, however, difficulty lies in the lack of qualitative or quantitative descriptions of geographic processes. The results of a model are validated by matching the output of the model with the real world, a good result being the ability to mirror that world *in silico*. Typically a model is validated by comparing the final simulated system state, with the real system at the same point in time. In order to validate the results of a simulation using the *nen* data model we need long-term empirical observations of the simulated processes. As with the data and literature on the RCEW used in the watershed application and described in Section 4, such process data is rarely if ever available. Without process observations, any simulation using the *nen* data model cannot be effectively validated.

5.7 Conclusion and future developments

The lack of appropriate data against which to validate process definitions and results of a *nen*-based model leads to questions of how we might go about observing and measuring processes in the field. Qualitative descriptions of processes, while available in certain cases, will always need to be quantified in some manner in order to provide a basis for comparison and formal analysis. Quantitative measurement devices also facilitate automation of analysis and validation. We do not know of any measurement approach that quantitatively records process information, which suggests there is a need for new data collection techniques that collect such information for comparison against model results. Furthermore, data theory needs to be developed, that is, new approaches to transforming real-world observations into something that can be analysed (Jacoby, 1991).

Currently the flux simulation environment is constrained to small models due to problems of computational complexity. To use this approach for models of larger spatial scale and of greater detail would require a significant rewrite of the software and consideration of advanced methods for accessing larger-scale computing resources. Alternatively, it should not be difficult to modify existing modelling software environments to implement the *nen* data model. However, given the potential of data models to shed new light on simulations, new open and flexible modelling platforms are needed that can easily incorporate new data models and new analytical and visualisation methods. Such a platform would provide a useful scientific environment for not only testing new models but testing new simulation methodologies.

In summary, describing the world as a set of processes requires a data model for modelling and simulating it as a set of processes. GIScience has thus far lacked a process-based data model for dynamic modelling, which has limited our analysis capabilities to information defining the state of a system at any instant of time rather than the processes that are specified in the model. The proposed data model, the *nen*, provides such an opportunity, resulting in new leverage gained in query and analysis of simulation results. These results have provided insight into the spatial dynamics of the simulated processes, and the data model will allow for the future exploration and testing of causal interactions among processes. The *nen* data model also provides new scope for visualisation and analysis of spatial processes, which is a goal of our continuing research.

References

Barabasi, A. (2002) *Linked: The New Science of Networks,* Cambridge, MA: Perseus Publishing.

Bevan, K. J. (2000) 'Uniqueness of place and process representations in hydrological modelling', *Hydrology and Earth System Sciences*, vol. 4, pp. 203–213.

Bittner, T. and Smith, B. (2003) 'Granular Spatio-temporal Ontologies', *AAAI Spring Symposium on Foundations and Applications of Spatio-Temporal Reasoning (FASTR),* Palo Alto, USA.

Casati, R., Smith, B. and Varzi, A. C. (1998) 'Ontological tools for geographic representation', in Gaurino, N. (ed.) *Formal Ontology in Information Systems*, pp. 77–85, Amsterdam, The Netherlands: IOS Press.

Dorogovtsev, S. N. and Mendes, J. F. F. (2002) 'Evolution of networks', *Advances in Physics*, vol. 51, pp. 1079–1187.

Dutay, J. C., Bullister, J. L., Doney, S. C., Orr, J. C., Najjar, R., Caldeira, K., Campin, J. M., Drange, H., Follows, M., Gao, Y., Gruber, N., Hecht, M. W., Ishida, A., Joos, F., Lindsay, K., Madec, G., Maier-Reimer, E., Marshall, J. C., Matear, R. J., Monfray, P., Mouchet, A., Plattner, G. K., Sarmiento, J., Schlitzer, R., Slater, R., Totterdell, I. J., Weirig, M. F., Yamanaka, Y. and Yool, A. (2002) 'Evaluation of ocean model ventilation with CFC-11: comparison of 13 global ocean models ventilation with CFC-11: comparison of 13 global ocean models', *Ocean Modelling*, vol. 4, pp. 89–120.

Fonseca, F. and Egenhofer, M. (1999) 'Ontology-Driven Geographic Information Systems', *7th ACM Symposium on Advances in Geographic Information Systems*, Kansas City, USA.

Hartshorne, C. (1998) 'The development of process philosophy', in Browning, D. and Myers, W. T. (eds.) *Philosophers of Process*, pp. 391–407, New York: Fordham University Press.

Jacoby, W. G. (1991) *Data Theory and Dimensional Analysis*. Sage University Paper Series on Quantitative Applications in the Social Sciences, no. 78, London: SAGE Publications.

Knutti, R., Fluckiger, J., Stocker, T. F. and Timmermann, A. (2004) 'Strong hemispheric coupling of glacial climate through freshwater discharge and ocean circulation', *Nature*, vol. 430, pp. 851-856.

Kuhn, W. (2001) 'Ontologies in support of activities in geographic space', *International Journal of Geographic Information Science*, vol. 15, pp. 613-631.

Marks, D. (2001) 'Introduction to special section: Reynolds Creek experimental watershed', *Water Resources Research*, vol. 37, no. 11, pp. 2817.

McMaster, R. and Usery, E. L. (eds.) (2005) *A Research Agenda for Geographic Information Science,* New York: CRC Press.

Peuquet, D. J. (2001) 'Making space for time: Issues in space-time data representation', *Geoinformatica,* vol. 5, pp. 11–32.

Raper, J. (2000) *Multidimensional Geographic Information Science,* New York: Taylor & Francis.

Reitsma, F. and Albrecht, J. (forthcoming). 'Implementing a new data model for simulating processes', *International Journal of Geographical Information Science.*

Rosenthal, S. B. (1999) 'Contemporary metaphysics and the issue of time: Re-thinking the "great divide"', *International Philosophical Quarterly,* vol. 39, pp. 157–171.

Smith, B. and Mark, D. M. (1998) 'Ontology and Geographic Kinds', *8th International Symposium on Spatial Data Handling (SDH'98),* pp. 308–320, Vancouver: International Geographical Union.

Thomasson, A. L. (2001) 'Geographic objects and the science of geography', *Topoi,* vol. 20, pp. 149–159.

Tomai, E. and Kavouras, M. (2004) 'From "Onto-GeoNoesis" to "Onto-Genesis": The design of geographic ontologies', *Geoinformatica,* vol. 8, pp. 281–298.

Tryfona, N. and Pfoser, D. (2001) 'Designing Ontologies for Moving Objects Applications', *Workshop on Complex Reasoning on Geographical Data,* Cyprus, [1 Dec 2001].

Wasserman, S. and Faust, K. (1994) *Social Network Analysis: Methods and Applications,* Structural Analysis in the Social Sciences, Cambridge: Cambridge University Press.

Worboys, M. F. (2001) 'Modelling changes and events in dynamic spatial systems with reference to socio-economic units', in Frank, A. U., Raper, J. and Cheylan, J-P. (eds.) *Life and Motion of Socio-Economic Units,* pp. 129–138, London: Taylor and Francis.

Yuan, M., Mark, D., Egenhofer, M. and Peuquet, D. (2005). 'Extensions to geographic representations', in McMaster, R. and Usery, E. L. (eds.) *A Research Agenda for Geographic Information Science,* pp. 129–156, New York: CRC Press.

Chapter 6

Comparing Map Calculus and Map Algebra in Dynamic GIS

Mordechai (Muki) Haklay

Department of Geomatic Engineering, University College London, England

6.1 Introduction

The integration of temporal information into Geographical Information Systems (GIS) has been the subject of extensive research for many years (Bergougnoux, 2000; Egenhofer and Golledge, 1998). This intense research effort stems from the inherent contradiction between GIS data models (be it raster or vector) and computer representations of dynamic processes. Due to their cartographic roots, data models in GIS have been designed to capture a static snapshot of reality (Albrecht, forthcoming). Thus, typical representations in GIS, where each layer is presented as a single file (such as those described in Tomlin, 1990), are geared towards describing the state of the study area at a single point in time. Over the years, representations that deal with temporal changes have been developed, especially within the context of spatial databases, where the changes can be handled at the feature level, instead of a monolithic handling of a whole layer (Worboys and Duckham, 2004).

Representations have been developed for dynamic phenomena where the challenges stem from the rate of updates and the need to visualise changes rapidly. Indeed, many solutions have been devised to deal with the dynamic aspect of GIS. In the vector model, for example, dynamic segmentation has been developed to allow the representation of changing events along static vector features. This representation is especially common in transport applications of GIS (Longley et al., 2001). In the raster model, dynamic modelling capabilities have been developed and implemented within packages such as PCRaster (Van Deursen, 1995) or IDRISI (Park and Wagner, 1997).

Despite these developments, the modelling of dynamic entities and processes in a GIS is still an active research issue (Couclelis, 2001; Laurini, 2001). The reason for this continued interest, as Laurini (2001) noted, is that there are many applications in which real-time dynamic representation is required. These applications range from environmental monitoring of pollutants to the management of a vehicle fleet. The recent advances in real-time location tracking, communication, digital mapping availability and the continued increase in computing power make these types of applications feasible, at least technically. As the technical challenges of

Dynamic and Mobile GIS: Investigating Changes in Space and Time. Edited by Jane Drummond, Roland Billen, Elsa João and David Forrest. © 2006 Taylor & Francis

implementing dynamic GIS diminish, researchers are now free to focus on the theoretical and conceptual challenges of the integration of temporal and dynamic aspects within GIS in novel ways.

This chapter focuses on Map Calculus (Haklay, 2004) and its potential applications in dynamic GIS. Map Calculus is an alternative to current representations in GIS, and is based on the use of function-based layers in a GIS (Haklay, 2004). A function-based layer is defined as the symbolic representation of a mathematical and spatial function. Map Calculus is best explained by comparing its core concepts to the current practice of representing surfaces in GIS in grids (rasters). The use of functional representation of layers existed in computer models in meteorology for many years (Goodman, 1985) and is being used in some global climate models, but it was not adopted in GIS and spatial analysis. The main strength of the new representation is the ability to treat analytical layers (layers that are based on manipulation of real-world observations) in their symbolic form, in a similar way to the manipulation of mathematical functions in software packages such as MATLAB. This can increase the GIS analytical toolbox and open up new directions in spatial analysis research.

In this chapter, the application of Map Calculus for dynamic GIS is examined and explained through the comparison with Tomlin's (1990) Map Algebra and Cartographic Modelling. The reason for this comparison is the link between Map Calculus and Map Algebra at the conceptual level, as explained in Haklay (2004), and the long use of Map Algebra and Cartographic Modelling (Tomlin and Berry, 1979) in environmental modelling and in dynamic GIS. Of course, dynamic GIS can be implemented in vector-based or object-based representations. However, the comparison of Map Calculus to these representations is beyond the scope of this chapter.

This chapter opens with a general comparison of Map Calculus and Map Algebra and Cartographic Modelling using an interpolation function and a simplified environmental model. Through these examples, the main principles of Map Calculus are explained and clarified. The following section moves to discuss the challenges of dynamic modelling in GIS, exploring the ways in which it is implemented in Map Algebra (Tomlin, 1990) and in PCRaster (Van Deursen, 1995) and outlining how such models can be implemented in a Map Calculus-based system. The chapter ends with conclusions and future directions for research.

6.2 Comparing map algebra and map calculus

6.2.1 Implementing spatial interpolation

The comparison of Map Calculus with Map Algebra provides a way to explain the main principles of Map Calculus, by allowing the reader to contrast them with the more familiar procedures of Map Algebra and Cartographic Modelling. For a detailed conceptual outline of Map Calculus see Haklay (2004). To make the comparison concrete, two common procedures in GIS are used here: the creation of an interpolated surface, using an Inverse Distance Weighted (IDW) function and the implementation of a spatial model through overlay functions. Naturally, these two examples do not reveal the full range of GIS operations that are available under Cartographic Modelling and Map Algebra (Tomlin, 1990) which include local, neighbourhood and zonal operations. However, within the confines of this chapter, the two examples set the scene for the discussion of dynamic models in the next section.

IDW is a common interpolation method and it is used widely within GIS. Like all interpolation functions, IDW operates on a set of sampled points (L1,L2,...Ln) and calculates the value for a new location L' by using the following equation:

$$L' = \frac{\displaystyle\sum_{i=1}^{n} \frac{1}{d_i^{\,p}} L_i}{\displaystyle\sum_{i=1}^{n} \frac{1}{d_i^{\,p}}} \tag{6.1}$$

where d_i is the distance from L' to the location L_i, and p is a power of the distance. Usually, the search radius is taken as a parameter of the function to limit the influence of remote data points. It is noteworthy that implementation of IDW function has been used for GIS research since its early days (for example Shepard, 1968).

In a GIS where Tomlin's Cartographic Modelling is implemented, IDW will be calculated in the following way. First, the user selects the spatial extent of the area for interpolation. Next, the user sets the spatial resolution (pixel size) of the grid that will be used to store the result of the IDW function. The next stage includes the main computational step – for each pixel, the computer takes the coordinates of the centre of the pixel, and uses them to calculate the IDW value for the cell. This is done by selecting data points from the search radius and including them in the calculations. The final value is stored in the pixel. Once all the values for all pixels have been calculated, the system writes the grid file and stores it on a mass storage unit – usually a hard disk – for future use. This process is represented in Figure 6.1(A).

In Map Calculus-enabled GIS, when the user requests the GIS to calculate the function, the system will register the manipulation in a symbolic form. If the point set is the layer "Height values", then the system will register a new layer as:

$$\text{IDW("Height values", } P_1 \ \dots \ P_n) \tag{6.2}$$

where P_1 to P_n are the parameters needed for this instance of the generic IDW formula. These will include search radius, the number of points that can be included in the computation, etc. The procedure for the definition of a function-based layer does not require any computation, but only the functionality to record the fact that the user made a decision to apply the function f to the data set x with parameter set p. Given the layer "Height values", with the field "Z" holding the actual values, search radius of 1000 units, power of 2, and a maximum of 12 neighbouring points, the internal representation of the layer can be:

```
Function -> "IDW"
Layer -> "Height Values"
Parameters -> ("Z",1000,12,2)
```

The stored definition of the layer can be used in other layers that are based on it, as explained in the next section. The computation of the function happens when the user requests the GIS to visualise the layer. When this happens, the GIS will calculate the value of the function for each pixel on the active display area of the screen. The GIS can use parameters, such as screen resolution and the current scale of the map, to minimise the amount of calculations. For example, if a user uses a common screen resolution of 1024 x 768, then the effective area of the map in a common GIS package (such as ArcGIS) is approximately 800 x 600 pixels, due to the elements of the graphical user interface which occupy the rest of the screen, such as the title bar, the status bar and toolbars. The active area requires about 480,000 calculations – not a major load on modern central processing units (CPUs).

Within the active area, each pixel's location can be calculated by using the current scale of the map, and the area on which the user is focusing. This will provide the definitions for the coordinates of each pixel. As the user zooms out or in, the scale of the map changes, and new calculations for the currently displayed area are carried out. Hence, to the system's user, a Map Calculus-enabled GIS behaves in the same way as any other GIS. This process is depicted in Figure 6.1(B).

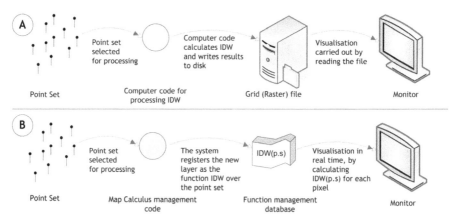

Figure 6.1. Computation of a function in a standard GIS (A) and in a Map Calculus-enabled system (B).

Another scenario for calculation occurs when the user wants to explore the grid with other (existing) grid layers. For this, Map Calculus-enabled GIS will have a separate interface that will allow the user to define the extent of the area to which the output is required and the resolution of the output grid. This grid will be produced in the same way as in standard implementations, by calculating the value for each pixel and producing a file.

6.2.2 Site suitability analysis

The major difference between Map Calculus and Tomlin's representation occurs when a set of layers is manipulated in order to complete an analytical task. For example, assume the problem that the user would like to solve is to create a suitability map for the construction of a wind farm in a study area of 10 km by 10 km. There are three wind farms in a given area, as well as five farm buildings. A data set of height values for sample points in the study area was assembled through a field survey. The suitability criteria is that the location must be within 1 km of an existing farm building, but over 1.5 km from an existing wind farm and in an area with a height over 500 m above sea level. The steps of the analysis are presented in Figure 6.2.

In a standard GIS representation it is common to use Tomlin's (1990) Map Algebra for such a task. Map Algebra provides a range of mathematical and spatial operations that can be carried out on a single layer (i.e. calculating slope) or between layers (i.e. adding cost surface to road network). This enables the user to construct sophisticated models by using these mathematical and spatial operations to merge layers. In each step of the process some operations are defined on grid layers, and, importantly, the output is a grid layer, too.

Thus, in its most generic form, the manipulation of grid layers in Map Algebra operates in the form:

```
OutputRaster = f(InRaster₁, … InRastern, P₁,  Pn)           (6.3)
```

where the function f will have 0 or more input layers (InRaster$_1$, … InRaster$_n$) and 0 or more scalar parameters (P$_1$, … P$_n$). In any function, there will be at least one input layer or parameter. The most notable aspect of this form is that each operation will result in an output layer, which will be a grid. For example, the IDW function that was described in Equation (6.3) can be represented in Map Algebra as:

```
IDWHeight = IDW("Height values", P₁ . Pn)                  (6.4)
```

Importantly, the difference between this form and the one presented in Equation (6.2) is that in Map Calculus, the computation ends in the definition of the layer, whereas in Map Algebra, the computation ends with the production of the grid layer *IDWHeight*.

Operationally, in Map Algebra, the analysis of the site suitability problem will require the creation of five or six grid layers that will be used during the process: the IDW grid, followed by a grid containing areas above the required height, two buffer grids and the final suitability grid. In some systems, Map Algebra operations are limited to binary operation, and thus the overlay will require two steps and a temporary grid. As noted, our study area is 10 km by 10 km, and therefore, the relevant area for the analysis can be calculated as follows: there are 5 farm buildings, and the new wind farm can be located within 1 km from an existing building. Thus, the area that is potentially suitable for the new wind farm is 15.7 km^2. Despite this, if the user selects pixel resolution of 5 m, the process includes up to 24,000,000 computations of which 83.3% are redundant. These excessive computations might lead to operational compromises such as a decision to increase pixel size, which reduces the overall accuracy of the model.

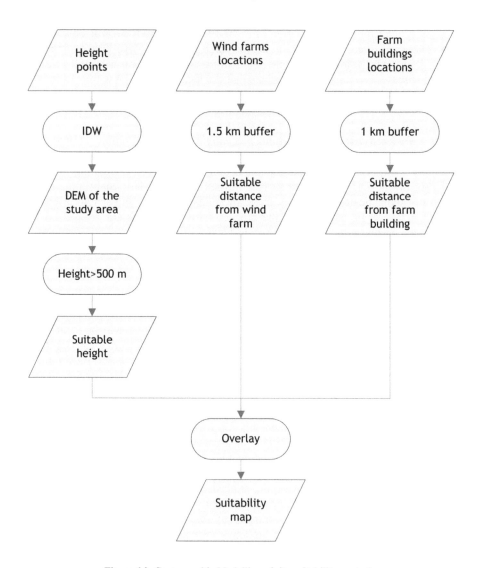

Figure 6.2. Cartographic Modelling of site suitability analysis.

It is also notable that, while the cost of digital storage has reduced dramatically in recent years, the management of multiple grid layers is still a technical and practical problem. While the latest version of ArcGIS has no limitations on the size of a grid layer, creating a grid of 23,000 x 23,000 cells of random floating point numbers will lead to an output file of 2.147 gigabytes (ESRI, 2004). Thus, a model with multiple grids can lead to very significant data volumes.

In Map Calculus, the use of symbolic representation of the layers guarantees very efficient storage. In the case of the earlier analysis, the process will include only one output and this may be produced only at the end of the modelling process. The GIS will operate as follows. First, the system will record that the user requested to perform the function IDW over the "height points" layer, as well as the buffers. The internal representation may look like (Table 6.1):

Table 6.1. Internal representation of layers in Map Calculus.

Layer ID	Layer definition
1	IDW("Height points","Z",100,12,2)
2	Buffer("Wind farms", 1500)
3	Buffer("Farms",1000)

As noted, each function includes reference to the input data set, and a list of parameters for this instance of the function. Hence, IDW holds four parameters as explained in the previous section. In the buffer functions, the parameter provides the distance in map units (metres in this case).

The next stage is to create the model. This can be represented in the symbolic form:

```
Locate(Value(IDW("Height points","Z",100,12,2))>500 &
Outside(Buffer("Wind farms", 1500)) &
Inside(Buffer("Farms",1000)))                        (6.5)
```

where *Locate* is a function that locates areas; *Value* provides a selection of output values from a surface function; and *Outside* and *Inside* are the definitions of the locations outside or inside the defined layer.

The final step is the visualisation of the map, possibly by creating an output grid file. In a Map Calculus-enabled GIS, the computation will include optimisation – before turning to calculate the more complex IDW function, the system will evaluate the other parts of the equation and find the area that incorporates layers 2 and 3. The resulting area will be the only part of the map for which IDW calculations will be carried out. The values will be compared to the value in the equation, and only areas above the required threshold will be marked on the final map.

This method of computation provides more precise results than it is possible to achieve with typical Map Algebra implementations, because of the compromise in pixel size that was explained in the previous section. This is especially relevant when a complex model is being computed, where it is more likely that users will choose a large pixel size to allow faster computation, as the selection of a small pixel size has a knock-on effect on the whole process. Another problem emerges in Map Algebra when data is computed with different pixel sizes, or when there is some shift in the origin of the grid, and therefore, the pixels from one grid do not match the pixels of the other grid accurately. In such cases, the binary operation needs to take into account the process of combining the two grids and this introduces errors into the model. All this is eliminated in Map Calculus, as the computation is carried out in the last step in a way that takes into account the full model and the specific aspects of each function that is being used in it.

6.3 Dynamic modelling in map algebra and map calculus

6.3.1 Spatiotemporal problem solving

In their original form, Cartographic Modelling and Map Algebra did not have explicit dynamic capabilities. Temporal aspects of geographical problems were translated into static representation. For example, assume that a goods delivery company is managing a vehicle fleet. The location of all vehicles is known, as they are equipped with satellite navigation systems and radio equipment. Information about the average speed in each road in the study area is provided in real time via traffic monitoring. As part of the system, the designers want to integrate the functionality to calculate the maximum travel time of each vehicle back to the warehouse. In this case, the Cartographic Model shown in Figure 6.3 applies.

The process in Figure 6.3 uses three input data sets – the location of the warehouse, the location of the vehicles and a map of the roads in the area. Following Tomlin's method (Tomlin, 1990, p. 143), the first step in the computation involves reclassifying the different roads according to their travel speed, by associating a lower value with roads on which it is possible to travel fast, and vice versa. Areas that are not part of the road network are classified as 'no data', to indicate that they should not be part of the calculations.

The map with the classified road network is then used to construct the travel-time map. For each location on the map, the distance from the warehouse is calculated. First, a network distance from the warehouse is calculated for each pixel, and information about the different classes of roads along the path is summarised. This information is then used to calculate the travel time by dividing the average speed for each type of road by the length of the segments belonging to this type on the path. The final value is then stored back in the pixel, representing the travel time from this point to the warehouse.

Once the travel time has been calculated, the location of the vehicles can be taken from their logs. Because it is assumed that the satellite navigation coordinates can give an inaccurate location of up to 30 metres, and therefore, the location can potentially be a location with "no data" value, a 50-metre buffer is created around each vehicle location to ensure that the observation is associated with a road.

The final step of the computation involves finding the pixel within the buffer with the highest time value. This procedure should be performed for each vehicle, and the output values will provide the requested answer.

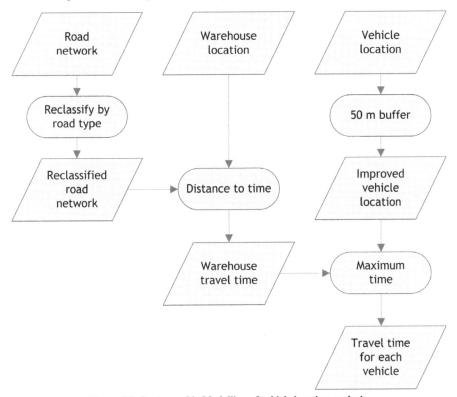

Figure 6.3. Cartographic Modelling of vehicle location analysis.

The Cartographic Modelling procedure requires the creation of four grid layers within the model. It can be assumed that there is no need to compute the road network connectivity model in each run of the model, but all the rest of the information should be computed in real time – the association of road segments with average speed, the location of the vehicles, and the shortest path of the vehicles to the warehouse. This is another class of Map Algebra functions:

```
Values = f(InRaster₁, … InRastern, P₁, … Pn)        (6.6)
```

where, as in Equation (6.3), the input is a set of grids and parameters, but the output is a set of values.

In Map Calculus, the model will be constructed from a set of building blocks, all presented as functions. The functions will include calculating a distance over a network, and calculating the nearest road segment to a given point. The layer definition will be of the form:

```
NetworkTime(Nearest("Vehicles","Roads"),
  "Warehouse","Roads","RoadTypeTable")                  (6.7)
```

where *NetworkTime* is a function that calculates the time on the network from the vehicles to the warehouse, and *Nearest* adjusts the location of the vehicles to the nearest segment on the road network. *NetworkTime* takes four parameters: origin, destination, network and a lookup table that translates the identification code of each road segment to average speed. The lookup table can be updated continuously from real-time data.

The fundamental difference between this model and the wind-farms siting case is that the current model needs to be rerun with every new input. In such situations, processing of multiple grids becomes a real hurdle in the provision of a timely response to the system's user. The main reason for the delay in response time is the redundant computations that do not contribute to the solution of the problem. This is a problem common to most Map Algebra and Cartographic Modelling implementations, where the algorithms perform the computations for all the pixels in the grid. In practice, it is unlikely that raster-based GIS will be used to solve this type of problem, and a vector-based GIS with a bespoke implementation of the *Nearest* function will be used instead of a generic buffering.

In a Map Calculus-enabled GIS the functions *Nearest* and *NetworkTime* can be optimised to ensure that they are not performing any computation that does not contribute to the final output. In this case, the optimiser can calculate first the *Nearest* function, which provides the location of the vehicle. This information will be used to calculate *NetworkTime* not for the whole network, but just for the parts of the network that contribute to the shortest path calculations for each vehicle.

The reason for the shortcoming of standard GIS implementations is that GIS is constructed as a toolbox where all operations are atomised and, therefore, designed in a generic way that cannot take into account the specific context in which they are utilised. In Map Calculus, the final model includes all the building blocks and the system can understand the semantics of the model. This ability will allow the creation of an optimiser that will reduce the number of calculations that are required to solve a specific problem.

6.3.2 Dynamic modelling

The example in the previous section dealt with spatiotemporal problem solving, where the GIS is required to represent information about a dynamic situation in the real world. As explained, the main challenge for the GIS is in handling temporal inputs and the need to process these inputs rapidly. Yet, the final output is static – it represents reality at a given snapshot in time when the inputs were sampled.

In contrast, in dynamic modelling within a GIS, the challenge is to represent a process in which both time and space influence the outcome. When such a process is modelled in a GIS, the ability of the GIS to integrate multiple data sources is invaluable, as it provides the framework for the model, such as the location of various physical features, the areas in which certain conditions apply, etc. At the same time, the representation of dynamic phenomena within a GIS is especially challenging, due to the static nature of its data models (Albrecht, forthcoming).

In recent years, there has been a growth in interest in the representation of dynamic models using Cellular Automata and Agent-Based Modelling (Albrecht, forthcoming; Couclelis, 2001). However, most of these models are loosely linked to a GIS, and operate in a separate computing environment.

Within GIS, PCRaster (Van Deursen, 1995) provides the clearest implementation of dynamic modelling which is tightly coupled with a fully functional GIS. PCRaster has been developed as a raster-based GIS with dynamic modelling capabilities and integrates a scripting language that was deliberately designed to allow domain experts, who are not necessarily GIS experts, to create their own models (Van Deursen et al., 2000). Noteworthy are PCRaster origins in physical geography with a focus on geomorphology and hydrology, although PCRaster has been used in other application areas. The system's database supports dynamic modelling by providing '...time series indexed on time and location, and by stacks of map layers representing the status of the model at different time steps' (Wesseling et al., 1996).

In PCRaster, a dynamic model will be programmed by setting a script that will have the following structure (Van Deursen, 1995).

```
timer BeginTime, EndTime, TimeStep;
initial InitiateModel;
dynamic ExecuteModel;
```

The timer statements inform the system about the simulation step by setting values for the beginning of the model run and the end time for the model run (thus, timer 1 20 1 will mean 20 iterations through the model). The initial statement provides instructions for setting up the model's environment, while the dynamic part will include the instructions that will be executed at each step of the model run.

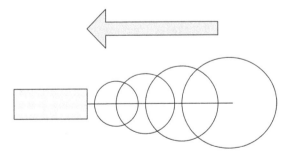

Figure 6.4. Model of a point-source pollution: the source (rectangle) moves in the direction of the arrow, and releases substance, spreading out as a function of time.

To compare the way in which a dynamic model can be executed in a Cartographic Modelling-based system and in Map Calculus, the next simple example will be used. A point-source pollution moves in a straight line, releasing a substance at a steady rate. The environment to which the pollution is released is stable and it spreads in all directions at the same rate. As a result, the pollutant will spread to a larger area, but at a reduced concentration as time passes. The model is depicted in Figure 6.4.

In a system like PCRaster, the model may be implemented in the following way (described here as pseudo-code, and not in PCRaster language):

```
timer 1, 20, 1;
initial
# Location - the location of the point-source
Location = StartLocation;

dynamic
PollutGrid = Spread(PollutGrid);
# SeedPollution - discharge at T0
PollutGrid = PollutGrid + Location*SeedPollution;
Report PollutGrid;
Location = Location(x+dx,y+dy);
```

The core of the process is implemented through a *Spread* function. This function takes a grid, in which there are several cells with a concentration of substance, and spreads the levels of concentration to the neighbouring cells, following predefined rules. In each step of the computation, the *PollutGrid* represents the state of the substance spread at a specific point in time. The last statement advances the computation to the next location of the pollution source.

In Map Calculus, this model will be presented in the following way. The concentration of the pollutant at any given point can be represented in a differential equation, which describes the concentration of the pollutant at a given location at time T. The definition of the function will be in the form:

```
Function -> "PointsourceSpread"
Layer -> "Start location"
Parameters -> (T0, InitialConcentration, Speed, Decay)
```

where *PointsourceSpread* is the generic function for the calculation of point-source pollution, and stores the algorithm to solve the differential equation; *Start location* describes the point from which the modelling will start; and the parameters describe the start time of the model (*T0*), the discharge concentration, the movement speed and the decay speed of the substance.

When the model is executed, the user is asked to enter the time over which they would like to view the dispersion model. This input, together with the stored parameters, allows the calculation of the concentration values for any location directly from the general equation. This can be done because the differential equation encapsulates the temporal dimension of the model and, therefore, can calculate the model's output for any unit of time that passed from *T0*.

Three aspects are emerging from the comparison of the two methods. They are: the way in which the state of the computation is carried in the system, the handling of the temporal dimension and the output volume.

In Cartographic Modelling, the state of the computation is stored in the grid and not as part of the logic of the model. The state of the system is the description of the environment at each time step. The computational steps of the model deal with the procedures that advance the system from one state to the next – adding more substance as the pollution source progresses, spreading the concentration and moving the source to its next location. These, however, do not reveal anything about the state of the environmental system. Indeed, the state of the system is being carried via the computation in the data structure *PollutGrid*. This grid is acting as the memory of the system. In Map Calculus, the state of the system is explicit. The functional representation holds the information about the environmental process that is being modelled, as well as the specific aspects of the instance – the time in which the discharge started, the level of concentration, etc.

As noted in the previous section, the majority of GIS implementations are based on generalised algorithms that do not take into account the specific aspects of the input or the operation. The ability of Map Calculus to take the function and the data set into account opens up the potential for more 'intelligent' algorithms that 'understand' the nature of the data and the manipulations that the user wishes to apply on them.

The second difference is in the handling of time. Note that in the Cartographic Modelling approach, the *Spread* function is oblivious to the specific time step – it works by spreading substance concentrations across neighbouring cells using a neighbourhood operator that is not influenced by the temporal dimension in an explicit way. This is a good demonstration of Couclelis and Liu's (2000) insight that '…models can predict the future to the extent that they are not about the future. We can indeed predict many aspects of what is to come because events are constrained by several different kinds of determination that are in themselves outside of time' – the implementation of *Spread* is clearly a-temporal. Because the computational steps do not encapsulate the temporal element, it is the data set that stores the state of the system at a given moment, as noted previously. Importantly, the user is forced to use a discrete view of time where the progression from one step to the next occurs in discrete units, each of them presented as a full step of computation.

By contrast, in Map Calculus the temporal aspect is an integral part of the differential equation. Time is expressed explicitly as the difference between T0 and the current time. While Couclelis and Liu (2000) still hold true, as the model itself is deterministic, it is clearly spatiotemporal and linked directly to the mathematical description of the model. Here, the handling of time is as a continuous element, and the user can request the model to produce output for any given point in time.

Finally, the difference in output is noteworthy. In Cartographic Modelling, in each computational step, the user must create an output grid (the *Report* statement in the script) and store it for future reference. The reason for this is that the internal representation of the system's state (*PollutGrid*) is written over in each step, and it is essential to store a description of each step, as otherwise the GIS will need to repeat the full computation from the first step to the place that the user wants to examine (see Wesseling et al. [1996] for an example of such an output). In Map Calculus, the calculation is performed in real time, and there is no need to store it, as the system can always produce an output by changing the current time value. Furthermore, it is easy to see that in a Map Calculus-enabled GIS, the user can instruct the system to produce values for a specific time frame automatically and to create animations which are valuable in understanding dynamic processes (Peterson, 1993).

6.4 Conclusion and future developments

Map Calculus provides a different way to manipulate layers in a GIS and the comparison of its principles to Map Algebra and Cartographic Modelling demonstrates its advantages over current representations in GIS. It provides an explicit representation of the spatial functions and their manipulation, provides a compact storage of function-based layers, and enables the interrogation of function-based layers at any required resolutions.

Within the context of dynamic GIS, Map Calculus allows easier linkage to rapidly changing inputs and the implementation of a dynamic model where the model is based on differential equations. These aspects will make the GIS more accessible to domain experts, as they can focus on the construction of the model and not on the finding of a way to translate the conceptual model to the constraints of the GIS. PCRaster is already offering such a translation (Van Deursen et al., 2000) but is limited by the structural constraints of Cartographic Modelling and Map Algebra. Map Calculus should be seen as the next step in making the GIS accessible to domain experts who are not familiar with GIS.

It is important to remember that Map Algebra and Cartographic Modelling have their own advantages. Of these, the simple and accessible data mode, which is based on the tessellation of the study area, holds a primary position. Indeed, the grid model is now used widely for dynamic models, especially in Cellular Automata models (Couclelis, 2001). Another advantage of these techniques is that the algorithms are simple to implement: the universal grid structure simplifies the processing and provides a clear conceptual framework. Even with the redundant computations, modern implementations provide reasonable response time and they are used for many applications of GIS.

At the same time, Map Calculus poses certain challenges. These include the reformulation of common GIS operators, creation of a function optimiser and consideration of optimal visualisation methods. Once these issues have been solved, Map Calculus can open up new avenues for spatial analysis and applications of Geographical Information Science.

Most importantly, Map Calculus is implemented only as a basic prototype. As such, its ability to handle and manipulate geographical data sets is very limited. A full evaluation of Map Calculus potential in geographical problem solving will be possible only through a full implementation of a system that will support it.

References

Albrecht, J. (forthcoming) 'Dynamic GIS', in Wilson, J. and Fotheringham, S. (eds.) *Handbook of Geographic Information Science,* London: Blackwell Publishers.

Bergougnoux, P. (2000) 'Editorial: A perspective on dynamic and multi-dimensional gis in the 21st century', *GeoInformatica,* vol. 4, no. 4, pp. 343–348.

Couclelis, H. and Liu, X. (2000) 'The geography of time and ignorance: dynamics and uncertainty in integrated urban-environmental process models', in *Proceedings, GIS/EM4 Conference,* [Online], Available: www.Colorado.edu/research/cires/Banff/upload/136 [20 Jun 2005].

Couclelis, H. (2001) 'Model frameworks, paradigms, and approaches', in Clarke, K. C., Parks, B. E. and Crane, M. P. (eds.) *Geographic Information Systems and Environmental Modeling,* pp. 34–48, New York: Prentice Hall.

Egenhofer, M. J. and Golledge, R. G. (1998) *Spatial and Temporal Reasoning in Geographic Information Systems,* New York: Oxford University Press.

ESRI (2004) *What is the maximum size a grid can be?* [Online], Available: http://support.esri.com/search/kbdocument.asp?dbid=14575 [25 Nov 2004].

Goodman, A. (1985) *Surface analysis: A structured bibliography,* Working paper 17, Department of Geography, Monash University, [Online], Available: http://www.deakin.edu.au/~agoodman/publications/biblio/surfacebiblio.html [1 Aug 2002].

Haklay, M. (2004) 'Map Calculus in GIS: A proposal and demonstration', *International Journal of Geographical Information Science (IJGIS),* vol. 18, no. 1, pp. 107–125.

Laurini, R. (2001) 'Real time spatio-temporal database', *Transactions in GIS,* vol. 5, no. 2, pp. 87–97.

Longley, P., Goodchild, M., Maguire, D. and Rhind, D. (2001) *Geographical Information Systems and Science,* New York: Wiley.

Park, S. and Wagner, D. F. (1997) 'Incorporating cellular automata simulators as analytical engines in GIS', *Transactions in GIS,* vol. 2, no. 3, pp. 213–231.

Peterson, M. P. (1993) 'Interactive cartographic animation', *Cartography and Geographic Information Systems,* vol. 20, no. 1, pp. 40–44.

Shepard, D. (1968) 'A two-dimensional interpolation function for irregularly spaced data', *Proceedings 23rd ACM National Conference,* pp. 517–524.

Tomlin, C. D. and Berry, J. K. (1979) 'A mathematical structure for cartographic modeling in environmental analysis', *Proceedings of the Annual Meeting of the American Congress on Surveying and Mapping and the American Society of Photogrammetry,* Falls Church, VA, pp. 269–283.

Tomlin, D. (1990) *Geographic Information Systems and Cartographic Modelling,* Englewood Cliffs: Prentice Hall.

Van Deursen, W. P. A. (1995) 'Geographical information systems and dynamic models: Development and application of a prototype spatial modelling language', *Netherlands Geographical Studies, 190,* Utrecht: Utrecht University.

Van Deursen, W. P. A., Wesseling, C. G. and Karssenberg, D. (2000) 'How do we gain control over GIS technology?', *Proceedings of the 4th International Conference on Integrating GIS and Environmental Modeling,* 2–8 September, Banff, Canada. Available: http://www.colorado.edu/research/cires/banff [20 Jun 2005].

Wesseling, C. G., Van Deursen, W. P. A. and Burrough, P. A. (1996) 'A spatial modelling language that unifies dynamic environmental models and GIS', in *Proceedings, Third International Conference/Workshop on Integrating GIS and Environmental Modeling,* [CD-ROM], Santa Fe,

NM, 21–26 January 1996. Santa Barbara, CA: National Center for Geographic Information and Analysis.

Worboys, M. and Duckham, M. (2004) *GIS – A Computing Perspective*, 2nd edition, Boca Raton: CRC Press.

Chapter 7

Constraints in Spatial Data Models, in a Dynamic Context

Peter van Oosterom

GIS-technology Section, Delft University of Technology, Delft, The Netherlands

7.1 Introduction

Constraints are important in every GI modelling process but until now have received only *ad hoc* treatment, depending on the application domain and the tools used. In a dynamic context, with constantly changing geo-information, constraints are very relevant; any changes arising should adhere to specified constraints, otherwise inconsistencies (data quality errors) will occur. In GIS, constraints are conditions that must always be valid for the model of interest. This chapter argues that constraints should be part of the object class definition, just as with other aspects of that definition, including attributes, methods and relationships. Furthermore, the implementation of constraints (whether at the front-end, database level or communication level) should be driven automatically by these constraints' specifications within the model. But, this is not possible yet, so this chapter will describe some implementation steps as interactively executed.

In certain applications some functions (linear programming in spatial decision support systems, survey least squares adjustment, cartographic generalisation, editing topologically structured data, etc.) partially support constraints. However, the constraints are not an integral part of the system and the constraint specification and implementation are often one and the same, and deep in the application's source code. The result is that the constraints are hidden in some subsystems (with other subsystems perhaps unaware of these constraints) and it may be very difficult to maintain the constraints in the event that changes are required. This is true for (G)IS in general, but is especially true for dynamic environments, with changing objects, where the support of constraints is required but presents a challenge. Example applications include cadastral or topographic data maintenance, Virtual Reality (VR) landscape design, and Web feature service.

7.1.1 Context

There are situations where certain types of constraints are well supported. Domain value constraints and referential integrity constraints in relational DBMSs (Date and Darwen, 1997) are standard functionalities. For example whenever one object refers to another via a foreign key, the DBMS checks that the referred object exits,

Dynamic and Mobile GIS: Investigating Changes in Space and Time. Edited by Jane Drummond, Roland Billen, Elsa João and David Forrest. © 2006 Taylor & Francis

otherwise the transaction or change will not be committed. Another more specific GIS example is the support of topological constraints, such as certain types of objects, which may not overlap. Topological constraints can be supported within the DBMS, by, for example, LaserScan Radius topology (2003) and Oracle spatial 10g (2003) with topology, or they can be supported at the 'middleware' level such as in ESRI (2002). Within the context of VR systems, constraints are often implemented as the behaviour of objects. An illustration is the constraint 'two trees (objects) cannot grow on the same location', which is realised (hard coded in the edit environment) by collision detection, a well-known computer graphics technique. Referential integrity, topological correctness and collision detection are just a few examples of constraint types, but the available solutions may only work in certain subsystems. Other subsystems may not be aware of these and may have different 'opinions' of correct data. So constraints must be implemented at various levels (or subsystems), including application (edit, simulate,...) level, data exchange (communication) level and database level.

   Although support for integrity constraints is patchy, there has been some research in this area. Primarily, integrity constraints are related to data quality (Hunter, 1996) and the source of errors (Collins and Smith, 1994) such as during data collection, data input, data storage, data manipulation, data output and the use of results. Cockcroft (1997) was one of the first researchers presenting a taxonomy of (spatial) integrity constraints. A contribution of the current chapter is a refinement of this taxonomy. Cockcroft (2004) advocated an integrated approach to handling integrity, based on a repository that contains the model together with the constraints. Cockcoft (2004) concluded that the constraints should be part of the object class definition, similar to other aspects of the definition. The repository is used both by the database and the application as a consistent source of integrity constraints. The current chapter continues these investigations into the possibilities of managing constraints in an integrated system-wide manner and adds data communication as an additional part of the system where constraints are important. It should be noted that much of the presented material is still a 'vision' and complete implementation is still in progress, though important parts have been proven.

7.1.2 Chapter overview

This chapter demonstrates the need for the integral support of constraints through four quite different cases: a VR system for landscape design (Section 7.2), cadastral data maintenance (Section 7.3), topographic data maintenance (Section 7.4) and a Web feature service (Section 7.5). All four applications deal with dynamic situations. The landscape design has an explicit temporal aspect, namely the simulation of tree growth. During both the initial design and the simulation these constraints should be met. In the case of the cadastral application, when parcels are changed, constraints have to be satisfied otherwise this could lead to inconsistencies, such as parcels overlapping or lacking an owner. Not further discussed, is in-car navigation using a topographic base map: if the moving point belongs to a car, a constraint could be that the point should always be on, or near, a road or related features, such as a parking lot. Based on the different constraints, experiences in the four cases (and the relevant literature), a classification of constraints is given in Section 7.6. Constraints can be related to the properties of an object itself and can also be based on relationships between objects. Constraints such as 'a tree must always be green' or 'the salary of a staff member should be higher or equal to the minimum salary' illustrate constraints based on properties of only one object. Examples of constraints considering relationships between two objects are 'a Yucca tree must never stand in water' and 'the salary of the boss must always be higher than the salaries of the other staff'. These constraints require formal description and definition. Section 7.7 discusses the formal specification of constraints within a (conceptual) model. The implementation of constraints, with focus on the DBMS, is described in Section 7.8. This chapter's last section concludes with the principal results and proposes further research directions.

7.2 Constraints in a landscape design VR system

With SALIX-2 (van Lammeren et al., 2002) a user can interactively introduce new objects (trees, bushes, etc.) to a 3D landscape. As is the case in reality, sometimes new objects have to be a certain distance from each other (for example, two trees have to be planted not closer that 3 m), from other objects or are even not in an area at all, for example a tree on a road (Louwsma, 2004).

7.2.1 SALIX background

Digitally supported landscape design contains intriguing challenges. These challenges have to do with modelling the changes in time of the architectural primitives (mainly trees and shrubs) and modelling the relation between architectural objects, their architectural primitives and their spatial configuration. Virtual Reality (VR) tools such as VR-construction sets and VR-viewers are widely available, and provide opportunities to experiment with a wide range of design proposals using a geo-database representation. Such (VR) geo-information systems offer a three-dimensional laboratory to experiment with landscape design proposals. SALIX-2 is a simulation program, exploiting these possibilities, developed for students of landscape architecture at Wageningen University (van Lammeren et al., 2002) (see Figure 7.1).

VR-scene manipulations make it possible to interact with a virtual scene object (Heim, 1998) such that an object (or its attributes) in the scene can be deleted or added. With SALIX-2 the underlying idea is a virtual environment for simulating the growth of plantation objects (bushes and trees). The students are able to plant bushes and trees interactively. Just as in the real world, one should be restricted from planting in particular areas. For that reason the system has to be provided with constraints related to the type of plantation and geo-information objects.

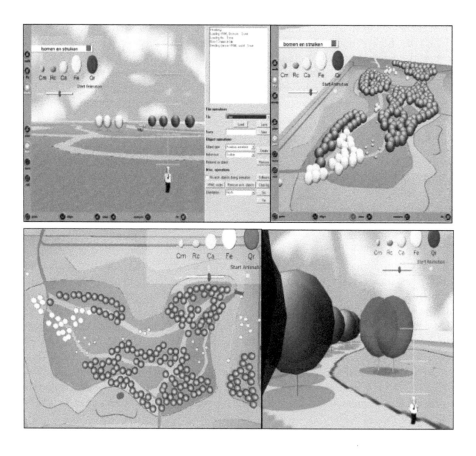

Figure 7.1. 3D scenes of SALIX-2: an interactive landscape modelling system. A constraint is violated in SALIX-2c (note the red, highlighted trees of Figure 7.1 on colour version following page 132).

7.2.2 Selected constraint examples

The SALIX-2 system currently maintains three classes of objects: trees, bushes and ground surfaces. The possible ground surfaces are water, paving, soft_paving, grass and bridge. There are five possible types of trees/bushes (CorAve, CorMAs, FraxExc, QueRob, RosCAn). Examples of rules for the position of objects in geo-VR environments can be: a tree must not overlap with water or a tree must be covered by a polygon with destination forest. For these constraints it is logical to represent a tree as a circle (an extended object) and not as a point (centroid). Table 7.1 shows examples of constraints for SALIX-2; see also Figure 7.2.

Table 7.1. Selected examples of relationship constraints for SALIX-2.

Type of relation	Constraints formulated with forced relations between objects
Direction	A bush always has to be placed south of a tree
Topology	Bushes always have to be disjoint or meet water A bush always has to meet or be disjoint with paved areas (also thematic constraint) (2 predicates)
Metric	Trees always have to be positioned > 1 metre from paving
Temporal	An oak always grows for 70 years
Quantity/ Aggregate (sum)	There must always be at least 10 trees on the specified ground surface
Thematic	A bush always has to meet or be disjoint with paved areas (note the mixed topological constraint)
Complex	The distance between trees inside water always is > 8 m AND the distance between the tree and the edge of the water always has to be < 0.5 metre AND the species must be a salix

7.2.3 Some lessons

The main lessons learnt with respect to the constraint support requirements of SALIX-2, the VR landscape modelling system are (Louwsma, 2004):

❑ constraints occur at different places, both in the VR user interface and data storage;
❑ when designing, immediate feedback to the user is important (see Figure 7.1, bottom); and
❑ simulation adds another 'dimension' to constraints, when creating an initial plantation layout everything may be correct, but after 5 years of simulated growth there may be conflicts, e.g. trees get too close.

7.3 Constraints in a cadastral application

In this section a cadastral data maintenance system (another application in which constraints play a major role) is discussed. Although cadastral systems also maintain important legal and administrative information, this section's focus is the spatial side of cadastre.

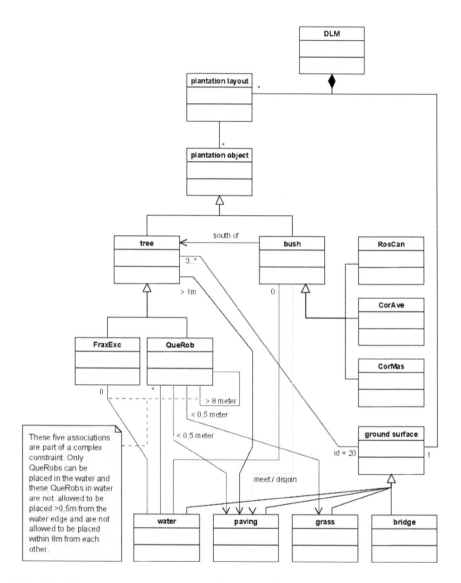

Figure 7.2. UML classs diagram representing the objects of interest and their constraints in SALIX-2 (see colour insert following page 132).

7.3.1 Dutch cadastral data

The Dutch cadastral map is based on a winged-edge topology structure (Van Oosterom and Lemmen, 2001); see Figure 7.3. The DBMS is considered very clean, topologically. Further, the model contains redundancy in the topological references:

both the (meaningless system) object_id reference to the left and right parcels and the (meaningful user) parcel_number references to the left and right parcels are stored and maintained. The topological consistency checks are hard coded and built into both the editor and the check-in software at the DBMS server side. However, the checks are currently not implemented within the DBMS itself (Ingres). The data set covers the Netherlands and contains history from 1997 to the present. The total number of current boundaries (polylines) is about 22,000,000 and the number of current parcels (topological faces) about 7,000,000. If all historic versions are counted, numbers roughly quadruple. There is a separate, but linked, subsystem containing the legal and administrative data.

Spatial types, topology references
(Boundary: polyline, or circular arc)

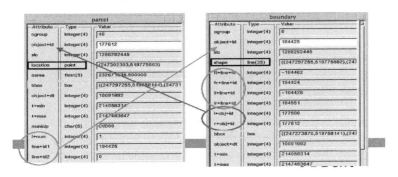

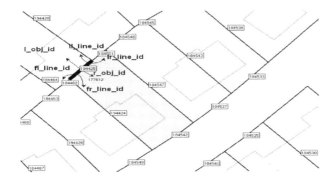

Figure 7.3. Winged-edge topology structure of the cadastral map.

7.3.2 Some examples of cadastral data constraints

Due to redundancies in the system and because, in general, topology references can be derived from the metric information, a large number of consistency checks can be defined for the cadastral model. Over 50 constraints have been defined, in a number of different categories.

In this section some example categories will be presented, accompanied by SQL select statements, which in the case of correct data should not find any objects. These statements could be considered the body of SQL assertions, with the 'create assertion' part skipped (see Section 7.8). (Discussion of constraints related to attribute value domain checks are also skipped, being trivial.) Five categories of cadastral constraints will be discussed.

1. *Metric checks.* The first example finds closed 'arcs' (but not circles), which can be detected by checking that the first and last (third) point defining the arc match, see CCVQ1 (Appendix 1 with the Cadastral Constraint Violation Queries). A second example constraint disallows straight 'arcs' (see Figure 7.4). Another example ensures every parcel has a reference point, which should be within the area of the parcel; this reference point should also be in the bounding box of the parcel, which is easily checked with the CCVQ2. The final example is that two different boundaries should not intersect, but should be disjoint or touch at their end points.

2. *Existence of topological references.* This can be compared to referential integrity checks in some administrative databases. A complication is that topological references can be signed (+ or -) in order to indicate proper orientation. The first constraint in this category checks whether the left (and right) parcel references from the boundaries do indeed exist (CCVQ3). The next example checks whether the winged-edge boundary-boundary reference (in this case the first left references) exists (CCVQ4). Then, starting from the parcel, a number of topological reference checks can be imagined. For example, as parcels can have island boundaries, these references also have to be correct. So, the reference from the parcel to the island boundary reference must exist. Further, as a parcel can have any number of islands (and the number of islands is encoded as an explicit attribute), it must be checked whether the correct number of parcel references are specified and if they all refer to existing boundaries. The final example in this constraint category checks whether the reference from the parcel to its outer boundary exists (CCVQ5).

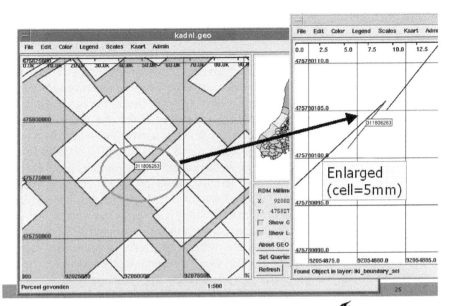

Closed 'circular arc' & geometric 1 mm gap   **ŤU**Delft

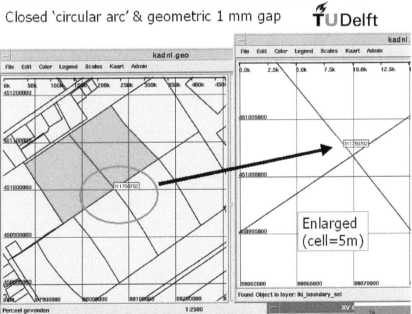

Straight line coded as circular arc   **ŤU**Delft

Figure 7.4. Some metric errors in the cadastral dataset (top: small gap between two boundaries, bottom: straight line encoded as circular arc).

3. The *correctness of a topological reference*, see Figure 7.5. A first example in this category is the check that two consecutive boundaries must have the same parcels on one side. In total there are eight combinations that have to be checked as each of the four winged-edge boundary-boundary references is signed, that is, the direction of the next edge may have to reversed (thereby switching the left and right hand sides). CCVQ6 checks the positive first left reference. A similar constraint in this category is that the end point of one boundary is the start of the next. As with the previous consistency check, there are again eight combinations which have to be checked; CCVQ7 shows the positive first left case again. Also in this category of constraints is the check as to whether the island boundary has the parcel at the correct side. Another constraint is whether the first coordinate of the island boundary lies within the bounding box of the parcel. Finally a check is given to see whether the outer boundary and parcel references back-and-forth are consistent (CCVQ8).

4. The fourth category of constraints to be considered is a *referential integrity check*, which determines whether two subsystems are consistent. The two subsystems are the geometric subsystem (LKI) and the administrative and legal subsystem (AKR). Every ground parcel in AKR should also be present in LKI (CCVQ9).

5. *Temporal constraints* ensure that the time intervals of two consecutive versions of an object do touch and assume no gaps or overlaps in the time dimensions of an object.

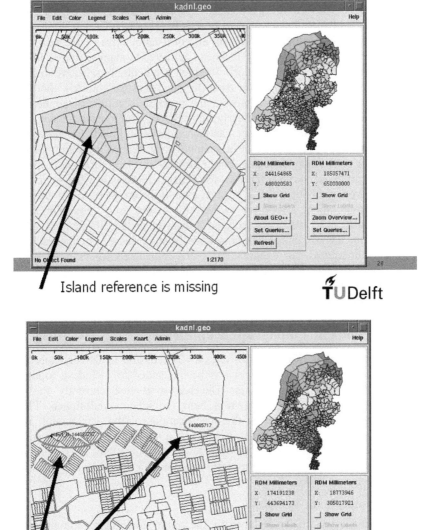

Figure 7.5. Some topology reference errors in the cadastral dataset (top: island reference is missing, bottom: parcel refers to wrong island).

7.3.3 Some lessons from cadastral data

The cadastral dataset is considered clean and is, by the nature of its structure, designed to avoid certain errors, such as overlapping parcels. During the conversion in 1977 from the old to the new version all consistency errors were resolved and removed. Further, the cadastral data has been delivered to many different customers and loaded into different systems, each of which is potentially sensitive to different errors. As the customers pay for the data, they will complain quickly about errors. However as the constraints discussed in the previous section and applied to the production data of 2004 have clearly illustrated, certain errors have, despite everything, been (re-)introduced (see van Oosterom et al. [2005] for more details). One important lesson has been that one should trust neither front-end nor middle ware alone for consistency checking, but implement checking throughout the whole system and particularly within the DBMS, which will contain what is considered valid data shared by multiple users. Further, even if the errors are not noticed in the production environment, they may be harmful in the users' environments; e.g. straight 'circular arcs'. Despite a thorough treatment of the different categories of constraints, not all possible constraints have been discussed. In the category of topological correctness, for example, it is not considered whether the complete domain is covered with parcels. This is an important type of topological constraint as there should be no gaps – in the cadastral case this is equivalent to an area without an owner.

7.4 Constraints in a topographic application

At first sight the types of constraints relevant to cadastral update and topographic update systems seem similar. But it has been decided to include a topographic application in this chapter's cases. The reason is that currently the Dutch topographic data maintenance system is being completely redesigned. The new design contains constraints within the specifications of the data model. Besides renewing the production environment (including a move from separate files to one geo-DBMS), the product itself is being renewed as TOP10NL. It is more object-oriented than map-oriented, contains nationwide unique identifiers to be delivered in GML-3 from late 2006.

7.4.1 Constraints in the TOP10NL

The constraints were initially designed for the conversion process, to make sure the new topographic production system only contained clean data. However, the same constraints will be used during future production editing (and this has been successfully tested in prototype versions of the future system) and geo-DBMS check-in. The constraints are specified in one source, which contains the complete model (or specification) of all object classes, attributes and relationships. The model is encoded in an XML-format developed by Vertis and the Topographic Service. In the remainder of this subsection fragments from this XML-format will be shown to illustrate a number of example constraints. The following five types of constraints were recognised (using Vertis/Topographic Service terminology):

1. Single entity, single attribute (thematic). This example shows a domain constraint specifying the fact that the width of water should be between one and 500 (metres) and the same XML encoding of this range domain type states that the default value is six (metres); see TMCX1 (Appendix 2 Topographic Model Constraints in XML). In the same category is the specification of valid values of the XML encoding for the railroad enumeration type, with allowed values 'verbinding' (connection) and 'kruising' (junction) given in TMCX2.

2. Single entity, multiple attributes (thematic). An example from this category checks a road object constraint that the 'NAAM_AWEG' (name a-road) attribute is filled and then the attribute 'WEGTYPE' (road type) must contain a specific value; in this case 'autosnelweg' (highway). Note the specific operator 'MVCONTAINS' which is used in the case when a multi-valued attribute contains specific value(s). The corresponding XML fragment is given in TMCX3.

3. Geometry (general rules, including minimum line length and minimum area). For example, if the width of a road is less than two metres, then the geometry type is line, otherwise the geometry type is area. The example from this category will not be further illustrated here in XML, because it is of the same type as the first category except that a constraint is specified for a geometric, not thematic, attribute.

4. Topology (several subtypes: covering_without_gaps, no_overlap, coincide, ...). As the model of the Topographic Service is not based on a topological structure, but consists of individual point, polyline and polygon features, topology constraints look different from those used for the cadastral data set, which was based on a topological structure. One could even state, in this case, that constraints are even more important, because there is no other facility supporting topological data quality. TMCX4 shows the constraint that two roads 'WEG_VLAK' (road area) may not overlap at the same height 'HOOGTENIVEAU' (height level). (Note the use of the topology operation 'AREAOVERLAPAREA' from the ESRI ArcGIS environment.)

5. Relationship. Every feature must have at least one specified source. The example TMCX5 checks that the return value of the operator 'BRONCOUNT' (source count) is greater than '0'.

7.4.2 Some lessons from the new topographic base data production system

The fact that the constraints are specified together with the model and used as a source for the realisation of different (sub)systems is a great leap forward. The same constraints are applied during the initial conversion from the old to the new system, during interactive editing (ArcGIS environment; see Figure 7.6) and at data storage (Oracle DBMS) during check-in.

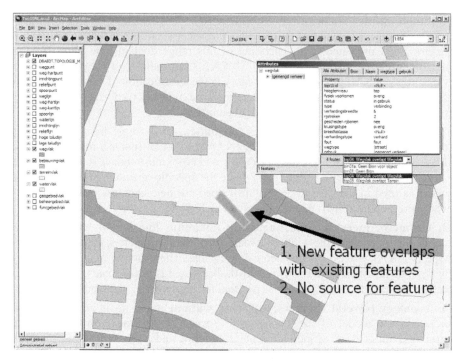

Figure 7.6. An error caught during editing of the topographic data.

Of course, things can always be further improved, for example:

1. the model itself could be specified in UML (and based on OGC/ISO TC211 standards);
2. instead of institutionally generated (XML) constraint encoding, perhaps a standard would have been better, e.g. OCL;
3. the exchange format is not (automatically) derived from the same source model (as used for the edit and storage subsystems); and,
4. the constraints are not yet included in the exchange format. Some could possibly be included in standard GML/XML encoding (for example the domains constraints) but more research is needed for the other types of constraints.

7.5 Constraints in a Web feature service

The last case to be presented also relates to the cadastral domain, but the context is quite different: an Internet GIS environment. This different context will reveal new insights into the important role constraints play in real-world implementations, and the current difficulties experienced in supporting them. With the availability of the standard OpenGIS Web Feature Service (WFS) (OGC, 2002) protocol, it is now finally possible to realise Internet-based geo-information processing environments, where clients cannot only view but also edit (update) data stored at multiple servers, and where products of one vendor can be combined with the server products of another. An evaluation of the WFS-Transaction protocol was carried out using a case study known as 'notary drafts cadastral parcel boundary'. The WFS protocol was analysed (Brentjens, 2004; Brentjens et al., 2004) and revealed, for more advanced edit scenarios, a number of possible improvements related to constraints.

7.5.1 Web feature services

Two classes of Web Feature Services are defined in the OpenGIS WFS specification: Basic WFS (for retrieving geographic features) and Transaction WFS (needed for editing geo-data) (OGC, 2002). The WFS protocol allows a client to retrieve geo-spatial (vector-) data encoded in Geography Markup Language (GML) from multiple Web Feature Services. GML is an XML encoding for the modelling, transport and storage of geographic information, including both the spatial and non-spatial properties of geographic features (OGC, 2003).

A 'Basic WFS' service implements the GetCapabilities, DescribeFeatureType and GetFeature requests. A client can request an XML-encoded capabilities document (containing the names of feature types that can be accessed via this service, the spatial reference system(s) and the spatial extent of the data, plus information about the operations that are supported) by sending the GetCapabilities request to the WFS service. The function of the DescribeFeatureType request is to generate an XML Schema document (XSD, 2004) with a description of the data structure of the feature types serviced by that WFS service. The GetFeature request allows for the retrieval of feature instances (with all or part of their attributes).

A 'Transactional WFS' offers the functionality for modifying geographic features as well, such as insert, update and delete. In order to do so, a Transactional WFS implements the Transaction request (a set of insert, delete and update actions that belong together). It could optionally implement the LockFeature and the GetFeatureWithLock request. When the transaction has been completed, a WFS will generate an XML response indicating the completion status of the transaction (and a list of newly generated feature identifiers assigned to the new feature instances). The purpose of the LockFeature request is to invoke a mechanism ensuring consistency by preventing simultaneous editing of these features by other users. A Lock element uses a filter to specify which feature instances should be locked. Finally, by using the GetFeatureWithLock instead of the GetFeature request, a client requests features be retrieved and locked simultaneously.

7.5.2 Notary drafts new parcel boundary

An example cadastral transaction is a notary who sketches a new parcel boundary as a parcel is split following a property transaction. The functional requirements for the interface between the Cadastral WFS server and the notary consist of the following requests (see Figure 7.7 top right): 1. Transfer whole parcel, 2. Merge two or more parcels, 3. Split parcel, 4. Get a reference cadastral map. The split of a parcel is the most interesting case as the required input is a parcel number; the result is a GML dataset with the parcel and context (if parcel does exist). The client action is to add one (or more) parcel boundary within the area of the parcel to be split and thereafter every implied part of the parcel is uniquely labelled with a text string (usually one letter 'A', 'B',..), (see Figure 7.7, bottom). Below is an example of the WFS protocol to pose a request to get two types of features (with the DescribeFeatureType request):

```
http://130.161.150.109:8080/geoserver/wfs/wfs?request=
DescribeFeatureType&typeName=DRAFT_BOUNDARY,DRAFT_PARCEL
```

Below the XML/GML fragment with geo-information is sent from the server to the client (or, in the case of an update, in the other direction):

```
<gml:lineStringMember>
 <gml:LineString>
   <gml:coordinates decimal="." cs="," ts=" ">
      106417204,448719275 106367184,448675614
   </gml:coordinates>
 </gml:LineString>
</gml:lineStringMember>
</gml:MultiLineString>
```

7.5.3 Evaluation of Transactional WFS

Though 'simple' editing proceeds well, a number of observations related to editing complex situations, such as handling topology (van Oosterom, 1997), can be made. First, concerning the integrity constraints in transactions, validation of (changes in) features should prevent a dataset containing features that violate topological rules or other restrictions. The WFS specification defines some operations and mechanisms that can be used for the validation of single features. It is not so easy however to enforce integrity constraints that concern combinations of features (as in the case of topologically structured data) or constraints that follow business rules. The server may check certain integrity constraints after the client posts a transaction and as a result the transaction may fail. However, the client does not know what the constraints are (except for the conditions implied by the GML application schema defining the individual feature types). It may be quite frustrating for a client to update data and then receive errors. An ongoing research topic is how and where to check integrity constraints (and other business rules) in WFS-based distributed systems.

One interesting question is whether it is possible (and meaningful) to translate constraints in the data model to constraints related to the structure of valid transactions. For example, a parcel split always implies at least deleting one old parcel, inserting a new boundary and two new parcel reference points on each side of the new boundary. How should constraints be related to operations in a valid transaction?

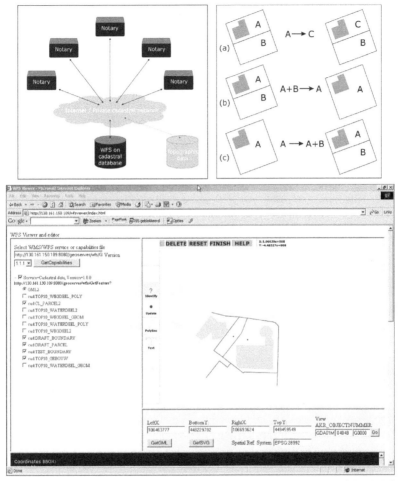

Figure 7.7. Case 'notary edits cadastral map via Internet' (top left: architecture of the Web feature server, top right: a number of typical cadastral edit operations, bottom: the edit Internet WFS client).

7.6 Classification of constraints

In order to better understand constraints and their use, it is important to classify them, including their spatial/dimensional aspects. Classification of the different types of (spatial) constraints reveals a complex taxonomy. Cockcroft (1997) presents a two-dimensional taxonomy of (spatial) constraints: the first axis is the static versus transitional (dynamic) distinction and the second axis is the classification into topological, semantic and user constraints. It is recognised that the transitional aspect of integrity constraints (allowed and valid operations; see also Section 7.5) is relevant, but in this chapter this is considered to be the 'other side of the same coin'. This section refines, based on the four presented cases, the second axis of Cockcroft's taxonomy by recognising five subaxes (or five different criteria) for the classification of integrity constraints:

1.   the *number of involved objects*/classes/instances;
2.   the *type of properties of objects and relationships* between objects involved: topologic (neighbourhood or containment), metric (distance or angle between objects), temporal, thematic or mixed;
3.   the *dimension* (related to the previous axis): 2D, 3D or mixed time and space, that is, 4D;
4.   the *manner of expression*: 'never may' (bush never may stand in water) or 'always must' (tree always must be planted in open soil); and
5.   the *nature of the constraint* can be: 'physically impossible' (tree cannot float in the air) or 'design objective' (bush should be south of tree).

With respect to the first subaxis of the constraint taxonomy, 'the number of involved objects', the following cases can be identified:

1.   one instance (restrictions on attribute values of a single instance);
2.   two instances from the same class (binary relationship);
3.   multiple instances of the same class (aggregate);
4.   two instances from two different classes (binary relationship); or
5.   multiple instances from different classes (aggregate).

Further, the fourth subaxis, 'the manner of expression' has only practical value for communicating the constraints between the users. Once the objects and the constraints are formally defined, the expressions 'never may' and 'always must' can be represented by one constraint, e.g. by the one that is more efficient from an implementation point of view. For example, the constraint 'a tree never may stand in water, or street or house' is equivalent to the constraint 'a tree always must be in a garden, or park' under the assumption that there are only five possible ground objects: water, street, house, garden and park.

Further consideration will be given to the second subaxis of the constraint taxonomy, that is constraints with respect to object attributes, spatial relationships and dimensions. A distinction is made between constraints:

1.   related to properties (attributes) or the state of objects, whether thematic, temporal or spatial (see subsection 7.6.1); and
2.   based on (spatial) relationships between objects (see subsection 7.6.2).

7.6.1 Constraints derived from the properties of objects

The categories of constraints are: thematic (non-spatial business logic-like), temporal, spatial (area, perimeter, length) and also mixed. *Thematic* information about objects can be found in the attributes of the objects, such as house (number of floors), road (maximum speed), grass (type). Thematic constraints ensure related attributes only get allowed values. *Temporal* property constraints specify allowed values for one or more of the time attributes; for example the start time (birth) of an object should be before the end time (death) of that same object. Other temporal property constraints may specify some valid values for time attributes; e.g. the date/time associated with an historic fact, should be somewhere in the past. *Spatial* constraints can be associated with spatial properties of one object such as size or shape. An example of size constraint can be 'a tree should not become higher than 30m', 'a canal must be at least 2 metres wide', and a constraint on shape could be 'a bush must be represented with a sphere'.

7.6.2 Constraints derived from spatial relationships between objects

The proposed (sub)classification of integrity constraints based on relationships between objects has the following components:

1. *thematic*, 'a parcel must always be owned by at least one person'. These relationship constraints are similar to the constraints for relationships found in business logic for non spatiotemporal systems.
2. *temporal*, 'the second object may only occur after the first object (adjacent in time)'. These relationship constraints are to be specified on the basis of frameworks for describing temporal relationships between objects. Peuquet (1995) and Kwon et al. (1999) describe the temporal relations between two time intervals. Given two time intervals, there are seven distinct ways in which these time intervals can be related (e.g. Before, Meets, Overlaps, Finishes, During, Starts, Equals). The temporal relations can be seen as relations between two objects with some time interval as existence time (with start and end time of existence as the boundaries of the time interval).
3. *spatial*, The formalisation of spatial relationship constraints is closely related to the formalisation of (spatial) associations between the objects. Thus when defining constraints the following subtypes exist:
   a. *topology* 'no trees and bushes inside water polygons; no trees and bushes inside paving polygons'. Topological constraints are to be constructed using frameworks for neighbourhoods (Egenhofer, 1989; Clementini et al., 1993). For example, if the boundaries of the two objects intersect but the interiors do not, the conclusion is that the objects meet. The constraint 'no bush in water' can be translated to 'no point-in-polygon' (assuming that a bush is represented as a point and water as a polygon), corresponding to the topology relationship 'not inside'.
   b. *direction* 'the trees should always be north of paving polygons, so people can walk in the sunshine'. Direction constraints are to be based on formalism for directional relations (Papadias and Theodoridis, 1997). The directional relations are defined as the position of an object with respect to

another object, as the directions can be given in degrees in the range of [0°, 360°] or verbally (Northeast, North, Northwest, West, Southwest, South, Southeast, East) with each expression standing for an interval of degrees. Algorithms are also developed to assign the right direction to an object.

c.   *distance* 'no trees inside the water, except if < 1 metre from edge or water bushes > 1 metre from paving (so the leaves do not overlap the paving)'. Distance constraints impose a constraint on a (Euclidian) distance between objects. They can be expressed in linear metres, or by more approximate linguistic terms such as 'closer than', 'further than' or 'interval distance'.

4.   *mixed*, such as quantity (or aggregate): 'the maximum number of 10 plantation objects in a specified area in the centre of the park'. It is common to mix these fundamental types of relationship constraints. Specifying a certain density of objects in a certain area is only implicitly related to spatial relationships. Knowing the distribution of objects in a certain area, the minimum distance between two objects can be computed and, eventually, the approach for metric constraints can be used. From a user point of view, however, a more intuitive approach will to specify a number per given area. This constraint can be given as a minimum, exact or maximum number of objects for an area (surface density). Examples of density constraints are 'maximum number of houses in a residential area' or 'minimum number of trees in an area'. Examples can be 'one tower, three benches and one statue must be placed in this garden'. This exact number of certain objects can be seen as a special case of a density constraint, because it can be defined as an exact number of certain objects for the whole area in the 3D model.

7.6.3 Dimensional aspects of constraints

The last aspect to be discussed here is related to dimension. In general all spatial relationship constraints can be specified for both 2D and 3D objects. Constraints can concern the 2D ground plane or the 3D objects (bushes and trees) that could be placed on the ground plane to create a spatial configuration. The rules concerning the ground plane find their origin in local policy and in the fact that some designations conflict with each other. The policy makers define area designations and note them in plans. For example, some areas get an urban designation, some rural and others are agricultural. These designations of the ground plane can easily be stored as an additional attribute of the separate polygons. However the policy makers' defined restrictions could still conflict: e.g. a road can never lie in water (except when a bridge or tunnel is built), a forest never lies on a major road and all houses should be reachable by a path or road. On the other hand, such conflicting constraints could be a source for strategic spatial decision making; and the rules for 3D objects can even be more complex, too.

7.7 Specifying constraints

Having seen the importance of constraints in different applications and presented a refined taxonomy, the next issue is how to specify the constraints. First of all, the specification of the constraints has to be intuitive for the user. The constraints have to be included in the object model. This model should be as formal as possible to be able to derive constraint implementations within the different subsystems (edit, store, exchange). Formal modelling is an essential part of every large project, but it is also helpful in small and medium-sized projects. Using a formal model permits communicating ideas with other professionals as well as describing clear, unambiguous views on implementation strategies. The Unified Modelling Language (UML), now a more or less 'default' state-of-the-art approach, will be used for object-oriented modelling (OMG, 2005a, Ch. 3). UML is a graphic language, which gives a wide range of possibilities for representing objects and their relationships. In general the language can be used for modelling business processes, classes, objects and components, as well as for distribution and deployment modelling. UML consists of diagram elements (icons, symbols, paths, strings), which can be used in nine different types of diagrams. The most appropriate diagram is the class diagram. It provides formalism for describing the objects/classes, with their attributes and behaviour, and relationships between these objects, such as association, generalisation and aggregation.

Despite their potential for formalizing objects and processes, UML class diagrams are typically not sufficiently refined to provide all the relevant aspects of constraints. Constraints are often initially described in natural language. Practice has shown that this results in ambiguities. In order to write unambiguous constraints, a non-graphic language is provided within UML for the further modelling of semantics (knowledge frameworks), namely the Object Constraint Language (OCL) (OMG, 2005b, Ch. 6). When an OCL expression is evaluated, it simply returns a binary value. The state of the system will not change when the evaluation of an OCL expression returns false. The advantage of using OCL is that – as with UML class diagrams – generic tools are available to support OCL (i.e. it is not GIS-specific); OCL has been used successfully in the context of GIS, an example is the IntesaGIS project with the GeoUML model specifying the 'core' geographic database for Italy (Belussi et al., 2004). The context of an invariant is specified by the relevant class; e.g. the object class 'parcel' is the context of the constraint 'the area of a parcel is at least 5 m^2. It is also possible within a constraint to use the association between two classes (e.g. every instance of the object class 'parcel' must have at least one owner, which could be depicted as an association with the class 'person'). OCL enables one to formally describe expressions and constraints in object-oriented models and other object modelling artefacts. Below are two examples in UML/OCL syntax (keywords in bold print):

```
context Parcel inv minimalArea:
        self.area > 5

context Parcel inv hasOwner:
```

```
self.Owner -> notEmpty()
```

Figure 7.2 shows the UML class diagram with the objects and the constraints (depicted as associations) used for SALIX-2 (introduced in Section 7.2 and Table 7.1). In principle there is no difference between a 'data model' relationship (association, aggregation, specialization) and a 'data model' integrity relationship constraint. Both are depicted as lines in the UML class diagram. From a high level conceptual (philosophical) point of view the difference may be very small. However, normal associations are often indented, in subsequent implementations, to be explicitly stored (in one or both directions), while the relationship constraints should not result in such an explicit storage, but in a consistency rule in the implementation environment. In order to make a difference between the two, normal relationships are depicted in black while integrity relationship constraints are depicted in colour. In the diagram notes can be used to explain the constraints on relationships and/or properties. These notes can contain either UML/OCL or natural language text.

7.8 Implementation of constraints

The specified UML (OCL) models (including the constraints), managed in a repository, should be the foundation for all subsystems, including the edit/simulation environment, the storage database (further described in the DDLs of the DBMS) and the data exchange subsystems. The edit/simulation application subsystem will not be discussed here. However, it is considered important to incorporate the integrity constraints from the model (automatically) in the applications (as has been done in the topographic application; see Section 7.4). With respect to data exchange, the eXtensible Markup Language (XML) can be used for the models containing the class descriptions at class level (XML schema document 'xsd') and for the data at object instance level ('normal' XML document with data 'xml'). XML documents also include the geometric aspect of objects (e.g. LandXML, GML, X3D). Further investigations are needed to incorporate integrity constraints in the XML schemas. The UML models (incl. the OCL) with constraints should be automatically translated to XML schemas. Note that this is different to encoding a UML model in an XML document according to the XML Metadata Interchange (XMI). In this section the implementation of integrity constraints will be illustrated with database examples. When the data are correctly stored in the DBMS all users will have access to the same consistent dataset.

Since SQL92 'general constraints' (assertions) are part of the standard and could be used to implement the OCL constraints. However, assertions are not supported in the currently available DBMSs and developers are referred to the use of triggers and procedures. Assertions may be considered as an intermediate step between UML/OCL and the actual implementation of constraints with triggers and procedures in the DBMS, therefore assertions will be discussed further in Subsection 7.8.1. This section will further present the implementation of constraints for the landscape modelling system SALIX-2 (see Subsection 7.8.2), in DBMS, using triggers (see Subsection 7.8.3).

Once the system is extended with support for constraints, the user should be informed about the kinds of constraints available. This can be a simple list of all the maintained constraints, or a more sophisticated attempt-alert approach in the interactive edit environment.

The apparent benefit of front-end implementation is direct interaction with the user. For example, if a user places a plantation object in the VRML scene, fast feedback of the validity of this placement can be realised if the constraints are also maintained in the visual environment (e.g. the VR component of the SALIX-2 system). However, the possibilities of changing the constraints within the VR-environment are limited, because currently the constraints are 'hard-coded' in the VR application code, as in many other edit/simulation environments. In future development environments it should be possible to automatically generate the part of the VR program application code that implements the constraints (as specified in UML/OCL).

Database implementation offers better management of constraints. If the constraints are stored in a database, they are stored in a central place, easily accessible and therefore easily adaptable. If in the VR/SALIX-2 example the application only connects to the database when saving or loading a plantation plan (such as with SALIX-2), there is not a connection during the interactive creation (editing) of the plantation plan. So the user only gets feedback when the plantation plan is saved, not when the plantation object is placed. An obvious and simple improvement is to automatically connect to the database after the user finishes a 'logical edit unit'. In this way immediate feedback is given, generated by the database, but presented in the edit environment.

Supporting integrity constraints in all subsystems is probably the optimal solution. That is, storing the constraints in a (model) repository on a central location and encapsulating this information in the different subsystems. Using this approach the feedback of the system will be significantly improved. However, if the constraints would be independently implemented more than once, this may lead to inconsistency between the subsystems. Therefore it is very important that the implementation of the constraints in the VR-environment is automatically derived and consistency is guaranteed.

7.8.1 Support for constraints in DBMS

In this subsection the support for constraints within the SQL92 is presented and compared to the actual functionality available in a number of DBMSs. According to the standard, three types of constraint are defined:

1. domain constraints, for example enumeration and range types;
2. general constraints (assertions) for any situation; and,
3. base table constraints, related to tables.

Domain constraints are relatively simple and will not be further discussed. The two other types are more interesting. With *general constraints*, also called *assertions*, and a rich set of spatial operators, many of the different types of constraints described in this chapter can be specified (according to the SQL 92 standard). The syntax of an assertion is:

```
CREATE ASSERTION <assertion_name> CHECK <constraint_body>
```

This syntax is quite simple and straightforward. When an attempt is made to commit the changes in a database, after a set of updates, inserts, and deletes, the assertion is checked and if the expression evaluates 'true' than the commit succeeds, otherwise it fails and the database remains in the old state. One could easily imagine automatic generation of these assertions from the UML/OCL invariants in ways similar to those by which database table definitions (DDL) can be derived from UML class diagrams.

To illustrate this a few examples will be given. First, an integrity constraint involving an aggregation 'the maximum height of the trees should be less than 10 (metres)':

```
create assertion size_is_ok check
  ((select max(height) from tree) < 10);
```

The same constraint can be formulated differently with the 'exists' construction in SQL, again specified as an assertion:

```
create assertion size_is_ok2 check
  (not exists (select * from tree where height >= 10))
```

Next, a different example showing a constraint involving a topological relationship 'there should be no tree standing in the water':

```
create assertion tree_not_in_water check
  (not exists (select * from tree, water
     where inside(tree.loc, water.polygon)));
```

It is clear from these examples that assertions are very powerful as any thematic, temporal, topological and geometric condition can be specified, between any number of tables. So, if assertion could be automatically derived from the UML/OCL models, this would conveniently implement constraints in the database. However, despite the fact the assertions are part of the SQL92 standard there is apparently no current DBMS (commercial or non-commercial) supporting their implementation (Oracle, DB2, Ingres, Informix, PostgreSQL, MySQL). The alternative might be *base table constraints* as in theory they are functionally equivalent to assertions. For example the previous constraint 'there should be no tree standing in the water' can be written as the following base table constraint:

```
create table tree (id integer, height integer, loc point,
   constraint tree_not_in_water check
     (not exists (select * from water
                where inside(loc, water.polygon)))));
```

However, again there is disappointment as the mainstream DBMSs do not support base table constraints with subselects. That leaves the question 'What types of database table constraints are supported?' Below in more detail are the four types of database table constraints that are supported in Ingres (and this is quite representative for other DBMSs):

1.   Unique constraint

```
create table ape(name char(10) unique not null, ....);
```

2.   Referential constraint

```
create table mary(id integer, ape_name char(10) references
ape(name))
```

3.   Primary key constraint

```
create table ape2(name char(10) primary key, ....);
create table mary2(id integer, ape_name foreign key (name)
references(ape2));
```

4.   Check constraint (this is the only one with some 'semantic' load)

```
create table nut(balance integer check (balance > 0),
spending integer);
  create table nut2(balance integer, spending integer,
    constraint not_too_much check (spending < balance));
```

Though useful, these four types of constraint are not powerful enough to support the (spatial) constraints of the examples. One last option available for general constraint implementation in DBMSs is the use of triggers (and often in combination with procedures). Below is a constraint, now implemented as a trigger in Oracle (and similar functionality is available in other DBMSs) that checks whether the value of 'a_value' is not above the allowed maximum.

```
create trigger not_too_much
after insert or update of a_value on a_table

 DECLARE
  total number;
 BEGIN
  select sum(a_value) into total from a_table;
```

```
if (total >= 100) then
 raise_application_error (
  num => -20000,
  msg => 'Cannot add/update "a_value", sum too big');
 end if;
END;
```

The advantage of this solution is that, although not pleasant to code, it really works. In practice CASE tools can generate some parts of the code; e.g. Oracle's CDM Ruleframe (see Subsection 7.8.3 for more details). With respect to the support of constraints in DBMS, as mentioned in Section 7.1, specific support for topology structures is improving (this can be compared to the built-in support of referential integrity constraints). As already mentioned some available solutions for managing topology constraints within the database are:

❏ Oracle 10g (2003) spatial includes some initial support for topological structures (DBMS checks topological consistency; e.g. is a loop closed, no crossing edges),

❏ LaserScan Radius topology is a (Oracle) DBMS solution (LaserScan, 2003, Louwsma et al., 2003) and

❏ also more 'middleware' types of solutions are available with support for topology constraints; for example the ESRI geo-database (ESRI, 2002).

7.8.2 Example assertions for the landscape design case

Though assertions are not directly supported in the available DBMSs, they are compact representations of integrity constraints and form convenient intermediate representations for generation from the UML/OCL model, for their final implementation within the database. A number of example constraints from the landscape design system, SALIX-2 will be presented as assertions. The tables of SALIX-2 that are used in the assertions are: prcv_treesrd_point (plantation objects of type trees and bushes) and prcv_gvkrd_poly (ground surface with description water, paving, soft_paving, grass, and bridge). The examples introduced in Section 7.2, Table 7.1. and illustrated in the UML class diagram of Figure 7.1., will be used. The first example is 'Bushes never lie inside water' (note the use of the Oracle 'sdo_relate' operator):

```
create assertion constraint_1 check (not exists (
select * from prcv_treesrd_point t, prcv_gvkrd_poly g
 where t.treetype in ('CorAve', 'CorMas', 'RosCan')
   AND g.descript = 'water'
   AND sdo_relate (g.geom., t.geom., 'mask=inside,
querytype=window')='TRUE'))
```

The second example is a metric constraint specifying 'Trees always have to be positioned > 1 metre from paving' (again, note the use of the Oracle spatial operator 'sdo_within_distance'):

```
create assertion constraint_3 check (not exists (
SELECT * FROM prcv_treesrd_point t, prcv_gvkrd_poly g
 WHERE t.treetype IN ('FraxExc', 'QueRob')
   AND g.descript IN ('paving', 'soft_paving')
   AND sdo_within_distance (g.geom., t.geom., 'distance=1')
='TRUE'))
```

These examples show how easily and naturally all kinds of constraints can be specified. The last example includes an aggregation function to specify the constraint 'There must always be at least three trees on a specified ground surface' (for this constraint the grass polygon with id 20 is used):

```
create assertion constraint_4 check ( (
SELECT count(t.treeid) FROM prcv_treesrd_point t,
prcv_gvkrd_poly g
  WHERE t.treetype IN ('FraxExc', 'QueRob')
    AND g.id=20
    AND
sdo_relate(t.geom,g.geom, 'mask=ANYINTERACT,querytype=window')=
'TRUE'
  ) >=3)
```

7.8.3 Triggers and procedures

The assertions are easy to specify, but as mentioned, not supported in mainstream DBMSs. Therefore triggers (and procedures) offer the only practical way to implement constraints. Triggers can be seen as small programs checking certain conditions and prompting alerts with respect to the conditions. Using triggers one can achieve the same effect as by defining assertions. A functionally correct, but not very efficient, way to implement the assertions would be to glue all Boolean expressions of the individual assertions into one large Boolean expression. During every commit to the database, this expression is checked (via a trigger and procedure) and if the result is false, the transaction fails. However, this is not an efficient implementation and in this subsection triggers and procedures will be used in a more 'customised' way. However, one has to develop the more specific code for the triggers (and the used procedures) oneself. The syntax of specifying a trigger is:

```
CREATE [OR REPLACE] TRIGGER <trigger_name>
BEFORE | AFTER
INSERT OR UPDATE [<column(s)>] OR DELETE ON <table_name>
[FOR EACH ROW [WHEN (condition)]]
<trigger_body>
```

A partial example ('bush must not stand in the water') of an implemented constraint in a DBMS using triggers and procedures is given in (Louwsma, 2004). Within the front-end application a new object is created and an insert statement is sent to the DBMS. The statement is checked for integrity constraints and feedback is given through DBMS outputs. In order to avoid the 'low level' hand-coding of constraints (or business rules), Oracle provides a development tool called Custom Development Method (CDM) for automatically generating this code for the DBMS (Muller, 2002; Boyd, 2001). The CDM RuleFrame is the business rules implementation framework of CDM. The rules consist of three parts (Muller, 2002):

1. a function that indicates when the rule should be validated;
2. a function that performs the actual validation, when the previous function indicates the need; and,
3. a handling procedure, that manages the communication with the outside world.

CDM RuleFrame does not check business rules at the moment the user performs insert, update or delete statements. Rather, CDM RuleFrame stacks the rules that have to be enforced and checks them only at the moment of commit. Stacking the rules and the business rule enforcement is performed in the Transaction Management component.

7.9 Conclusion and future developments

7.9.1 Results obtained

Constraints have not attracted much attention in GIS, despite the need for them in dynamic GIS applications as illustrated in the four different cases where constraints were analysed. Each case had its own language specifying the constraints (natural text, XML format, SQL assertions). It was concluded that formalisation is needed and that (spatial) constraints must be specified in UML/OCL. This chapter also proposed a classification (taxonomy) of the constraint types relevant to a dynamic GIS environment based on: 1) number of involved objects/classes; 2) properties of objects and/or relationships between objects: metric, topological, temporal, thematic or mixed; 3) dimension (2D, 3D or 4D); 4) manner of expression 'never may' or 'always must'; and, 5) nature of the constraint 'physically impossible' or 'design objective'.

From the single UML/OCL model specification, implementation of different parts of the system should be automatically derived although current environments are not yet capable of supporting this. The implementation should be: 1) front-ended, allowing to generate direct feedback to the users during editing; 2) during data definition, thus making sure only valid data are stored and accepted; and 3) during the communication or data exchange in the case of loosely coupled clients and servers, thus making sure the client is aware of the constraints. Database assertions do seem quite close to the UML/OCL invariants and it is therefore rather disappointing that these assertions are not yet implemented in mainstream DBMSs. Currently, the implementation of the more complex constraint types, which includes nearly all spatial constraints, has to be realised by triggers and procedures.

7.9.2 Further investigations

In real interactive applications an end-user who is a non-programmer, such as a landscape designer, must be able to change, delete or make new constraints.

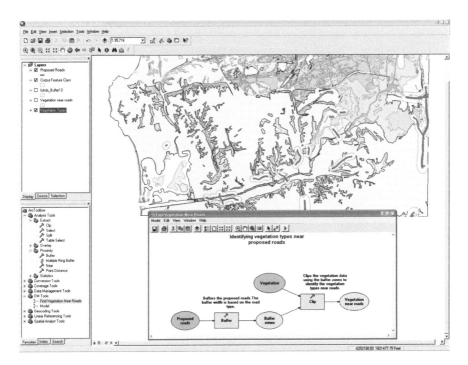

FIGURE 1.1 Simple 2D cartographic model of the impact of proposed new roads on vegetation in San Diego County, USA.

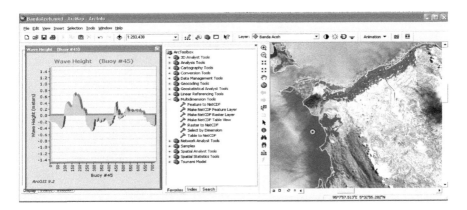

FIGURE 1.6 ArcGIS map and chart derived from a NetCDF simulation file of Banda Aceh, Indonesia.

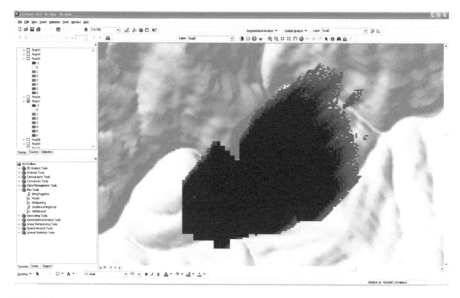

FIGURE 1.8 Visualisation of one step of a dynamic fire simulation mode.

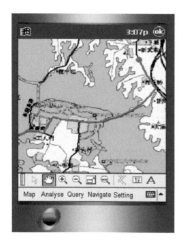

FIGURE 2.2 (a) The flood publishing software offers five functions, namely, map viewing, flood analysis and simulation, flood information querying, navigate to different monitoring sites and system calibration. Colour Figure 2.2 (b) Inundation map.

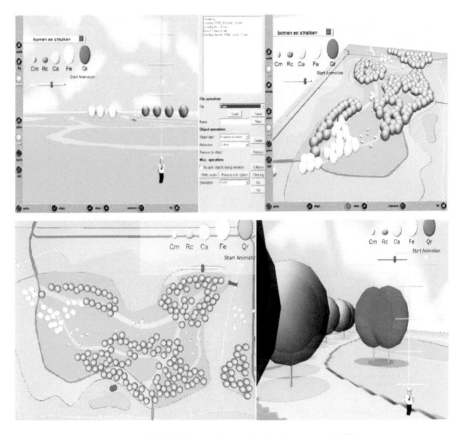

FIGURE 7.1 3-D scenes of SALIX-2: an interactive landscape modelling system.

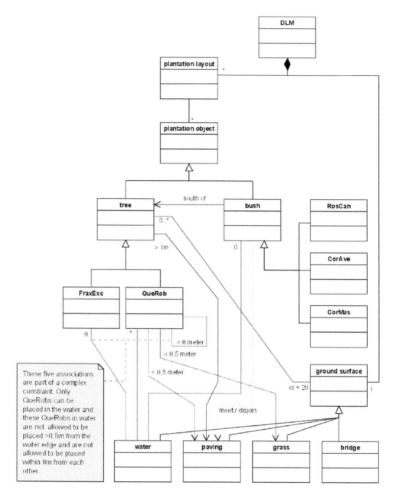

FIGURE 7.2 A constraint is violated in SALIX-2c (note the red, highlighted trees of Figure 7.1).

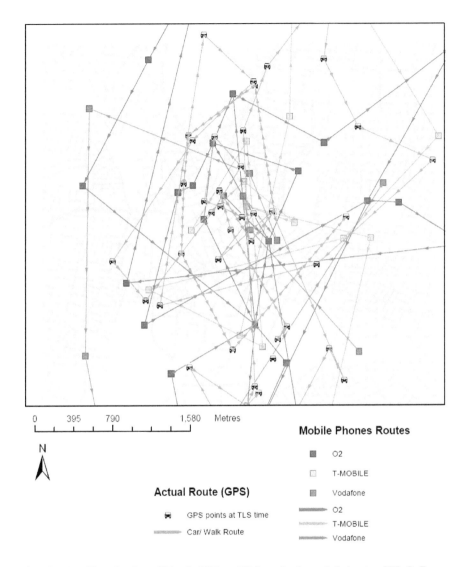

FIGURE 11.5 Visualization of Test 2, GPS vs. TLS tracks through Leicester, UK. © Crown Copyright/database right 2004. An Ordnance Survey/Edina Digimap supplied service.

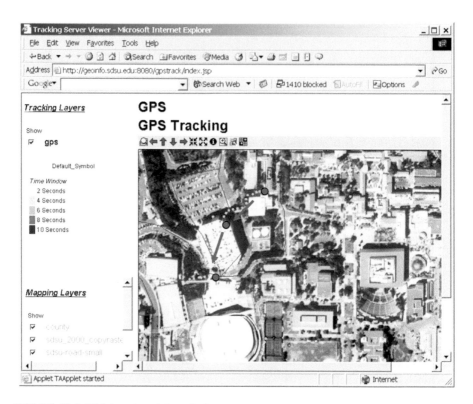

FIGURE 12.5 Web-based real time GPS tracking services (the moving red dot indicates a person with the Mobile GIS device sending GPS location back to an Internet map server in real time).

FIGURE 12.6 Land cover changes detection by Mobile GIServices (Tsou, 2004). The different coloured polygons indicate the different types of landcover changes. (First published in Cartography and Geographic Information Systems, vol. 31 (3), p. 164.)

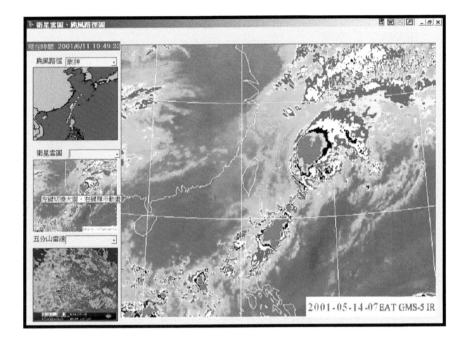

FIGURE 12.8 Typhoon Information Display System for Taiwan. (Top box on left shows the predicted path of typhoons, middle box shows the satellite image, and bottom box shows local radar information. Users can click on the left-mouse button for a full-size display or click on the right button to display animation.)

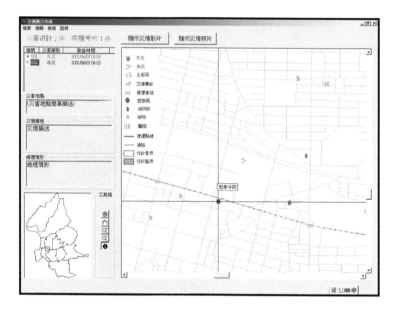

FIGURE 12.9 Damage Condition Display System in Taiwan that can track multiple disaster recovery and rescue tasks. (Top table on left shows ID of each disaster and its symbol - red for fire, blue for flood. Second table describes the disaster location and street names; third table describes the disaster situation and the damage level; fourth table shows the rescue and recovery actions taken by local government. Additional functions provide video and photos of the disaster taken at different locations. This screen shot shows that there are two major disaster events: one processed and one still awaiting action.)

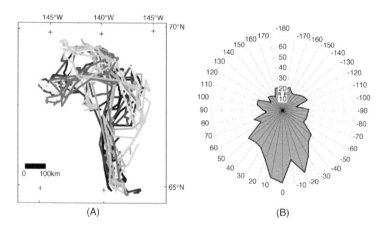

FIGURE 14.1 Exploration of geospatial lifelines. (A) Mapping the geospatial lifelines of moving point objects in a static map ignores completely the temporal aspect of motion and leads to confusing representations, as illustrated here with the tracks of only a dozen caribou migrating by the Beaufort Sea during two seasons. (B) The turning angle distribution of the same group of caribou illustrates the directional persistence in their motion (0° for straight on).

Therefore finding an easy way to interactively specify constraints is required. This is closely related to modelling. For example looking to the UML class diagram with all relevant object classes and their (restricted) relationships.

The user should get visual 'feedback' during editing; e.g. red or green highlighted areas during insert. These highlights should be derived from constraints, the instance geometry in a DBMS and be based on spatial functionality such as buffers and overlays. An investigation should take place to see whether the highlighted areas can be created inside the DBMS (and presented as views) or using specific GIS software.

This research can be extended to 2½D or 3D referenced objects. The objects of interest are currently often limited to point objects, polyline objects and polygon objects, but should be extended to volume (polyhedron) objects. However, data types and operations for 2½D and 3D geometry in DBMS are not yet available, but will be indispensable for implementing 2½D and 3D constraints concerning 2½D or 3D objects.

One issue not yet mentioned in this chapter is that, with a number of user-specified constraints, conflicts may arise; therefore, a good mechanism has to be developed to check for this. When users can change the constraint definitions of an existing application, then a conflict check should take place. This is of concern for existing software. For example topology constraints can be implemented in the ESRI (2002) geo-database software, but there is no check as to whether these conflict with each other. This can result in unwanted system behaviour; e.g. 'an infinite loop' correcting one error, which results in another error. There will never be a correct situation satisfying both rules (integrity constraints) as they are conflicting.

Finally more research is needed for topics such as: the automatic translation of UML/OCL to XML schema (as used in the exchange); the automatic translation of UML/OCL models to the edit environment (as in the case of the topographic data maintenance application); the relationship between consistent data and operations (as illustrated in the Web feature service case) and extending the constraints to space-time/simulation. The ultimate goal would be to support an environment in which it is possible to delete, change and add new constraints and automatically rebuild new versions of the different subsystems, during edits, storage and data exchange. Good support for constraints in a dynamic GIS environment is essential for data quality control and decisions based on these data.

Acknowledgements

All persons and organisations involved in the case studies are acknowledged: VR landscape design (Jildou Louwsma, MSc student, Sisi Zlatanova, TUD supervisor and Ron van Lammeren, Wageningen University, research supervisor), Dutch Cadastre (Christiaan Lemmen and Peter Jansen), Topographic Service (Nico Bakker), Notary drafts parcel boundary (Thijs Brentjes, MSc student, Marian de Vries, TUD supervisor and Tom Vijlbrief, Cadastre supervisor). Elfriede Fendel is acknowledged for the careful proofreading of the first version of the manuscript.

Many thanks to Roland Billen and Jane Drummond for their editorial support. The author expresses his gratitude to the research program 'Sustainable Urban Areas' at Delft University of Technology for making this publication possible.

References

Belussi, A., Negri, M. and Pelagatti, G. (2004) 'GeoUML: A Geographic conceptual model defined through specialization of the ISO TC211 standards', in *Proceedings 10th EC GI & GIS Workshop, ESDI State of the Art,* Warsaw, Poland, (23–25 Jun 2004).

Boyd, L. L. (2001) 'CDM RuleFrame – the business rule implementation framework that saves you work', *ODTUG 2001, Business Rules Symposium,* Oracle Corporation, iDevelopment Center of Excellence.

Brentjens, T. J. (2004) *OpenGIS Web Feature Services for Editing Cadastral Data; Analysis and Practical Experiences,* MSc Thesis Geodetic Engineering, Delft University of Technology.

Brentjens, T., de Vries, M., Quak, W., Vijlbrief, C. and van Oosterom, P. (2004) 'Evaluating the OpenGIS Web Feature Services protocol with the case study 'Distributed Cadastral Transactions'', in Fendel, E.M. and Rumor, M. (eds.) *Proceedings of the 24th Urban Data Management Symposium,* Chioggia (27–29 Oct 2004), ch. 25, pp. 11–22.

Clementini, E., Di Felice, P. and van Oosterom, P. (1993) 'A small set of formal topological relationships suitable for end-user interaction', in *SSD'93: the Third International Symposium on Large Spatial Databases,* Singapore, pp. 277–295, LNCS no. 692, Berlin: Springer-Verlag.

Cockcroft, S. (1997) 'A taxonomy of spatial data integrity constraints', *GeoInformatica,* 1,4, pp. 327–343.

Cockcroft, S. (2004) 'The design and implementation of a repository for the management of spatial data integrity constraints', *GeoInformatica,* vol. 8, no. 1, pp. 49–69.

Collins, F. C. and Smith, J. L. (1994) 'Taxonomy for error in GIS', in Congalton, R. G. (ed.) *Proceedings International Symposium on Spatial Accuracy in Natural Resource Databases "Unlocking the Puzzle",* pp. 1–7, Williamsburg: American Society for Photogrammetry and Remote Sensing.

Date, C. J. and Darwen, H. (1997) *A Guide to the SQL Standard,* 4th edition, ch. 14, pp. 197–218, Boston, Massachusetts: Addison-Wesley.

Egenhofer, M. J. (1989) 'A formal definition of binary topological relationships', *Proceedings of 3rd International Conference on Foundation of Data Organisation and Algorithms,* Paris, pp. 457–472.

ESRI (2002) *Working with the Geodatabase: Powerful Multi-User Editing and Sophisticated Data Integrity,* ESRI.

Heim, M. (1998) *Virtual Realism,* New York: Oxford University Press.

Hunter, G. J. (1996) 'Management issues in GIS: Accuracy and Data Quality', in Hunter, G. J. (ed.) *Proceedings of Conference on Managing Geographic Information Systems for Success,* Aurisa: Melbourne, Australia, pp. 95–101.

Kwon, Y-M., Ferrari, E. and Bertino, E. (1999) 'Modelling spatio-temporal constraints for multimedia objects', *Data & Knowledge Engineering,* vol. 30, pp. 217–238.

van Lammeren, R., Ogrin, D., Marusic, I. and Simanic, T. (2002) 'Virtual Reality in the landscape design process', *Proceedings International Conference on Landscape Planning in the Era of Globalisation,* pp. 158–165.

Laser-Scan (2003) *Technical Product Description - Topology Users Guide,* Cambridge, UK.

Louwsma, J. (2004) *Constraints in Geo-Information Models Applied to Geo-VR in Landscape Architecture,* MSc Thesis Geodetic Engineering, Delft: Delft University of Technology.

Louwsma, J., Tijssen, T. and van Oosterom, P. (2003) 'Topology under the microscope', *Geoconnexion,* vol. 2, no. 6, pp. 29–30.

Muller, S. (2002) *CDM RuleFrame Overview: 6 reasons to get framed!,* Oracle Corporation, iDevelopment Center of Excellence.

OGC (2002) Vretanos, P. (ed.) *Web Feature Service Implementation Specification*, version 1.0, Reference number: OGC 02–058, OpenGIS Consortium Inc.

OGC (2003) Cox, S., Daisey, P., Lake, R., Portele, C. and Whiteside, A. (eds.) *OpenGIS Geography Markup Language (GML) Implementation Specification*, version 3. Reference number: OGC 02–023r4, OpenGIS Consortium Inc.

OMG (2005a) Unified Modeling Language: Superstructure, version 2.0, formal/05-07-04, August 2005, [Online], Available: http://www.omg.org/cgi-bin/doc?formal/05-07-04.

OMG (2005b) Unified Modeling Language - Object Constraint Language (UML-OCL) 2.0 Specification Version 2.0, ptc/2005-06-06 June 2005, [Online], Available: http://www.omg.org/docs/ptc/05-06-06.pdf.

van Oosterom, P. (1997). 'Maintaining consistent topology including historical data in a large spatial database', *Proceedings of Auto-Carto 13*, Chrisman, N. (ed.), Bethesda: ACSM & ASPRS, pp. 327–336.

van Oosterom, P. and Lemmen, C. (2001) 'Spatial data-management on a very large cadastral database', *Computers, Environment and Urban Systems*, vol. 25, no. 4–5, pp. 509–528.

van Oosterom, P., Tijssen, T. and Penninga, F. (2005) 'Topology storage and use in the context of consistent data nanagement', *GISt report no. 33*, Delft, [Sep 2005].

Oracle (2003) *Oracle Spatial Topology and Network Data Models, 10g Release 1 (10.1)*, Author: Oracle.

Papadias, D. and Theodoridis, Y. (1997) 'Spatial relations, minimum bounding rectangles and spatial data structures', *International Journal of GIS*, vol. 11, no. 2, pp. 111–138.

Peuquet, D. J. (1995) 'It's about time: a conceptual framework for the representation of temporal dynamics in GIS', *Annals of the Association of American Geographers*, vol. 84, no. 3, pp. 441-461.

XSD (2004) XML Schema, [Online], Available: http://www.w3.org/XML/Schema, [14 Jul 2005].

Appendix 1. Cadastral constraint violation queries (CCVQ) in SQL

```
    /* CCVQ1 */
SELECT object_id, numpoints(shape),
    anypoint(shape, 1), anypoint(shape, 2), anypoint(shape, 3)
FROM xfio_boundary
WHERE numpoints(shape)=3 and interp_cd=3 and tmax = 0 and
   ogroup = 6 and (anypoint(shape, 1) = anypoint(shape, 3));
/* CCVQ2 */
SELECT object_id from lki_parcel
WHERE inside(location, geo_bbox) != 1;
    /* CCVQ3 */
SELECT l_obj_id FROM lki_boundary
WHERE l_obj_id not in (select object_id from lki_parcel);
    /* CCVQ4 */
SELECT object_id, fl_line_id FROM lki_boundary
   WHERE abs(fl_line_id) not in (select object_id from lki_boundary);

    /* CCVQ5 */
SELECT object_id, line_id1 FROM lki_parcel
WHERE abs(line_id1) not in (select object_id from lki_boundary);
    /* CCVQ6 */
SELECT s.object_id, s.fl_line_id
FROM xfio_boundary s, xfio_boundary r
```

```
WHERE s.fl_line_id > 0 and s.fl_line_id=r.object_id and
   s.tmax=0 and s.ogroup=6 and r.tmax=0 and r.ogroup=6 and
   s.r_obj_id <> r.l_obj_id;

   /* CCVQ7 */
SELECT s.object_id, s.fl_line_id
FROM xfio_boundary s, xfio_boundary r
WHERE s.fl_line_id > 0 and s.fl_line_id=r.object_id and
   s.tmax=0 and s.ogroup=6 and r.tmax=0 and r.ogroup=6 and
   (anypoint(s.shape, 1) <> anypoint(r.shape, 1));

   /* CCVQ8 */
SELECT s.object_id, s.line_id1
FROM xfio_parcel s, xfio_boundary r
WHERE s.line_id1 > 0 and s.line_id1=r.object_id and
   s.tmax=0 and s.ogroup=46 and r.tmax=0 and r.ogroup=6 and
   (s.object_id <> r.r_obj_id);

/* CCVQ9 */
   SELECT count(*),municip FROM mo_object
WHERE pp_i_ltr='G' and x_akr_objectnummer not in
   (select x_akr_objectnummer from lki_parcel)
GROUP BY municip;
```

Appendix 2. Topographic model constraints in XML

```
TMCX1:
   <Domein>
        <Naam>dWaterBreedte</Naam>
        <Registreren>J</Registreren>
        <Beschrijving>Breedte voor Waterdeel</Beschrijving>
        <Type>Range</Type>
        <DataType>int</DataType>
        <CodedValueData/>
        <RangeData>
                <Minimum>1</Minimum>
                <Maximum>500</Maximum>
                <Default>6</Default>
        </RangeData>
        <SplitRegel>Duplicate</SplitRegel>
        <MergeRegel>DefaultValue</MergeRegel>
   </Domein>

TMCX2:
   <Domein>
        <Naam>dSpoorTypering</Naam>
```

```
        <Registreren>J</Registreren>
        <Beschrijving>Typeringen voor Spoorbaandeel</Beschrijving>
        <Type>CodedValue</Type>
        <DataType>int</DataType>
        <CodedValueData>
                <Code>44</Code>
                <Value>verbinding</Value>
                <Default>J</Default>
        </CodedValueData>
        <CodedValueData>
                <Code>45</Code>
                <Value>kruising</Value>
                <Default>N</Default>
        </CodedValueData>
        <RangeData>
                <Minimum/>
                <Maximum/>
                <Default/>
        </RangeData>
        <SplitRegel>Duplicate</SplitRegel>
        <MergeRegel>DefaultValue</MergeRegel>
    </Domein>

TMCX3:
    <AttribuutRegel>
        <Nummer>att007a</Nummer>
        <VervolgNummer/>
        <Categorie/>
        <Beschrijving>Als Naam_Aweg is ingevuld,
                dan WegType moet 'autosnelweg' bevatten</Beschrijving>

        <FoutMelding>WegType bevat niet autosnelweg</FoutMelding>
        <TriggerNiveau>1</TriggerNiveau>
        <VervolgOperator/>
        <FeatureKlasse>EDT_WEG_VLAK</FeatureKlasse>
        <AlsAttribuut>NAAM_AWEG</AlsAttribuut>
        <AlsOperator>!=</AlsOperator>
        <AlsWaarde>""</AlsWaarde>
        <DanAttribuut>WEGTYPE</DanAttribuut>
        <DanOperator>MVCONTAINS</DanOperator>
        <DanWaarde>|autosnelweg|</DanWaarde>
    </AttribuutRegel>
TMCX4:
```

```
<AttribuutRegel>
     <Nummer>top04</Nummer>
     <VervolgNummer/>
     <Categorie/>
     <Beschrijving>Indien Wegvlak overlapt met Wegvlak
             dan moet HOOGTENIVEAU verschillend zijn</Beschrijving>

     <FoutMelding>Wegvlak overlapt Wegvlak</FoutMelding>
     <TriggerNiveau>1</TriggerNiveau>
     <VervolgOperator/>
     <FeatureKlasse>EDT_WEG_VLAK</FeatureKlasse>
     <AlsAttribuut>OBJECTID</AlsAttribuut>
     <AlsOperator>AREAOVERLAPAREA</AlsOperator>
     <AlsWaarde>EDT_WEG_VLAK</AlsWaarde>
     <DanAttribuut>HOOGTENIVEAU</DanAttribuut>
     <DanOperator>!=</DanOperator>
     <DanWaarde>FEATURE2.HOOGTENIVEAU</DanWaarde>
  </AttribuutRegel>

TMCX5:
  <AttribuutRegel>
     <Nummer>brn01</Nummer>
     <VervolgNummer/>
     <Categorie/>
     <Beschrijving>Iedere feature moet een Bron
hebben</Beschrijving>
     <FoutMelding>Geen Bron</FoutMelding>
     <TriggerNiveau>1</TriggerNiveau>
     <VervolgOperator/>
     <FeatureKlasse>EDT_WEG_VLAK</FeatureKlasse>
     <AlsAttribuut>OBJECTID</AlsAttribuut>
     <AlsOperator>&gt;=</AlsOperator>
     <AlsWaarde>0</AlsWaarde>
     <DanAttribuut>OBJECTID</DanAttribuut>
     <DanOperator>BRONCOUNT&gt;</DanOperator>
     <DanWaarde>0</DanWaarde>
    </AttribuutRegel>
```

Part III

Display and Visualisation

These three chapters forming 'Part III – Display and Visualisation' of 'Dynamic and Mobile GIS: Investigating Changes in Space and Time' examine the challenges of visualisation within the constraints of mobile GIS, particularly the small-screen of a typical PDA, but also the limitations of using only GPS as a positioning system. It is noted that in all three chapters it is the visualisation of transportation networks that is under consideration.

In Chapter 8, Malisa Plesa and William Cartwright remind us that the 240 x 320 pixel resolution and 16-bit colour display of the typical PDA restrict the appearance of a displayed map. So, although inexperienced users express a preference for 'realistic' visualisation, photorealism, for example, is very difficult to achieve within the PDA environment. A series of tests, including a navigation task, were given to ten subjects. These tests required representing a downtown area realistically (using oblique photography) and non-realistically and determining their usability.

The tests showed that the non-realistic map was more effective. For example, the slight tonal variations required to display photography could not be shown on the PDA, limiting its utility for the display of realistic imagery. The non-realistic map was able to utilise a series of colours that achieved the maximum contrast and legibility available from this form of map delivery. As well as the results obtained showing that non-photorealistic graphics are more effective than photorealistic graphics in the context of 3D mobile cartography, non-realistic representations are easier to develop than realistic scenes, so are more suited to delivery on mobile devices.

Plesa and Cartwright's test, although involving transport networks, was for pedestrians; the representation of buildings and other neighbourhood features was as important as that for the road network. The extra rapid decision making required for vehicle navigation, compared to pedestrians, will almost entirely limit the display to the road network. Suchith Anand, Mark Ware and George Taylor, in Chapter 9, consider the generalisation of transport networks specifically for Mobile GIS with PDA display. Transportation networks are considered ideal candidates for using 'schematisation' to ease the interpretation of information by generalisation. Schematic maps are diagrammatic representations based on linear abstractions of networks. The authors of Chapter 9 have looked at the generalisation techniques that can be applied to generate schematic maps from large-scale digital geographic data, and among their conclusions is that, to be effective, a route map must show all turning points on a road.

Route finding using mobile GIS not only requires an adequate representation of the transport network, but also adequate positioning. Reliable and accurate determination of the current user position is considered a prerequisite not only for car navigation (automotive) applications but also for a wide variety of mobile GIS applications. The algorithm, described by Britta Hummel in Chapter 10, is not specifically for car navigation applications and so can be integrated into any mobile GIS requiring positioning. But the challenge, in navigation along a mapped network, is not just to say where you are on the Earth's surface (which GPS can do adequately), but where you are in the map.

Hummel presents a map matching method that exclusively relies upon information from a standard GPS receiver. But, the method exploits the vehicle position and orientation history, information about road network topology, driving restrictions and the assumed driving direction for each road element, to allow correct positioning on the network.

This part of the book, although ostensibly addressing visualisation in the dynamic and mobile GIS environment in an unrestricted manner, has concentrated on the visualisation of the transportation network and positioning within it. The reader might feel that there are other visualisation challenges for mobile GIS, and indeed there are, but given the way mobile GIS will be applied (i.e. in dynamic circumstances), the visualisation of the transport network provides a real visualisation challenge, and is a necessary first step to whatever follows.

Chapter 8

An Evaluation of the Effectiveness of Non-Realistic 3D Graphics for City Maps on Small-Screen Devices

Malisa Ana Plesa and William Cartwright

RMIT University, Australia

8.1 Introduction

When we think of navigating through a city we need to determine where we are, where we wish to get to, and mentally calculate how to get there. We use landmarks, signs and other relevant built and natural environmental cues to assist our wayfinding. But we generally rely upon maps to provide an overview of the space we wish to move through and as a tool to assist navigation. Once, these were paper products, designed, delivered and consumed under the 'umbrella' of a print mindset. But, now, paper maps are just one of the choices in the plethora of locational and wayfinding tools available. We are offered maps on mobile telephones, Personal Digital Assistants (PDAs), on television monitors, in kiosks and even as illuminated billboards.

However, in many instances the way in which the maps are designed for use with contemporary information delivery tools, and especially those that deliver geographical information via small-screens, is no different to the designs for their paper map cousins. 'Just' applying the 'rules' of paper maps, or maps on large-screen computer graphics, cannot ensure a usable mapping product for navigation and wayfinding. A different approach needs to be explored.

The use of Non-Photorealistic Rendering (NPR) for 3D cartography displays on mobile devices was explored. A research 'gap' was identified, and the following questions were found to be unanswered:

1. Are non-photorealistic computer graphics more effective than photorealistic graphics for the delivery of three-dimensional, spatial information on mobile devices?
2. What is the potential of non-photorealistic computer graphics, combined with three-dimensional cartography, for the representation of spatial information on mobile devices?

One way to address these questions is to create and test a prototype built on Döllner and Walther's (2003) theory. This prototype consisted of two 3D maps: one non-photorealistic map that utilised similar design strategies as those outlined by Döllner and Walther (2003), and one aerial photograph that provided an example of a

Dynamic and Mobile GIS: Investigating Changes in Space and Time. Edited by Jane Drummond, Roland Billen, Elsa João and David Forrest. © 2006 Taylor & Francis

photorealistic map. They were created as static views and the extent of each map focused on the same urban area, situated just south of the Melbourne CBD, Australia. The prototype was delivered on a Personal Digital Assistant (PDA) and tested *in situ* with a representative target population.

This chapter provides a background to the application of non-realistic computer graphics to mobile devices and describes the results from a research project formulated to find an answer to the above questions.

8.2 Photorealism vs. non-photorealism

Since the introduction of computer graphics, the ultimate goal has been to achieve photographic realism (Schumann et al., 1996; Durand, 2002; Gooch and Gooch, 2001). Computer graphics has subsequently realised a need for the generation of abstract imagery, which has led to a rapidly growing interest in NPR (Markosian et al., 1997; Goldstein, 1999; Gooch and Gooch, 2001). Non-Photorealistic Rendering is focused on emphasising the most relevant components of an image, while suppressing unimportant details. This helps to overcome the expense related to the creation of realistic imagery, while still providing users with the same information. This can be seen as something akin to the bird's-eye views shown in some early map-like artefacts that were produced to show cityscapes. Since photorealism has become achievable, little research has been undertaken into alternative methods of information display, and the benefits of photorealism to user understanding currently remain unclear (Schumann et al., 1996; Markosian et al., 1997; Ferwerda, 2003; Gooch and Gooch, 2001).

Non-photorealism is predominantly focused on user understanding. It effectively communicates subtle information while also highlighting important details. Research suggests that the human mind is able to complete abstract information, and that realism does not directly influence human image interpretation (Duke et al., 2003). Non-photorealism represents a form of functional realism, whereby knowledge relating to the properties of objects are provided to the user, allowing them to make reliable visual judgements (Gooch and Willemsen, 2002). Photorealism provides the same visual response or stimulation as the original scene. In contrast, functional realism provides users with the same visual information, allowing them to use imagery to help them complete real-world tasks (Ferwerda, 2003).

8.3 An NPR technique for mobile city models

Döllner and Walther (2003) presented a non-photorealistic rendering technique for 3D city models. Principles derived from cartography, cognition and non-photorealism form the basis of this technique, which may hold a number of advantages over photorealistic representations. Less graphical and geometric detail is required for the construction of these models in order to produce favourable results. The potential of non-photorealistic models for small-screen delivery was outlined, and it was found that these models had a number of theoretical advantages over photorealism in this context.

Most studies concerning 3D map design for mobile devices focus on the development and use of photorealistic imagery and do not attempt to explore alternative methods of display. It is argued that more focus needs to be directed at determining the most effective technique to display this information on handheld devices, as current research does not appear to address this issue. Contemporary cartography has followed a similar path to that of computer graphics, but as computer graphics realises the need for NPR in some applications, 3D cartography is still primarily interested in realism. Döllner and Walther (2003) have identified a possible need for non-photorealism in city models and have devised an NPR technique. While Döllner and Walther's (2003) technique is theoretically feasible, there was an interest to test its effectiveness in a real-world situation. This research involved the development of a prototype that simulated their technique. The prototype was then used to compare the technique with a realistic representation before testing in the field with real users. By doing this, it was hoped that a clearer understanding relating to effective 3D map design for small-screen devices could be established.

8.4 Developing the prototypes

A prototype was built to demonstrate how Döllner and Walther's (2003) expressive rendering of city models could be applied to a map for display on a small-screen, and to compare this with the use of a photorealistic map. Expressive rendering involved manipulating an initial image produced with a CAD package so that it begins to resemble a hand-drawn sketch. Lines defining the edges of buildings are made to change thickness along their length, colours are chosen from basic hues and the general impression of the image appears more uneven that a typical CAD output. Expressive rendering has as its aim to produce simple images that impart a more general image of a building, rather than a hard-edged precise CAD drawing.

This prototype contained a 2.5D map of an urban area, encompassing both pan and zoom functionality. It was designed to be delivered on a small-screen device for evaluation. The development of the prototype involved four stages:

1. Data collection;
2. Base map production;
3. Map design; and
4. Preparation for delivery.

The map coverage of the prototype included the southern part of the city of Melbourne's CBD and the eastern portion of the Southgate Arts and Leisure Precinct. This area was selected because of its popularity as an entertainment destination for both tourists and Melbournians. The area features a variety of land cover types, including parkland, varying building densities and forms, different types of pedestrian access routes, and the Yarra River, which is ideal to demonstrate the concepts governing the map design.

A 1:4,000 scale photograph provided a large-scale image of the area, clearly showing the footprints and position of buildings. Because of the non-realistic nature of the map, precise building heights were not necessarily required. Instead, building heights were measured in the field using a clinometer. In order to keep the simplicity of the map, building heights were generalised so that buildings with similar heights in the real world would appear to have the same heights on the map.

Due to the 2.5D nature of the map, it was deemed important to select the most appropriate viewing angle. It was found that constructing a simple 3D model could eliminate this problem. A 3D model can be easily viewed from an infinite number of points, allowing for a quick method of comparison to aid in the selection process. This also makes revision easier, as the map does not need to be recreated each time. The amendments can be applied to the model, and the viewing angle can be adjusted accordingly. This was completed using *AutoCAD*. It was used to digitise the aerial photograph and extrude building footprints.

The map was designed according to the NPR principles devised by Döllner and Walther (2003). Döllner and Walther devised a rendering technique for city models that combined the principles of cartography, cognition and non-photorealism. Non-photorealistic city models are by no means a 'new' revelation in cartography, and there is ample historic evidence of their existence long before computer cartography. These traditional city models have become somewhat dated with the development of computer technology, which is now able to create images and models that mimic reality. Even though photorealism is now attainable, it is still computationally expensive and unfit for many of today's display mediums.

The creation of the prototype provided a valuable insight into the different needs and methods required for photorealistic and non-photorealistic city models. Even though the photorealistic map used in the prototype was not created from scratch, some useful observations have been made and can be used to compare these different needs and methods.

The first stage of production for the non-realistic map involved digitising the vertical aerial photograph and extruding building footprints, resulting in a basic 3D wireframe model. In order to achieve maximum realism, a photorealistic city model requires a considerable amount of time and skill in the production stage. While a non-photorealistic model requires more time in the design stage, this helps to set clearly defined goals for production, making it a relatively simple process.

The non-realistic map developed for the prototype utilised NPR software to produce the final map. The use of such software provides a level of automation in production that is not available to the creation of realistic models. This software also makes modifications somewhat easier. A non-expert developer could return to the initial model and easily apply design changes before using the NPR software to output a new design. In addition, it is believed that a non-realistic map would require less frequent updates, as real-world changes to building façades and the landscape would not compromise its currency as much as it would a realistic map.

The NP renderer, *Penguin* from McNeel, was used to add NPR effects to 3D models created in *Rhino* and *AutoCAD* (McNeel, 2003). *Penguin* runs inside these programs to provide seamless rendering without the need for exporting or starting over. It features two rendering modes: sketch (which simulates hand drafting) and cartoon (which utilises a limited number of shades and edges to generate 'illustration' drawings). Its ability to render both axonometric and perspective scenes, the user-controlled anti-aliasing option, and its customisable render output were thought to be important factors for designing 3D maps for small-screen delivery. These features also make it quick and simple to replicate the rendering process, and the end result of the rendering could be customised so that it resembled the city models created by Döllner and Walther (2003).

Figure 8.1 presents a section from the final non-photorealistic map as created in *AutoCAD* and rendered with *Penguin*.

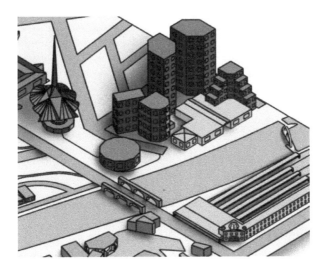

Figure 8.1. The non-photorealistic map as created in AutoCAD and rendered with Penguin.

An oblique aerial photograph of Melbourne was obtained to provide a photorealistic representation for comparison. It was taken from Melbourne's North East facing a South Westerly direction. This angle provides a clear view of many of the landmarks within the area.

The prototype was simulated on a handheld device. Macromedia *Flash,* a multi-media software development tool, was used to create the simulation. Each map was imported into *Flash* at a resolution to fit the largest scale displayed. This was done to avoid pixellation. Zoom and pan functionality were incorporated. A labelling function was also included, enabling users to view feature labels for roads and points of interest. This functionality was also applied to the realistic map, which was simulated in *Flash* in a similar manner to allow for effective comparison. This was teamed with a 'tutorial' map of the world, which was used as skills exercise for participants prior to the commencement of the evaluation stage of the project.

Macromedia provide a *Flash* Player for handheld devices running the operating system *Pocket PC*. *Flash Player 6 for Pocket PC* enables publishing *Flash* content on PDAs. A stand-alone player allows for the creation of 'projectors', executable files that can be viewed on Pocket PC devices without the need for a *Flash* player. A Compaq *iPAQ* Pocket PC H3700 PDA, a standard, colour display running Microsoft *Pocket PC 2002* was used to deliver the product. Screen shots taken from the prototype, showing both the non-realistic and realistic maps, are shown in Figure 8.2.

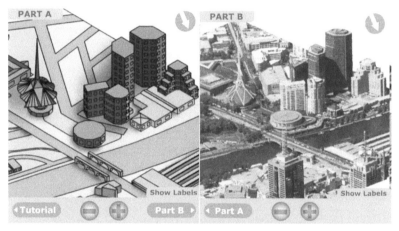

Figure 8.2. PDA screen shots: the non-realistic map (left) compared to the realistic map.

8.5　　Testing and evaluating the prototype

The evaluation procedure required test participants to use each of the maps displayed on a mobile device while undertaking realistic navigational activities. After using each map to undertake a set of real-life scenarios, users had enough usage time to develop opinions relating to each map. Upon conclusion of the evaluation session, they were debriefed so that data about their likes and dislikes could be recorded. This data was used to establish a greater understanding of user preferences, which assisted in developing recommendations for 3D map design for small screens.

8.5.1 Evaluation method

The method used to test the prototype and gather data was similar to that of a usability study: the only difference was that it was concerned with the usability of the maps, rather than the interface. Rubin (1994) has outlined a Comparison Test, which compares two or more designs. Data relating to each design is gathered, and their results are compared to establish which design shows most potential. For comparison tests, the best results are achieved by using extremely different alternatives because it forces test participants to really consider why one design is better than the other, and the reasons why this is so (Rubin, 1994). These factors influenced the test design, test execution and data analysis stages of this study.

8.5.2 Selection of test participants

If reliable results are to be achieved, it is important to conduct user testing with real users. This provides an example of how the prototype would be used in reality, while identifying the exact problems that users have while using it (Nielsen, 1993). Studies undertaken by Dumas and Redish (1999) suggest that factors such as relevant experience and motivation matter more than demographic factors if satisfactory results are to be gained from a usability study. Users' experience differs along three main dimensions (Nielsen, 1993): computer experience, experience with the system and knowledge of the domain. These differences must be considered when selecting test participants.

It was decided that the intended users of the prototype created for this study should possess the following characteristics (Figure 8.3):

- Developed map-reading skills;
- Proficient in computer use; and
- Experience using handheld devices.

Test candidates were university students who knew the City of Melbourne, but did not have an intimate knowledge of the study area. All of the participants were aged between 18 and 25. Male was the dominant gender, with only three females participating in the evaluation. All candidates volunteered to participate in the evaluation.

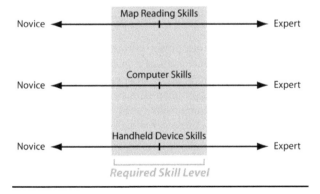

Figure 8.3. Desired skill level of test participants required for this study.

Extensive computer experience was a common factor among all participants; they each had at least five years' experience, and half had used personal computers for more than ten years. They had also been exposed to a wide range of map products, and all had used at least two different types of digital mapping applications. Most participants considered their map reading expertise to be 'competent', with the remaining four specifying 'reasonable' and 'expert' capabilities. Only three participants had used a PDA prior to undertaking the evaluation, although none of them had previously used one to operate a mapping application. Half had had experience using 3D (or 2.5D) maps, including visualisations, bird's-eye views and city plans.

8.5.3 How many to test?

In deciding how many participants were required for this study, factors associated with time and accuracy were considered. There needed to be enough participants to uncover any major issues associated with the prototype, while also fitting into the specified time frame. A study undertaken by Virzi (1992) revealed that 80 per cent of usability problems are uncovered with four or five test participants, including the most severe problems. This also suggested that more participants were less likely to detect additional problems.

A typical usability test includes six to twelve test participants in two or three subgroups (Dumas and Redish, 1999). This number generally creates a balance between time and information gained. The time spent testing is only one of the factors to consider when deciding on the number of participants, as time allocated to recruiting and analysing results is also increased with additional participants.

It was decided that ten participants would be sufficient for this study. These would then be broken up into two subgroups of five participants.

8.5.4 Test materials

Test materials needed to be developed to communicate with the user and collect the required data. These needed to be completed prior to the pilot test so that they could be evaluated and modified accordingly. Rubin (1994) outlined the most common test materials used in usability studies. The ones that were incorporated into this study are:

- ❑ Orientation script – a preset introduction that is read to participants prior to the commencement of a usability test.
- ❑ Background questionnaire – designed to gather data relating to their age, gender, education, map-using expertise and computer experience. Close-ended questions were used, requiring participants to choose their answer from a preselected list of options.
- ❑ Data logging sheet – divided into four parts, one for each scenario.
- ❑ Task scenarios.
- ❑ Post-test questionnaire, used to gather preference data once users have gained some perspective relating to the product (Dumas and Redish, 1999). This provides an indication of the product's perceived strengths

and weaknesses, by calling on participants' opinions and feelings (Rubin, 1994).

The post-test questionnaire used in this study consisted of three parts. Part A and part B, respectively, corresponded to the non-photorealistic map and the realistic map. They required scaled responses, utilising both Likert scales and semantic differential scales. For a Likert scale, users are presented with a statement (e.g. 'I found the non-realistic map easier to use.') and asked to rate their degree of agreement by marking the appropriate checkbox. Sematic differential scales list two opposite terms (e.g. legible and illegible) along some dimension and require users to indicate their opinion by placing their rating somewhere along that scale. Part C was designed to gather preference data by providing participants with open-ended questions. These questions required the participants to list the reasons why they did or did not prefer a particular map (e.g. 'Please state the reasons why you preferred the map you selected in Q2.'), and also provided them with an opportunity to make any final comments or suggestions. Questions were kept brief, specific and free of bias. They were also linked directly to the objectives of the study, and asked for information that cannot be directly observed by other means (Rubin, 1994).

8.5.5 Testing procedure

Thomas (2003) identified 'kinds of meaning' that are derived from interpretation methods. The 'kinds of meaning' established from this study include comparison, evaluation and generalisation. The non-photorealistic map was compared with the photorealistic map, and their strengths and weakness evaluated according to the test results. Raw test results were compiled to ascertain whether patterns in the data would emerge.

Observations were recorded onto a data logging sheet and route sketches helped to paint a picture of the common routes taken, and the areas where participants encountered problems. The routes taken by each participant were redrawn for each task so that common routes could be identified. This was compared to results gained from questionnaires and debriefing.

Qualitative and quantitative data was derived from questionnaire responses. Qualitative data concentrated on subjective satisfaction, and was concerned with uncovering participant preferences. This was done so that likes and dislikes could be seen to help determine why or why not participants preferred a particular map.

The debriefing session helped to clarify any issues witnessed through observation and questionnaire responses. It also provided an opportunity to delve into extraneous details, such as opinions relating to the map's orientation and level of detail. Debriefing comments were tabulated to identify patterns and summarised to extract meaning.

The test consisted of three stages: introduction, test tasks and debriefing.

A pilot test was carried out prior to the actual experiment to identify the approximate test duration and any problems that may occur with its design. Pilot tests are important because they allow issues to be identified and to rectify them before the test (Rubin, 1994; Dumas and Redish, 1999).

Participants were introduced to the study through an orientation script. They were then required to fill out a brief background questionnaire to gather basic data relating to their experience and preferences. Participants were also required to undertake a simple tutorial, designed to familiarise them with the map's functionality.

A within-subject testing method was employed so that each test participant got to use each map. This method controls user variability, which is a problem often encountered with between-subject testing (Nielsen, 1993). Half of the users began with the non-realistic map, while the other half started with the realistic map. This was done to ensure that bias did not enter into the results when users transferred the skills they had already obtained in the first task.

The test consisted of two parts (one part focusing on each map), with each part containing two scenarios and a related task. Scenarios were employed because they added a real-life context to the tasks. Users were asked to 'think aloud' while undertaking each task. This procedure required the test participant to continuously think out loud while using the prototype. By incorporating such a protocol, an understanding of the participants' view of the prototype could be gained through their verbalised thoughts (Nielsen, 1993).

This debriefing stage required participants to fill out the post-test questionnaire. Each participant was asked four questions, and their responses were recorded. The debriefing session is a valuable addition to any usability study, as it provides a final chance to clarify ideas and gather information about participants' performance and opinions (Rubin, 1994).

8.6 Results from the evaluation

8.6.1 Observations

Observations were recorded during the evaluation session. Due to the fact that this evaluation session was the first time any of the participants had used a PDA for navigational purposes, it took them a while to get started. This involved panning around the map, adjusting the zoom level, and turning the feature labels on and off. They needed to orientate the map in order to begin navigating. The 'think aloud' protocol was used, and it revealed that all participants relied on nearby landmarks to find their orientation and plan the route they were going to take. The river was also a significant feature that participants used to determine their direction of travel. All participants used the labels facility on the map to find the location of the points of interest they were to navigate to before deciding on the best route to take.

Halfway through the test, participants were required to use the alternative map. Participants who began with the non-realistic map were now confronted with a realistic map and vice versa. Even though both maps had the same orientation, participants still needed to re-establish their current location on the map. Most needed a moment to adjust the zoom level and pan around the map to become familiar with it. During this changeover period, the 'think aloud' protocol produced many comments regarding the participants' opinions of the two different maps.

Participants freely voiced their thoughts about the differences they had noticed between the non-realistic map and the realistic map.

8.6.2 Post-test questionnaire

Participants filled out the post-test questionnaire immediately after completing the final task. The three parts to the questionnaire were concerned with obtaining data relating to participants' opinions regarding the non-realistic map, the realistic map and a comparison of the two.

According to Likert scale responses, all participants agreed that the non-realistic map was clearer than the realistic map, while eight indicated that the realistic map appeared to be cluttered. Eight participants also revealed that the non-realistic map was better suited to their needs, and seven indicated that the realistic map was not easier to use. During this stage of the questionnaire, eight individuals indicated that they favoured the non-realistic map, while two confirmed that it was the realistic map that they preferred. Half believed that the information provided by the realistic map was useful, and eight agreed that the non-realistic map was more aesthetically pleasing. In relation to effective design for small screens, seven participants believed that the non-realistic map was more appropriate, while only one thought that the realistic map was more suitable.

Semantic Differential Scales provided a quick summary of participants' views relating to the two maps. The non-realistic map was rated positively by all participants in the clarity, usefulness, usability, functionality and legibility sectors. Nine participants also rated the non-realistic map positively for appropriateness and innovation. A negative rating with a value of 1 was given by a participant expressing their dislike of the non-realistic map, and another negative rating of one suggested that this participant found the non-realistic map to be unappealing. Definite variability was witnessed in the semantic differential scale responses regarding the realistic map. Each scale seemed to attract widespread responses. The realistic map was generally rated towards the positive side of the scale for usefulness, usability, functionality and appropriateness. Negative responses dominated for clarity, appeal, legibility and innovation. Likeability was split between participants, with four ratings on both the positive and negative sides of the scale, and two neutral ratings.

The final two questions required participants to indicate their personal preferences. Eight participants preferred the non-realistic map and found it to be more useful. In contrast, two participants favoured the realistic map for both questions.

In the extended answer section of the post-test questionnaire, the non-realistic map attracted many positive responses. Most of these responses praised the non-realistic map for its clarity. Candidates noted that this made the map uncluttered, simpler and more legible. Participants also favoured its symbolic attributes, as this made it easier to differentiate between objects, to locate features quickly and to navigate easier. Its innovative and aesthetic attributes also attracted positive comments. One negative comment was received: this participant stated that the non-realistic map only provided the shape attribute of objects, while ignoring their other features.

Negative comments dominated those in reference to the realistic map. Most criticism related to the way information was presented. Participants described the map as cluttered, pixellated and confusing. Many indicated that it showed too much information, leading to reduced legibility and causing it to be difficult to use. Positive comments related to personal preferences and the fact that the map included more detail. One participant stated that the realistic map was valuable because users could view realistic features on the map and easily identify them in the real world.

The section provided for additional comments attracted a variety of responses. These can be summarised as positive comments relating to both maps, as well as some suggestions for methods to improve the maps.

8.6.3 Debriefing

A debriefing session took place immediately after participants had finished completing the post-test questionnaire. By this stage, the test monitor had prepared four questions to ask the participant. These questions were generally concerned with gathering more preference data, as well as touching on areas requiring further clarification.

The questions requiring a response that compared non-realistic and realistic maps helped to expand on the information gained in the written response section of the post-test questionnaire. Participants spoke about which map they preferred and why, as well as providing additional positive and negative comments relating to each map. Some also spoke of methods that could be incorporated to improve the maps.

The orientation of the maps was an issue that needed to be clarified. Both of the maps were presented with an unconventional orientation, and it was felt that participant input relating to this concern would be valuable. Most of the participants remarked that the map was not orientated to north; in fact, only one participant did not notice this. This was established through observation and the 'think aloud' protocol employed during the evaluation session. The responses relating to orientation were varied – some participants did not seem to mind either way, while others were strongly opposed to maps that are orientated unconventionally. Such variability in comments suggests that this issue requires further research to establish the most effective method of orientation.

Other questions touched on the type of information participants used from the maps, as well as what methods they used to navigate and identify features. Aside from the labels provided on each map, participants generally relied on factors such as building height, building form and spatial arrangement for identification purposes. They also used the location of prominent landmarks to find their way around.

Questions regarding map scale and zoom functionality were also asked. Responses to these questions varied between participants. One participant found the lack of scale bar to be a problem while navigating, while another participant identified a need for more zoom levels.

8.7 Discussion

8.7.1 General observations

The creation and delivery of the prototype, as well as participant information gained from the evaluation, provided a useful insight into the differences between realistic and non-realistic maps delivered on mobile devices. The information acquired relates to that of map development, map delivery, user preferences and the appropriate level of realism required for mobile navigation. Observations within these categories have helped to meet the objectives of this research, and in doing so have provided answers to the research questions.

Observations relating to the appearance of each map on the mobile device show that the non-realistic map was more effective. The 240 x 320 pixel resolution and 16-bit colour display of the PDA restricted the appearance of each map. Colour limitations caused the realistic map to lose colour variation, resulting in a flat composition. Slight tonal variations could not be shown on the PDA, limiting its utility for the display of realistic imagery. Colour for the non-realistic map was chosen for aesthetic appeal, as well as to avoid these colour drawbacks on the PDA display. The non-realistic map was able to utilise a series of colours that achieved the maximum contrast and legibility available from its delivery medium. Pixellation was also reliant on the selected zoom level. In order to keep file size down, only one image of each map was used in the *Flash* file.

The test results showed that the non-realistic map was more widely accepted and understood by test participants than the realistic map. Of the ten participants, eight preferred the non-realistic map, while only two selected the realistic map.

The post-test questionnaire required participants to rate each map on a semantic differential scale to indicate their opinion about its clarity, usefulness, usability, functionality, aesthetic appeal, legibility, appropriateness, innovation and likeability.

Figure 8.4 illustrates the number of positive and negative ratings given by participants for each map. The non-realistic map attracted far more positive responses than the realistic map. While the non-realistic map was stronger across all categories, responses given for the realistic map were quite neutral in their positive to negative ratio, achieving a fairly consistent result across all rating values. The attitudes towards the realistic map were indifferent, and there were no significant advantages related to any of the categories covered.

Written responses, as well as information gained from the 'think aloud' protocol and the debriefing session, provided a valuable insight into what participants thought about each map. Overall, the majority of participant reactions and comments regarding the non-realistic map were positive. In contrast, the realistic map attracted more criticism and negative responses.

Participants described the non-realistic map as clear, simple, uncluttered and legible. For many, these attributes made it easier for them to use. The symbology employed allowed them to differentiate between classes of features, and in doing so sped up the feature identification process. One participant stated that the non-realistic map provided just enough detail for them to find their way to where they needed to go. The 'think aloud' protocol revealed that participants using the non-realistic map tended to make reference to significant landmarks more then they did when using the realistic map. Landmarks stand out clearly on the map. Participants also tended to view the non-realistic map from the second zoom level, suggesting that it still remains clear when viewed at smaller scales. Participants also commended the non-realistic map for its aesthetic appeal and innovation.

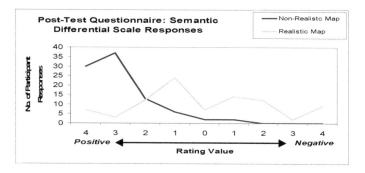

Figure 8.4. Semantic Differential Scale responses.

The non-realistic map also received some criticism. One participant indicated that the buildings on the map appeared to be spaced further apart then they were in reality. This participant also criticised the ground plane because it gave the false impression that the area was flat. Others suggested that more detail on building façades would aid in feature recognition, and build their confidence when trying to identify features.

Participants found the realistic map useful for visual references. They could locate features on the map and could associate them with the corresponding features in reality. The detailed colours and shades presented on buildings were also favoured by one participant. Many participants commented on the clarity of the Arts Centre on the realistic map because it managed to stand out effectively from other buildings.

The most common criticism made by participants regarding the realistic map was its cluttered appearance. This confused participants as it affected clarity and legibility, while also making it difficult for them to extract the information that they required. Some stated that it presented 'too much detail', distracting them from the essential information. This plethora of detail was also found to be irrelevant for the map's purpose, and caused unused information to permanently clutter the small-screen display. It also obscured paths, which made route planning difficult for some participants. Image quality was found to be an issue: pixellation reduced clarity by adding to the image clutter and affected the legibility of feature labels. Participants seemed to prefer the zoom level that presented the largest scale when using the realistic map. This suggests that image clutter reduced usability at smaller scales so users zoomed in to the largest possible scale to narrow their field of view to specific features.

The overwhelming positive response to the non-realistic map indicated that this technique is definitely a viable option for displays on small-screen devices. The criticism regarding the realistic map suggests that users do not want or need that level of detail to perform navigational activities using a mobile device. In fact, the realistic map hindered some participants by making it difficult for them to locate the information they were seeking. More investigation into appropriate design needs to be undertaken, but the non-realistic map certainly exceeded the realistic map in terms of user preference and understanding.

8.7.2 How much realism is required for navigation?

The amount of realism users said they wanted prior to the evaluation session differed greatly from what they preferred after using each map. Prior to the test, more than half of the participants stated that they would prefer a 3D map that exhibited a high degree of realism. In contrast, only two participants preferred the realistic map after the evaluation session.

Participant responses to the question asking them what degree of realism they would prefer in a 3D map, and their map preference taken from their responses to the post-test questionnaire are compared in Figure 8.5.

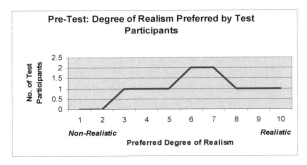

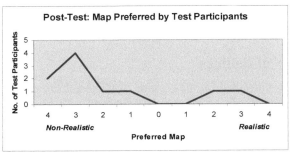

Figure 8.5. Background questionnaire responses vs. post-test questionnaire responses.

While participant preferences differed before and after the test, it is also possible that their perception of realism differed. Is everyone's perception of realism the same, and is it possible to measure realism? In his paper 'Three Varieties of Realism in Computer Graphics', Ferwerda (2003) discussed three varieties of realism: physical realism, photorealism and functional realism. While the first two categories have clearly defined visual characteristics, the boundaries that define functional realism remain relatively unclear. If one user can employ a functionally realistic image to perform a task, does that mean everyone can use the same image to perform the same task equally effectively? Ferwerda (2003) devised a method to measure functional realism by assessing the relationship between accuracy and fidelity. Although this may be a useful tool for developers when assessing their products, users are still unable to voice their opinions about the degree of realism they want because there is nothing governing the scale. During the testing stage, users did not realise that they did not want more realism until they were given the opportunity to use two different maps. Perhaps user desires would become clearer if functional realism was segmented into further categories that correspond to how closely the image depicts reality before user testing takes place.

The spatial information required by the user to undertake the tasks required in the context of the product needs to be identified. In the case of the prototype, its purpose was to act as a navigational aid for pedestrians while walking in the area. The principal task undertaken with such a product involves the user identifying their current location, finding the whereabouts of their destination, and then determining the route between the two points. The non-realistic map provided just enough information for the user to undertake these operations, making it easy for them to extract the required information. In contrast, the realistic map provided much more visual information, most of which was irrelevant to users at the time. This had a negative effect as it distracted users from the information they were looking for, and hindered them while they attempted to extract only what they required.

The creation of a realistic map does not utilise the principles of selection and generalisation because it presents the world as it is. Realistic detail is not necessary or desired when the purpose of a map is to help users get from one place to another. This is further complicated when the information is presented on a small screen because the restricted viewing area and the capabilities of the device do not accommodate the display of intricate detail.

8.8 Summary of findings

Data must be very detailed for the creation of photorealistic models, whereas non-photorealistic models can be effectively developed with less detailed data. Non-photorealistic models are much more time and cost efficient, and the final product still provides all of the information required by users.

The design of non-photorealistic maps is closely related to conventional map design. Even though the city models developed for evaluation were constructed using digital tools, they still utilised many traditional map design and production processes. By employing conventional processes, the design of non-photorealistic maps is more familiar to the cartographers who create them, and the final product is more familiar to the users who exploit them.

Three-Dimensional non-photorealistic maps are more effective than photo-realistic maps when delivered on small-screen devices. Non-photorealistic maps can be designed to utilise colours able to achieve the maximum contrast and legibility available from their delivery medium.

Non-photorealistic maps are more widely accepted and understood by users than photorealistic maps.

Non-photorealism is driven by clearly defined goals that work to provide users with just enough information. It also presents the information in a symbolic way, which ensures that the same information is communicated to all users. This reduces image clutter and accelerates users' ability to extract the information required.

When users say they want a map to be more realistic for navigation, they actually imply that they want a map that will effectively enable them to complete a specific navigational task. This does not mean that the map needs to be a point-by-point re-creation of reality, but rather a functional tool that will enable them to successfully carry out an intended task.

The results obtained show that non-photorealistic graphics are more effective than photorealistic graphics in the context of 3D mobile cartography. Non-realistic representations are generally easier and quicker to develop than realistic scenes, and are decidedly more suited for delivery on mobile devices. Not only did the non-realistic map exceed the realistic map on a technological level, but it was also more widely accepted and understood by target users. It was found that the non-realistic map provided a clear and highly usable product that fulfilled their wayfinding needs. The realistic image was found to be cluttered and incoherent. It presented too much information.

Non-photorealistic graphics offer the potential to improve displays of 3D spatial information on mobile devices. This research was an exploratory step into determining the potential of using non-photorealism in this context. There exists the need for further investigation. Non-photorealism appears to be the better alternative for small mobile displays. Further research needs to be undertaken so as to determine the most effective technique. This could also focus on available technology, but instead of using it to create complex scenes, as in photorealism, it could have a prime focus on the automation of the technique, as well as providing new formats to optimise delivery.

8.9 Conclusion and future developments

The study reported in this chapter focused on the design of 3D city maps for delivery on small-screen devices. It compared the creation, delivery and use of non-photorealistic representations in comparison to photorealistic representations, and aimed to determine which method was more effective.

It described a method for introducing non-photorealistic rendering, an area of much interest to computer graphics and 3D mobile GIS developers alike. Also, it identified many advantages associated with non-photorealistic imagery, and validated its usefulness over photorealism when small mobile devices are used. It demonstrated that non-photorealistic graphics are better than photorealistic graphics for the display of 3D maps on mobile devices, where the image provided advantages over photorealism for navigation.

A number of recommendations can be made to improve the current technique and extend this research:

(a)   While this technique identified that non-photorealism has potential, it is important to extend this knowledge by developing and testing fully working 3D models. The development and testing of a prototype that utilises true 3D non-photorealistic and photorealistic maps would provide better insight into the advantages and disadvantages associated with each technique. It could also lead to identifying appropriate interaction methods, which would differ considerably to the basic zoom and pan functionality applied to the 2.5D maps used in this study.

(b)   Currently, there is no formal method that can be used to measure realism. A suitable scale that clearly defines differing levels of realism needs to be devised. The development of such a scale would ensure that there are no unclear boundaries between what users say they want and what they actually need.

(c)   It might be beneficial to develop a range of maps at differing levels of realism, or maps with an adjustable level of realism.

(d)   Three-dimensional non-photorealistic maps possess attributes that could be delivered in both vector and raster formats. It is important to identify the most appropriate display method so as to fully optimise its suitability for small-screen delivery.

(e)   If non-photorealistic graphics do have advantages over photorealistic graphics for the delivery of mobile maps, they may also have benefits in other areas of cartography. The application of non-photorealistic representations in other areas needs to be explored.

The many benefits of non-photorealism identified illustrate the cartographic potential of non-realistic images for small-screen displays. Further research is now needed that focuses on the most appropriate methods for the development and delivery of non-photorealistic maps, so that this technique can be refined.

References

Döllner, J. and Walther, M. (2003) 'Real-time expressive rendering of city models', *Proceedings Seventh International Conference on Information Visualization (IV '03)*, London, England, 16–18 July. IEEE Computer Society, pp. 245–250.

Duke, D. J., Barnard, P. J., Halper, N. and Mellin, M. (2003) 'Rendering and Affect', *Computer Graphics Forum*, vol. 22, no. 3, pp. 359–368.

Dumas, J. S. and Redish, J. C. (1999) *A Practical Guide to Usability Testing*, Revised Edition, Exeter: Intellect Books.

Durand, F. (2002) 'An invitation to discuss computer depiction', *Symposium on Non-Photorealistic Animation and Rendering (NPAR 2002)*, Annecy, France, 3–5 June, pp. 111–124.

Ferwerda, J. A. (2003) 'Three varieties of realism in computer graphics', *Proceedings of the SPIE - The International Society for Optical Engineering,* Santa Clara, CA, USA, 21–24 Jan, pp. 290–297.

Goldstein, D. (1999) 'Intentional non-photorealistic rendering', *Computer Graphics,* vol. 33, no. 1, pp. 62–63.

Gooch, A. A. and Willemsen, P. (2002) 'Evaluating space perception in NPR immersive environments', *2nd International Symposium on Non-Photorealistic Animation and Rendering (NPAR '02).* Annecy, France, 3–5 June, pp. 105–110.

Gooch, B. and Gooch, A. (2001) *Non-Photorealistic Rendering,* Natick, MA: A K Peters.

Markosian, L., Kowalski, M. A., Trychin, S. J., Bourdev, L. D., Goldstein, D. and Hughes, J. F. (1997) 'Real-time nonphotorealistic rendering', *Proceedings of the ACM SIGGRAPH Conference on Computer Graphics,* Los Angeles, CA, USA, 3–8 August, pp. 415–420.

McNeel (2003) 'Penguin - Non-Photorealistic Renderer', [Online], Available: http://www.penguin3d.com/, [18 August 2004].

Nielsen, J. (1993) *Usability Engineering,* Boston: Morgan Kaufmann.

Rubin, J. (1994) *Handbook of Usability Testing,* New York: John Wiley & Sons.

Schumann, J., Strothotte, T., Raab, A. and Laser, S. (1996) 'Assessing the effect of nonphotorealistic rendered images in CAD', in Bilger, R., Guest, S. and Tauber, M. J. (eds.) *Proc. Computer Human Interaction (CHI'96),* New York: ACM Press, pp. 35–42.

Thomas, M. R. (2003) *Blending Qualitative and Quantitative Research Methods in Theses and Dissertations,* Thousand Oaks, CA: Corwin Press.

Virzi, R. A. (1992) 'Refining the test phase of usability evaluation: How many subjects is enough?', *Human Factors,* vol. 34, no. 4, pp. 457–468.

Chapter 9

Generalisation of Large-Scale Digital Geographic Datasets for MobileGIS Applications

Suchith Anand[1], J. Mark Ware[2] and George Taylor[2]

[1] Centre for Geospatial Science, University of Nottingham, England
[2] Faculty of Advanced Technology, University of Glamorgan, Wales

9.1 Introduction

This chapter builds upon the display and visualisation theme of this part of the book and focuses on the automatic production of schematic maps on demand for small-screen mobile devices using a simulated annealing technique. Mobile GIS applications derive benefits of map generalisation by rendering relevant information legible at a given scale by filtering the required information as well as enhancing the visualisation of the large-scale data on small-screen display devices. With the advent of high-end miniature technology as well as digital geographic data products like OSMasterMap® and OSCAR® it is desirable to devise proper methodologies for map generalisation specifically tailored for MobileGIS applications. Schematic maps are diagrammatic representations based on linear abstractions of networks. Transportation networks are the key candidates for applying schematisation to help ease the interpretation of information by the process of cartographic abstraction (Avelar, 2002). Generating schematic maps is an effective means of generalisation of large-scale digital datasets for display on small-screen display screens and is primarily aimed at enhancing visualisation and also making such maps user friendly for interpretation. Hence the relevance of schematic maps in mobile applications and their automated production underpins the theme of this part of the book.

The remainder of this chapter is set out as follows. Section 9.2 provides some background information on Mobile GIS. Section 9.3 looks into map generalisation requirements from a MobileGIS perspective. Section 9.4 introduces schematic maps and gives a short review of previous automated solutions to the problem of schematic map generation. Section 9.5 outlines the key generalisation processes involved in the production of schematic maps. Section 9.6 contains a description of the simulated annealing-based schematic map generator algorithm that forms the basis for this chapter. A prototype implementation of this algorithm is described in Section 9.7, and some experimental results are presented. The chapter concludes in Section 9.8 with a summary of the results and a discussion of future work.

Dynamic and Mobile GIS: Investigating Changes in Space and Time. Edited by Jane Drummond, Roland Billen, Elsa João and David Forrest. © 2006 Taylor & Francis

9.2 Mobile GIS

Mobile GIS refers to the use of geographic data in the field on mobile devices, such as networked PDAs. MobileGIS applications act according to a geographic trigger, such as input of a place name, postcode, position of a GPS user, location information from mobile phone network, etc. The main components of a MobileGIS application are a global positioning system (GPS) receiver, a handheld computer (e.g. a PDA), and a communication network with GIS acting as the backbone (Figure 9.1).

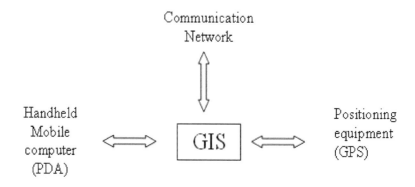

Figure 9.1. The basic components of MobileGIS application.

Mobile GIS is a relatively new technology, but with the availability of digital geographic datasets its application potential has increased tremendously. There is a huge amount of available geographic information that can be re-purposed for mobile GIS applications; together with the ability to filter and personalise content by reference to a user's physical location, this will provide compelling business and research opportunities in this emerging field. This work looks into how suitable map generalisation techniques can be applied to generate schematic maps from large-scale digital geographic data to enable more effective means of map interpretation on small-screen display devices.

9.3 Map generalisation – Mobile GIS perspective

The process of simplifying the form or shape of map features, usually carried out when the map is changed from a large scale (i.e. more detailed) to a small scale (i.e. less detailed), is referred to as generalisation. This necessitates the use of operations such as simplification, selection, displacement and amalgamation of features that takes place during scale reduction (Ware et al., 2003).

Through the introduction of OSMasterMap®, the Ordnance Survey has now made available a seamless digital map database of the UK. The OSMasterMap® data features are digital representations of the world. All real-world objects are represented as explicit features and each identified by a unique TOID (Topological Identifier). The features have survey accuracy ranging from ±1.0 m in urban areas to ±8.0 m in mountain and moorland areas (OS, 2005).

The key benefits OSMasterMap® has over the previous large-scale digital geographic dataset OSLandline®, as summarised by ESRI (2005), include providing a single, consistent seamless national digital base map; improved topological structure thereby increasing functionality and flexibility for map display; improved speed, accuracy and simplicity of derived data capture through the new data structure of point, line and polygon features; ease of integrating other datasets thereby adding value to the geometry of features by taking advantage of unique TOID referencing. With the large-scale use and application of mobile devices it is now possible to deliver digital geographic information for mobile GIS applications. OSMasterMap with its advantages provides immense opportunities for MobileGIS applications. Also the need to deliver the required map information on small display screens of devices, such as PDAs, necessitates the application of appropriate map generalisation techniques that are specifically tailored for this purpose.

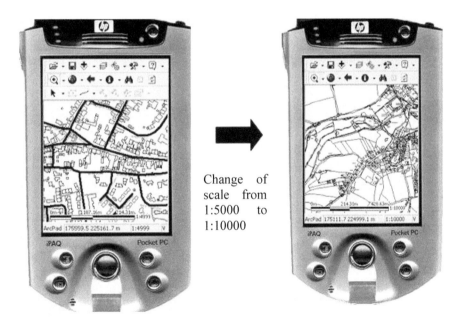

Change of scale from 1:5000 to 1:10000

Figure 9.2. In order to verify the suitability of OSMasterMap data for small-screen devices, the data for the St David's area in Wales was loaded in ESRI's ArcPad and tested on an HP iPAQ PocketPC h5400 series for display at various scales to find out the extent of spatial conflicts between features and data volume (Figure 9.2). There is explicit proof of graphic conflict during scale changes and the dataset needs to be tailored for small-screen devices specifically for MobileGIS by applying suitable map generalisation techniques. For example, it is necessary to apply scale-based symbolisation as well as applying suitable generalisation operators like simplification, displacement, amalgamation, etc.

To understand the demands for mobile applications, the general user requirements of small display devices (PDAs in this case) have been studied. In comparison to contemporary desktop computers which have processing power in the range of 4GHz, memory of 512Mb and storage capacity around 80 Gb, the processing capability of PDAs is much lower in the range of 400 MHz and their memory capacity is in range of 64 Mb. This highlights the issues associated with processing and storage of large-scale voluminous datasets in thin client mobile devices. Also the low display resolution of 240 x 320 pixels as well as the smaller display area of $50cm^2$ of PDA screens make it necessary that the final output image is generalised as per appropriate small display cartographic specifications to give maximum clarity and readability. The basic criteria are easily readable font, recognisable symbols, mutually exclusive colour at each level of information and the comprehensive use of area colour with few geometric details of objects (GiMoDig Project, 2003). In summary, PDAs have different form factors such as display resolution, varying numbers of display lines, horizontal or vertical screen orientation and hardware specification when compared to contemporary desktop computers. Hence GIS applications that are to be used in PDAs need to be tailored appropriately. The application of suitable automated map generalisation techniques will help in filtering redundant data enabling faster and more efficient rendering, as well as in noise reduction in the rendered image and enhancing the essential details.

A suitable cartographic display specification was developed to represent OSMasterMap data on small-screen devices and tests were carried out at a wide range of display scales (Anand et al., 2004). It was found that there is graphic conflict between features during scale reduction and since the display screen is comparatively small the problem becomes much more apparent. Once the same dataset was displayed as per the developed cartographic specification, better graphic representation was obtained (Figure 9.3). For example it can be seen in Figure 9.3 that the low display resolution and smaller display area of PDA screens makes it necessary to apply the small display cartographic specification to give maximum clarity and readability to the output map.

9.4 Schematic maps

The way people construct and interact with geographical maps has to be regarded as a valuable clue to the properties of the underlying mental structures and process for spatial cognition. Geographical maps are described as spatial representation media that play an important role in many processes of human spatial cognition (Berendt et al., 1998). A schematic map is a diagrammatic representation based on linear abstractions of networks. Typically transportation networks are the key candidates for applying schematisation to help ease the interpretation of information by the process of cartographic abstraction. Schematic maps are built up from sketches, which usually have a close resemblance to verbal descriptions about spatial features (Avelar, 2002). The London Tube map is one of the well-known examples of a schematic map.

Figure 9.3. OSMasterMap® data (Ordnance Survey © Crown Copyright. All rights reserved, 2005) displayed in an HP iPAQ using ESRI's ArcPad. The figure shows how appropriate symbolisation can enhance readability and usability of maps. Image on the left explicitly showing poor visualisation and image on the right displayed at the map specification guidelines for 1:5000 scale showing better data visualisation.

Generating schematic maps involves reducing the complexity of map details while preserving the important characteristics. When performed manually, this is a time-consuming and expensive process. The application of GIS tools has led to the realisation that the efficiency of the cartographer could be increased through the automation of some of the more time-consuming generalisation techniques. Contemporary GIS software contains tools for automating processes like line simplification that allow basic generalisation to be performed. Although these algorithms go some way to help in the automated production of schematic maps, there is lot of work to be done on developing fully automated schematic map generalisation tools. Differing geometric and aesthetic criteria are used to design a schematic map keeping in mind the common goals of graphic simplicity, retention of network information content and presentation legibility (Avelar et al., 2000).

Agrawala and Stolte (2001) in their work present a set of cartographic generalisation techniques specifically designed to improve the usability of route maps. These techniques are based on cognitive psychology research, which has shown that an effective route map must clearly communicate all the turning points on the route, and that precisely depicting the exact length, angle and shape of each road is much less important. They show how these techniques are applied in hand-drawn maps and demonstrate that by carefully distorting road lengths and angles and simplifying road shape, it is possible to clearly and concisely present all the turning points along the route. Avelar (2002) presents the automatic generation of schematic maps from traditional, vector-based, cartographic information. By using an optimisation technique, the lines of the original route network are modified to meet geometric and aesthetic constraints in the resulting schematic map. The algorithm preserves topological relations using simple geometric operations and tests.

Due to their abstracting power, schematic maps are an ideal means for representing specific information about a physical environment. They play a helpful role in spatial problem-solving tasks such as way finding. Schematic maps provide a suitable medium for representing meaningful entities and spatial relationships between entities of the represented world. While topographic maps are intended to represent the real world as faithfully as possible, schematic maps are seen as conceptual representations of the environment (Casakin et al., 2000). When generalising, the cartographer tries to maintain the topology of the features as accurately as possible, i.e. the cartographer might sacrifice absolute accuracy in order to maintain relative accuracy (João, 1998). As discussed earlier, the key characteristic of mobile devices is their limited processing capacity, memory and available display area. This necessitates that suitable operations are carried out to filter redundant data from the voluminous large-scale digital datasets to help reduce data volume as well as enhancing visualisation and readability of the final output. Schematic maps are an effective way of achieving this outcome.

Though schematic maps have found successful application in underground tube map design, Morrison (1996) in his study of public transportation maps in western European cities demonstrates that schematic maps are not suitable for surface transport maps like bus networks. This highlights the problem of developing techniques that are specific for generating schematic maps of surface transportation networks.

9.5 Key generalisation processes for schematic maps

A schematic map is a diagrammatic representation based on linear abstractions of networks. Typically transportation networks are the key candidates for applying schematisation to help ease the interpretation of information by the process of cartographic abstraction. Schematic maps are built up from sketches which usually have a close resemblance to verbal descriptions of spatial features (Avelar, 2002). The best example of modern-day schematic map is the London Tube map originally designed by Harry Beck in 1931. An electrical engineer, he based his design on a circuit diagram and used a schematic layout. The map locally distorted the scale and shape of the tube route but preserved the overall topology of the tube network (LTM, 2004). Morrison (1996) describes the appropriateness of using schematic maps for different modes of transport.

The basic steps for generating schematic maps are to eliminate all features that are not functionally relevant and to eliminate any networks (or portions of networks) not functionally relevant to the single system chosen for mapping. All geometric invariants of the network's structure are relaxed except topological accuracy. Routes and junctions are symbolised abstractly (Waldorf, 1979).

Elroi (1988) refined the process by adding three more graphic manipulations. Lines are simplified to their most elementary shapes. Line simplification algorithms such as the Douglas–Peucker algorithm, can be applied to road datasets to remove unwanted detail and produce a simplified version of the network (Figure 9.4).

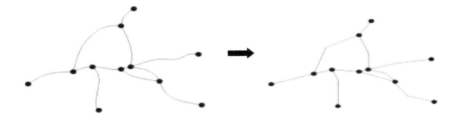

Figure 9.4. First step in the schematisation process is line simplification, which can be achieved using an algorithm such as that of Douglas and Peucker (1973).

Also lines are re-oriented to conform to a regular grid, such that they all run horizontally, vertically or at a 45-degree diagonal. Finally, congested areas are increased in scale at the expense of reducing scale in areas of lesser node density.

Graphic legibility is an important criterion and is achieved using appropriate display styles for the point, line, area features, etc., as per the small display cartographic specification guidelines. This will enhance the readability of the generated schematic map as well as improving usability. Other factors that need to be taken into consideration are fixing the aspect ratio of the resulting image to make the effective use of map space when trying to fit and display on a small-screen display device of 240 x 320 pixel resolution (Agrawala, 2001).

As the first step in the process is line simplification, algorithms like the Douglas–Peucker algorithm can be applied to road datasets to remove unwanted detail and produce a simplified version of the network. When generating schematic maps from large-scale datasets for navigation systems, the basic user inputs are the initial and final destinations. Based on this the system will have to generate an appropriate schematic map depicting the turning point information with turning directions coupled preferably with map labels for each road and the distance to be travelled on that road. The local landmarks on the route from the PoI (Points of Interest) dataset can also be displayed, enhancing the navigational usability of the generated schematic map. This is especially important if the system is to be used for generating tourist maps. Also, by enabling different levels of scale for the schematic, the global properties of the route can be conveyed to the user. Factors, auch as optimal aspect ratio of the resulting image to make effective use of the map space when trying to fit on a small display device of 240 x 320 pixel resolution, have to be taken into account. Enabling support for vertical and horizontal scrolling will add more flexibility to the user (Agrawala, 2001).

9.6 Schematic map generation using simulated annealing

This work is concerned with the problem of effective rendering of large-scale digital geographic datasets on small display devices by developing appropriate optimisation techniques for generating schematic maps. At present, schematic maps are produced manually or by using graphic-based software. This is not only a time-consuming process, but requires a skilled map designer. The challenge of replacing an experienced cartographer with a computer that can make the same decisions to produce a schematic map is compelling. Also there are no cartographic guidelines to help the design of schematic maps. Automatic generation of schematic maps may improve results and make the process faster and cheaper. It would also help in extending the use of schematic maps to transportation systems of cities around the world (Avelar and Muller, 2000).

Simulated Annealing (SA) (Kirkpatrick et al., 1983) is a probabilistic heuristic optimisation technique used for finding good approximate solutions to the global optimum of a given function in a large search space. SA has been used as an optimisation tool in a wide range of application areas, including routing, scheduling and layout design (e.g. Cerny, 1985; Elmohamed et al., 1998; Chwif et al., 1998), including automated cartographic design (Zoraster, 1997; Ware et al., 2003). In this chapter, the schematisation process is considered as an optimisation problem. Given an input state (a network layout), an alternative state can be obtained simply by displacing one or more of the network vertices. The search space being examined is therefore the set of all possible states of a given input linear network. Each state can be evaluated in terms of how closely it resembles a schematic map. However, finding the best state by exhaustively generating and evaluating all possible states is not possible, as for any realistic data set the search space will be excessively large (i.e. there are too many alternative layouts). SA offers a means by which a large search space can be searched for near optimal solutions. A standard SA algorithm, which is adopted for use in this work, is shown in Figure 9.5.

At the start of the optimisation process SA is presented with an initial approximate solution (or state). In the case of the schematic map problem, this will be the initial network (line features, each made up of constituent vertices). The initial state $M_{initial}$ is then evaluated using a cost function; this function assigns to the input state a score that reflects how well it measures up against a set of given constraints. If the initial cost is greater than some user defined threshold (i.e. the constraints are not met adequately) then the algorithm steps into its optimisation phase. This part of the process is iterative. At each iteration, the current state $M_{current}$ (i.e. the current network) is modified ($M_{modified}$) to make a new, alternative approximate solution. The current and new states are said to be neighbours. The neighbours of any given state are generated usually in an application-specific way. A decision is then taken as to whether to switch to the new state or to stick with the current. Essentially, an improved new state is always chosen, whereas a poorer new state is rejected with some probability P, with P increasing over time. The iterative process continues until stopping criteria are met (e.g. a suitably good solution is found or a certain amount of time has passed).

```
input: M_initial, Schedule, Stopconditions

set M_current equal to M_initial
set T to T_initial (from Schedule)
evaluate M_current
while notmet(Stopconditions)
                select Vertex at random
                generate random Displacement
                displace Vertex
                evaluate M_modified
                if M_modified is better than M_current
                        M_modified becomes M_current
                else
                                    -ΔE/ T
                        P = e
                        M_modified becomes M_current with probability P
                endif
                update T according to Schedule
endwhile
```

Figure 9.5. Shows the Simulated Annealing (SA) algorithm used as optimisation process for producing schematic map. SA is presented with an initial approximate solution and then evaluated using a cost function. If the initial cost is greater than some user-defined threshold then the algorithm steps into its optimisation phase. At each iteration, a vertex is chosen at random in the current state and subjecting it to a small random displacement. The new state is also evaluated using the cost function and a decision is then taken as to whether to switch to the new state or to stick with the current. An improved new state is always chosen, whereas a poorer new state is rejected with some probability. The iterative process continues until stopping criteria are met.

At each iteration the probability P is dependent on two variables: ΔE (the difference in cost between the current and new states); and T (the current *temperature*). P is defined as:

$$P = e^{-\Delta E / T}$$

T is assigned a relatively high initial value; its value is decreased in stages throughout the running of the algorithm. At high values of T higher cost new states (large negative ΔE) will have a relatively high chance of being retained, whereas at low values of T higher cost new states will tend to be rejected. The acceptance of some higher-cost new states is permitted so as to allow escape from locally optimal solutions.

9.7 Experimental results

Prototype software for producing schematic maps for transportation network data has been developed. The software makes use of the simulated annealing optimisation technique described in Section 9.6. The schematic software is currently implemented as a VBA script within ArcGIS. This technique has been used previously to control operations of displacement, deletion, reduction and enlargement of multiple map objects to help resolve spatial conflict arising due to scale reduction (Ware et al., 1998).

A brief summary of the schematisation process is given below:

Define constraints – these are the constraints that are to be met by the derived schematic map. The current software caters for three constraints: (i) topology – ensures that original map and derived schematic map are topologically consistent; (ii) angular – if possible, edges should lie in horizontal, vertical or diagonal direction; and (iii) minimum edge length – if possible, all edges should have a length greater than some minimum length.

Simplify lines – input data will typically contain redundant vertices. These are removed by application of a suitable line simplification algorithm (in our case the Douglas–Peucker algorithm).

Evaluate and optimise – evaluate the simplified input map (against constraints) and if required make use of simulated annealing optimisation to refine. The simulated annealing part of the process is iterative. At each iteration, the current map is modified slightly (in our implementation this involves displacing a single vertex) and re-evaluated. A decision is then taken as to whether to keep the new map or revert to the previous. Essentially, an improved map is always retained, whereas a poorer map is rejected with some probability p, with p increasing over time. The process continues until stopping criteria are met (e.g. a suitably good map is generated or a certain amount of time has passed).

The tests are applied to real datasets and schematic maps are automatically generated in response to a selected set of constraints from large-scale digital geographic dataset (OSCAR® road dataset in this case). The topology of the network is preserved during the schematisation process. This approach provides promising results in the production of automated schematic maps. Examples are shown in Figures 9.6 and 9.7. These maps are subsequently displayed within the ArcPad application on an HP iPAQ PDA. Example output is shown in Figure 9.8. Also aesthetic improvement of the resulting schematic map is achieved using appropriate display styles for the point, line and area features, etc., as per the small display cartographic specification guidelines, which will enhance usability of the generated schematic map.

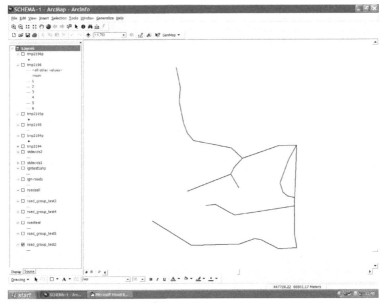

Figure 9.6. Pregeneralised data OSCAR® road dataset (Ordnance Survey © Crown Copyright. All rights reserved, 2005).

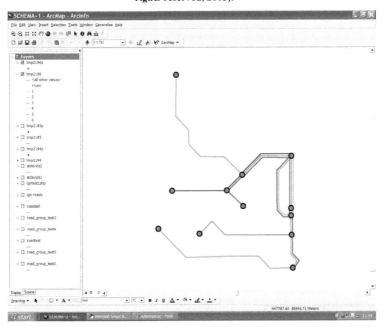

Figure 9.7. Schematic map of the same roads shown in Figure 9.6 generated by the simulated annealing software and symbolised automatically.

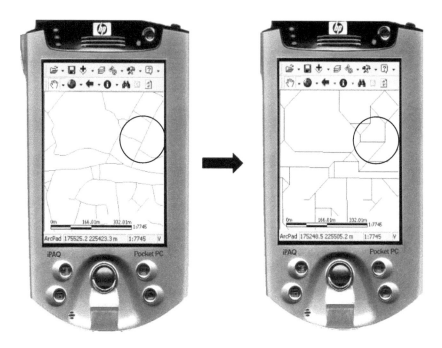

Figure 9.8. OSCAR® road dataset (Ordnance Survey © Crown Copyright. All rights reserved, 2005) displayed in an HP iPAQ using ESRI's ArcPad. The image on the left is before applying schematisation and the image on the right is displayed after applying schematisation. There are areas for future improvement as highlighted where certain sections, practically straight on the left, have sharp bends on the right. This again is dependent on the amount of schematisation applied.

9.8 Conclusion and future developments

This chapter looks into the development of automated means of generating schematic maps from large-scale digital geographic datasets that are tailored for mobile GIS applications. A prototype Simulated Annealing technique has been used to derive a schematic map with reduced linear information from the detailed OSCAR® dataset. The key theme of this chapter is to demonstrate the practical application of the simulated annealing technique in the automatic generation of optimal on the fly schematic maps from large-scale geographic datasets specifically tailored for mobile GIS applications. A simulated annealing algorithm and implementation for generating schematic maps based on a number of user-defined constraints is presented. Results show the algorithm to be successful in producing schematic maps from large-scale transportation network data.

Future work will concentrate on refining the technique through the use of additional constraints and also the analysis of the extent to which the predefined road classifications in the OSCAR® dataset are affected during the schematisation process with respect to the original map. Also it is intended to do further work on the automated application of the appropriate small display specific symbology for the generated schematic map based on the referenced display scale to enhance visualisation and usability.

Acknowledgement

The authors express thanks for the Ordnance Survey for providing the data used in this work.

References

Agrawala, M. (2001) *Visualizing Route Maps*, Stanford, CA: Stanford University.

Agrawala, M. and Stolte, C. (2001) 'Rendering effective route maps: improving usability through generalization', *Proceedings of SIGGRAPH 2001*, Los Angeles, pp. 241–249.

Anand, S., Ware, J. M. and Taylor, G. E. (2004) 'Map Generalization for OSMasterMap Data in Location Based Services and Mobile GIS Applications', *Proceedings of 12th International Conference on Geoinformatics*, Gävle, Sweden, pp. 54–60.

Avelar, S. and Muller, M. (2000) 'Generating Topologically Correct Schematic Maps', *Proceedings of 9th International Symposium on Spatial Data Handling*, Beijing, pp. 1472–1480.

Avelar, S. (2002) *Schematic Maps On Demand: Design, Modeling and Visualization*, Zurich: Swiss Federal Institute of Technology.

Berendt, B., Barkowsky, T., Freksa, C. and Kelter, S. (1998) 'Spatial representation with aspect maps', *Spatial cognition: An Interdisciplinary Approach to Representing and Processing Spatial Knowledge*, Berlin: Springer-Verlag, pp. 157–175.

Casakin, H., Barkowsky, T., Klippel, A. and Freksa, C. (2000) 'Schematic Maps as Wayfinding Aids', *Lecture Notes in Artificial Intelligence-Spatial Cognition 1849*, pp. 59–71.

Cerny, V. (1985) 'Thermodynamical approach to the travelling salesman problem: An efficient simulation algorithm', *Journal of Optimization Theory and Applications*, 45(1), pp. 41–51.

Chwif, L., Barretto, M. R. P. and Moscato, L. A. (1998) 'A solution to the facility layout problem using simulated annealing.' *Computers in Industry Archive*, 3 6(1-2), pp. 125–132.

Douglas, D. H. and Peucker, T. K. (1973) 'Algorithms for the reduction of the number of points required to represent a digitized line or its caricature', *The Canadian Cartographer*, 10, pp. 112–122.

Elmohamed, S., Fox, G. and Coddington, P. (1998) 'A Comparison of Annealing Techniques for Academic Course Scheduling', *Proceedings of 2nd International Conference on the Practice and Theory of Automated Timetabling*, Syracuse, NY, USA, pp. 92–114.

Elroi, D. S. (1988) 'Designing a Network Linemap Schematization Software Enhancement Package', *Proceedings of the Eighth Annual ESRI User Conference*, Redlands, CA.

ESRI (2005) 'A roadmap to success with OSMasterMap', [Online], Available:www.esriuk.com/pdf/media/OSMasterMap.pdf [09/08/05].

GiMoDig (2003) 'Geospatial info-mobility service by real-time data-integration and generalization', [Online], Available: http://gimodig.fgi.fi/ [20/07/05].

João, E. (1998) *Causes and Consequences of Map Generalization*, London: Taylor and Francis.

Kirkpatrick, S., Gelath, C. D. and Vecchi, M. (1983) 'Optimization by simulated annealing', *Science*, 220, pp. 671–680.

LTM (2004) 'Transport in London Archives', [Online], Available: http://www.ltmuseum.co.uk/ [04/05/05].

Morrison, A. (1996) 'Public transportation maps in western European cities', *Cartographic Journal*, vol. 33, no. 2, pp. 93–110.

Ordnance Survey (2005) 'OSMasterMap & OSCAR Technical Information, Ordnance Survey UK website', [Online], Available: http://www.ordnancesurvey.co.uk [04/07/05].

Waldorf, S. P. (1979) *Schematic Navigational Map Design*, Lawrence: University of Kansas.

Ware, J. M., Jones, C. B. and Thomas, N. (2003) 'Automated cartographic map generalisation with multiple operators: a simulated annealing approach', *The International Journal of Geographical Information Science*, vol. 17, no. 8, pp. 743–769.

Zoraster, S. (1997) 'Practical results using simulated annealing for point feature label placement', *Cartography and Geographical Information Systems*, vol. 24, no. 1, pp. 228–238.

Chapter 10

Map Matching for Vehicle Guidance

Britta Hummel

University of Karlsruhe, Germany

10.1 Introduction

The past years have revealed a dramatically increasing interest in the use of mobile Geographical Information Systems (GIS) in various automotive applications. Car navigation systems employ digital maps to guide the driver to the desired destination. Next-generation driver assistance systems will use enhanced maps in order to present precise navigation hints (including speed limits, locations of gas stations, restaurants, etc.), and for assisted vehicle control. Furthermore, methods for autonomous enhancements of existing maps using video and lidar sensors are currently under development. An exhaustive overview of potential applications for upcoming mobile GI systems can be found in Chapter 2.

All applications share the need for a robust assignment of the measured vehicle position to a road segment in the digital map. This process is called *map matching*. Since the emergence of the field in the 1970s (French, 1989), considerable progress has been reported (cf. the surveying articles of Bernstein and Kornhauser, 1998; White et al., 2000; Quddus et al., 2003; and Lakakis et al., 2004). However, users of navigation systems still encounter some erroneous map matching results.

On the contrary, manufacturers increasingly aim for a simple architecture of navigation systems. Hence, a growing number of navigation systems do not rely on multiple vehicle sensors— such as a combination of DGPS, odometer and gyro— but instead restrict themselves to GPS only. This is necessarily true for the newly evolving PDA (Personal Digital Assistant) navigation systems, which are identified in Chapter 1 as key devices for next-generation GIS, as well as for low cost in-car navigation solutions.

In this chapter a robust map matching algorithm is presented which exclusively relies upon information from a standard GPS receiver (however, the integration of data from an integrated GPS dead-reckoning unit is straightforward). Mobile phone-based location systems can equally serve as input. The achievable accuracy of mobile phone location is examined in detail in Chapter 11.

In contrast to standard map matching techniques, the whole *vehicle path* is estimated for each time step within an iterative, statistically optimal Bayesian estimator (a Bayesian estimator using a different formulation has been developed by Scott and Drane, 1994). The algorithm is suitable for all maps using the standard road segment representation as piecewise linear links. An introduction to digital maps and GPS is given in NCHRP (2002), for example. The map database used by the authors is off-the-shelf and frequently used in today's navigation systems. Errors of up to 40 metres with respect to ground truth data have been encountered.

Section 10.2 derives the Bayesian classifier for matching one single GPS position and orientation datum to the map. Incorporation of the classifier into a Hidden Markov Model in Section 10.3 accounts for the vehicle position and orientation history, information about the road network topology, driving restrictions and the assumed driving direction.

The vehicle path estimation proves to be robust even for challenging inner-city scenarios, some of which are shown in Section 10.4. A further improvement of the quality of current navigation systems for platforms without access to in-vehicle sensors (i.e. odometer or gyro) is anticipated and this is discussed in the final section (10.6).

The reliable and accurate determination of the current user position is considered a prerequisite not only for automotive applications but also for a wide variety of mobile Geographical Information Systems. The algorithm described is not specifically tailored to automotive applications and can thus be integrated into any mobile GIS requiring positioning information.

10.2 Bayesian classification of GPS data

Map Matching can be formulated as a stochastic classification task: The measured GPS position and orientation vector $\mathbf{x} = (x;\ y;\ \phi\ )\mathsf{T}$ is to be assigned to the road element $\mathbf{k}_i$ with highest *a posteriori* probability:

$$\hat{i} = \mathrm{argmax}_i\ p(\mathbf{k}_i|\mathbf{x}) \qquad\qquad (10.1)$$

The map represents a road element as a line segment defined by its start and end vertex. Figure 10.1(a) illustrates the classification task. A situation that clearly justifies the use of both position and orientation information is given in Figure 10.1(b). While standard position-based map matching procedures would erroneously assign the encircled position data to road element $\mathbf{k}_3$, the orientation data assist in their correct assignment.

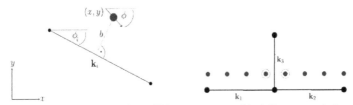

Figure 10.1. Dots (not nodes) indicate GPS measurements, black line segments denote road elements from the map. *(a)* **Classification task: Given a GPS measurement** *(x; y;* ϕ *)T* **the probability of being located on an arbitrary road element** k_i **has to be estimated. Orthogonal distance** *b* **and orientation difference thereby serve as criteria.** *(b)* **Example of erroneous assignment to road element** k_3 **for the circled GPS measurements if only positional information is used.**

The assumption of uniformly distributed *a priori* probabilities for the road elements $p(k_i)$, together with Bayes formula (cf. Duda et al. [2001], for example), yields:

$$\operatorname{argmax}_i p(k_i|x) \overset{\text{Bayes Theorem}}{=} \operatorname{argmax}_i \frac{p(x|k_i)\, p(k_i)}{p(x)}$$

$$\overset{p(X)\,\text{const.}}{=} \operatorname{argmax}_i p(x|k_i)\, p(k_i)$$

$$\overset{p(k_i)\,\text{uniformly distributed}}{=} \operatorname{argmax}_i p(x|k_i) \quad . \tag{10.2}$$

The class conditional probability $p(x_j|k_i)$ of the vehicle state measurement **x** when traversing road element k_i is modelled by two, statistically independent, random variables:

❑ The Euclidean distance *b* between vehicle position **x** and k_i is modelled as zero-mean, normally distributed random variable *B* with standard deviation σB.

❑ The angular difference δϕ between vehicle orientation ϕ and the orientation of the road element ϕ*i* is modelled as zero-mean, normally distributed random variable **F**. The standard deviation is σF.

The values for the standard deviations σB and σF have to account for the uncertainties in both the map and the GPS receiver data. The GPS orientation information becomes less reliable at lower speeds; therefore, σF is chosen to be inversely proportional to the measured GPS speed.

Equation 10.1 can now be rewritten as the following Mahalanobis distance:

$$
\begin{aligned}
\hat{i} &= \operatorname*{argmax}_i p(\mathbf{x}|\mathbf{k}_i) \\
&= \operatorname*{argmax}_i p_B(b(\mathbf{x},\mathbf{k}_i)) \cdot p_\Phi(\delta\phi(\mathbf{x},\mathbf{k}_i)) \\
&= \operatorname*{argmax}_i \frac{1}{\sqrt{2\pi}\sigma_B} \cdot \exp\left(-\frac{b(\mathbf{x},\mathbf{k}_i)}{2\sigma_B^2}\right) \cdot \frac{1}{\sqrt{2\pi}\sigma_\Phi} \cdot \exp\left(-\frac{\delta\phi^2(\mathbf{x},\mathbf{k}_i)}{2\sigma_\Phi^2}\right) \\
&= \operatorname*{argmin}_i \frac{b^2(\mathbf{x},\mathbf{k}_i)}{\sigma_B^2} + \frac{\delta\phi^2(\mathbf{x},\mathbf{k}_i)}{\sigma_\Phi^2} .
\end{aligned}
$$

$$(10.3)$$

This yields the desired classifier for a single time instant. Figure 10.2 illustrates the properties of the classifier for one particular road element. After assigning the current vehicle state to a road element, the vehicle position and orientation estimates are updated accordingly. The updated position is determined by the orthogonal projection of the GPS position on the assigned road element. The updated orientation equals the orientation of the road element.

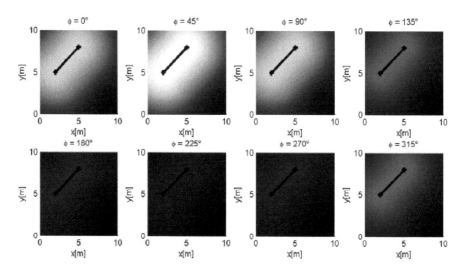

**Figure 10.2. Properties of the classifier. The black line denotes a road element with an angle of 45°
with respect to the x axis. The likelihood p(x/ki) is brightness-coded with respect to different
vehicle orientations. The highest likelihood is observed for a vehicle orientation of 45°, the lowest
for a vehicle driving just in opposite direction, i.e. 225°.**

10.3 Incorporation of position history and network topology

Up to now, the proposed classifier exclusively uses position and orientation information from the current time step. Additionally, all information concerning the topology of the road network is discarded. A considerable increase of classification robustness can be achieved by including the following features in the classifier:

Position history and orientation history: using all previously measured position and orientation data will lead to a significant reduction of the impact of gross measurement errors.

Road network topology: considering the relations among different road elements will inhibit impossible consecutive map matchings (i.e. a transition from road element k_i at time t to road element k_j at time $t+1$, although k_i and k_j are not connected).

The features described can be fully incorporated in the map matching process by the Hidden Markov Model described in the next section.

10.3.1 Hidden Markov Model (HMM)

An important class of Markov Models can be represented by a stochastic finite state machine, with state transitions and outputs being described by probability distributions. A Hidden Markov Model is defined by the five-tuple: state space, set of possible observations, transition probabilities, emission probabilities and initial state distribution. Duda et al. (2001), for example, provide an introduction to Hidden Markov modelling. Figure 10.3 depicts the proposed model. Each road element k_i constitutes one element of the state space. The emission probabilities $p(\mathbf{x}|k_i)$ correspond to the classification rule from Equation 10.2. The transition probabilities $p(k_j|k_i) = a_{ij}$ represent the road network topology: Two elements have a non-zero transition probability only if they share at least one vertex. No state transition is preferred: $a_{ij} = 1/s_i \quad \forall j$.

We can now formulate the optimum estimate for the path. $\hat{\mathbf{i}} = \left(\hat{i}_T, \hat{i}_{T-1}, \ldots, \hat{i}_1 \right)$ for an observed input sequence $\mathbf{X}_T, \mathbf{X}_{T-1}, \ldots, \mathbf{X}_1$ using the chain rule as:

$$\hat{\mathbf{i}} = \operatorname{argmax}_{\mathbf{i}} \left(\prod_{t=2}^{T} p(\mathbf{x}_t | \mathbf{k}_{i_t}) \right) \cdot p(\mathbf{k}_{i_t} | \mathbf{k}_{i_{t-1}}) \cdot p(\mathbf{x}_1 | \mathbf{k}_{i_1}) .$$

$$(10.4)$$

The Viterbi algorithm is used for a minimum cost computation of the best path. It iteratively computes the statistically optimal sequence of state transitions for a given sequence of vehicle states.

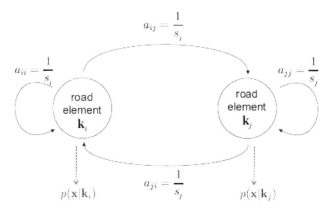

Figure 10.3. First-order Hidden Markov Model. Circles denote the model states, thin arrows denote state transitions. Road elements ki and kj are assumed to be connected. The dashed arrows indicate the output probabilities.

10.3.2 Extended HMM

The Hidden Markov Model is further augmented by considering the roads' driving restrictions (i.e. one-way streets) and, moreover, the assumed driving direction of the vehicle. Both are incorporated by the following model enhancements:

- The elements of the state space are enhanced by a flag denoting the driving direction. One road element can thus yield one (for one-way streets) or two elements in the state space.
- The transition probabilities between two state space elements are set to a very small value for contradictory driving directions, reflecting probability for doing a U-turn.

Figure 10.4 depicts the proposed model extension. Traversing a road element opposite to its allowed driving direction is no longer permitted. Additionally, paths with contradictory driving directions are assumed very unlikely.

10.3.3 Detection of erroneous map topology

The maximum *a posteriori* probability given by Equation 10.2 can directly be used as a measure of goodness of the classification result. A very low value indicates a coarse GPS measurement error, as already indicated by low horizontal/vertical dilution of precision (HDOP/VDOP) values within the GPS receiver protocol, or a modelling error. Modelling errors refer to an erroneous map topology due to missing road elements. Hence, the algorithm inherently provides a means for detecting erroneous map data.

In the case of a detected model error the Hidden Markov Model is reset by discarding all previously acquired position data.

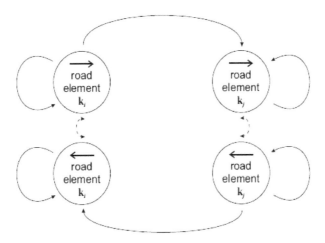

Figure 10.4. Extended Hidden Markov Model depicting the case where road elements k*i* and k*j* are bidirectional. Emission and transition probabilities have been omitted. Only state transitions with non-zero transition probability are shown. A dashed transition arrow indicates a low transition probability (reflecting the U-turn probability).

10.3.4 Revised position estimate

The proposed classifier assigns GPS data to the most likely road segment of the digital map. This allows for a subsequent update of the vehicle state estimate. Within this contribution the updated position is determined by the orthogonal projection of the GPS position on the assigned road element. The updated orientation equals the orientation of the road element. Another possibility of computing the updated position estimate using the vehicle speed data from the GPS sensor for dead-reckoning is described in Ochieng et al. (2003).

10.4 Examples in a complex urban environment

The proposed map matching has been successfully tested on an experimental vehicle in the inner city of Karlsruhe, Germany. All tests are run on standard hardware (Pentium 4, 2 GHz processor). The digital map is commercially available and frequently used in today's navigation systems. A standard low-cost GPS receiver without differential corrections is used with an estimated standard deviation of position and orientation measurements of 10-15 metres and 15°. Data is acquired at 1 Hz. The processing time of the algorithm is 0.01 seconds per GPS datum. Real-time performance can thus be achieved on systems up to hundred times slower, for example on PDA processors. Several test runs were performed with a total amount of more than four hours of online testing (equalling about 15,000 measured GPS data points) in dense urban area. Examples of the computed path results are shown in Figure 10.5.

Few intermediate misclassifications occurred for 0.4% of the data points due to severe deviations of the GPS measurement (up to 80 metres with respect to the correct road element) or due to occasionally coarse map digitisation (up to 40 metres deviation from ground truth data). All of those cases were based on the following configuration: The vehicle was standing still close to an intersection and the GPS points were slowly drifting away from the true position. It is believed that those few cases will elegantly be circumvented by preferring self-transitions over transitions to any follow-up road elements for low vehicle speeds within the emission probabilities of the Hidden Markov model.

All intermediate misclassifications are completely compensated by the algorithm leading to a completely error-free posterior vehicle path estimate! Figure 10.6 illustrates how an intermediate misclassification in a very complex road configuration was automatically corrected by the HMM towards the correct path as soon as enough measurements corroborated the belief in the correct path. One exceptional case leading to one erroneous path estimate was observed which is analysed in Figure 10.7.

The travelled route contained three map topology errors, referring to missing road elements. All three cases have been successfully classified as *modelling error* by the algorithm (cf. Section 10.3.3).

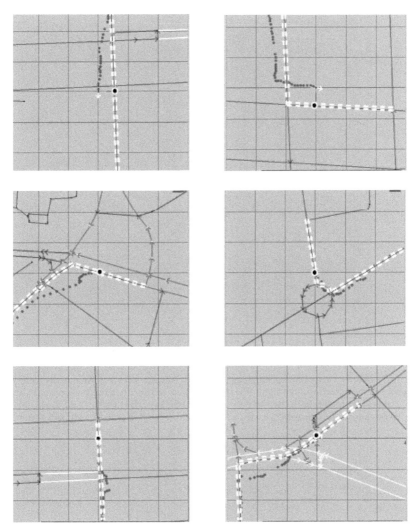

Figure 10.5. Map matching results in complex situations. One element of the background grid covers a 50 m area. The small dots correspond to measured raw GPS positions. The computed path is marked by white dashed double lines. The black dot with white surrounding corresponds to the map matched vehicle position for the current time step. Despite severe deviations between GPS measurements and the road elements from the map, the correct path (according to the classification by a human observer) has been successfully found in all situations.

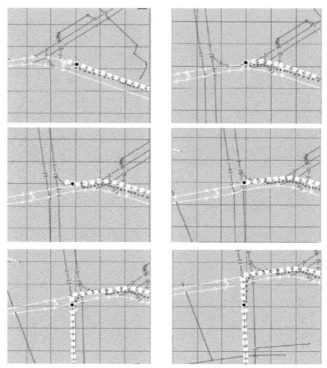

Figure 10.6. Sequence of map matching results in a complex situation. An initially correctly assigned path (top left and top right) was intermediately misclassified due to erroneous GPS and map data (middle left), but was corrected towards the correct path as soon as enough measurements had corroborated the belief in the correct path (middle right and bottom).

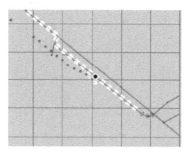

Figure 10.7. Erroneous map matching result. The vehicle was erroneously assigned to the lower, parallel running road element although the measured vehicle orientation didn't indicate any right turn. This is due to the fact that the positions of approximately 20 follow-up position measurements yielded a significantly larger error for the upper, correct element, leading to a larger overall error. That case could only be corrected if the standard deviation of the orientation would be set to a very small value compared to the standard deviation of the position. This decision is not justified because of the often large orientation deviations of the road elements in the map compared to ground truth.

10.5 Conclusion and future developments

A map matching method has been presented which exclusively relies upon information from a standard GPS receiver. The method exploits the vehicle position and orientation history, information about road network topology, driving restrictions and the assumed driving direction for each road element. A Hidden Markov Model has been formulated leading to a statistically optimal, iterative Bayesian estimation procedure. In contrast to conventional map matching, the whole vehicle path is estimated at each time step. An initialisation procedure as required by other methods is not needed. The method has proven to be robust even on challenging inner-city scenarios. Real-time performance has been shown. An improvement of navigation quality for platforms without access to vehicle sensors (odometer or gyro, respectively) is anticipated by the method. This is especially the case for handheld navigation systems. Furthermore, the studies revealed that additional vehicle sensors are not necessary in order to yield robust map matching results.

The current contribution focused on *global* map matching, i.e. determining the most likely road segment within the network. However, *local* map matching, i.e. determining the lateral vehicle pose within the road segment still remains an open challenge (an approach using Differential GPS is described in Du et al. [2004]). Current research focuses on video sensor-based estimation of the number of lanes and the subsequent estimation of the vehicle ego pose, namely lateral offset and orientation with respect to the road segment. Further work concerns methods for an automated extension of the digital map by the estimated attributes.

References

Bernstein, D. and Kornhauser, A. (1998) 'An introduction to Map matching for personal navigation assistants', in *The Transportation Research Board - 77th Annual Meeting*, Washington, D.C.

Du, J., Masters, J. and Barth, M. (2004) 'Lane-Level Positioning for In-Vehicle Navigation and Automated Vehicle Location (AVL) Systems', in *Proceedings of the Intelligent Transportations Systems Conference*, pp. 35–40.

Duda, R. O., Hart, P. and Stork, D. G. (2001) *Pattern Classification*, New York: John Wiley & Sons.

French, R. L. (1989). 'Map Matching Origins, Approaches and Applications', in *Proceedings of the Second International Symposium on Land Vehicle Navigation*, pp. 91–116.

Lakakis, K., Savvaidis, P., Ifadis, M. and Doukas, D. I. (2004) 'Quality of Map-Matching Procedures Based on DGPS and Stand-Alone GPS Positioning in an Urban Area', in *Proceedings of the FIG Working Week, TS 29/3*.

NCHRP (2002) *NCHRP Synthesis 301: Collecting, Processing, and Integrating GPS Data into GIS*, Washington, D.C.: National Academy Press.

Ochieng, W. Y., Quddus, M. A. and Noland, R. B. (2003) 'Map-matching in complex urban road networks', *Brazilian Journal of Cartography (Revista Brasileira de Cartografia)*, 55(2), pp. 1–18.

Quddus, M. A., Ochieng, W. Y., Zhao, L. and Noland, R. B. (2003) 'A general map matching algorithm for transport telematics applications', *GPS Solutions Journal*, vol. 7, no. 3, pp. 157–167.

Scott, C. A. and Drane, C. R. (1994) 'Increased Accuracy of Motor Vehicle Position Estimation by Utilising Map Data, Vehicle Dynamics and Other Information Sources', in *Proceedings of the Vehicle Navigation and Information Systems Conference*.

White, C. E., Bernstein, D. and Kornhauser, A. L. (2000) 'Some map matching algorithms for personal
     navigation assistants', *Transportation Research Part C,* vol. 8, no. 1–6), pp. 91–108.

Part IV

Motion, Time and Space

Part IV of the book 'Dynamic and Mobile GIS' focuses on the study of mobility and the use of devices, such as mobile phones and mobile GIS, for tracking the movement of people, for disaster management and for environmental monitoring. Pablo Mateos and Peter Fisher in Chapter 11 start by arguing that mobile phone location (or its technological successors) might become a new spatial reference system for the new millennium, when it will be possible to achieve spatiotemporal resolution of less than 20 metres and less than 10 seconds. The authors propose that this new spatial reference system should be called the 'new cellular geography' (in contrast to the development of the postcode to become the 'new geography' in the 1990s). Mateos and Fisher believe that mobile phones can potentially measure mobility patterns of people through the analysis of the 'spatiotemporal signature' of their mobile phones. However, such measurement is currently limited by problems associated with the poor spatiotemporal *accuracy* of the technology. Chapter 11 therefore presents an evaluation of the spatiotemporal accuracy of mobile phone location (using 2004 data for the UK). This is in order to determine current appropriate scales of application of mobile phone location as an automated method to measure and represent the mobility of people in contemporary cities. In addition, the authors propose that the work and data presented in Chapter 11 can also provide a baseline against which future enhancements of mobile phone location methods (e.g. 3rd Generation mobile phones or voice-over Wi-Fi technology) can be compared with.

Ming-Hsiang Tsou and Chih-Hong Sun in Chapter 12 suggest that mobile GIS is one of the most vital technologies for the future development of disaster management systems because it extends the capability of traditional GIS to a higher level of portability, usability and flexibility. The authors argue that an integrated mobile and distributed GIService, combined with an early warning system, is ideal to support disaster management, response, prevention and recovery. The chapter proposes the term 'mobile GIServices' to describe a framework that uses mobile GIS devices to access network-based geo-spatial information services. Chapter 12 proposes that, with the progress of new wireless communication technology and GPS techniques, mobile GIServices will help to monitor real-world dynamic changes and provide vital information to prepare and prevent natural hazards or human-made disasters.

Cristina Gouveia et al. in Chapter 13 propose the creation of an Environmental Collaborative Monitoring Network (ECMN) that relies on citizens using either mobile phones or mobile GIS in order to carry out environmental monitoring. The chapter explores the use of mobile computing and mobile communications, together with sensing devices (such as people's own senses like smell and vision), to support citizens in environmental monitoring activities. The authors suggest that this environmental collaborative monitoring network can form a framework that not only supports public participation but also promotes the use of data collected by citizens. The chapter evaluates the possibility to create a mobile collaborative monitoring network where each node is a citizen, willing to participate within environmental monitoring, but with no predefined location. The authors argue that mobile communication and computing are crucial developments in this respect, as they may be used to link citizens and therefore create new opportunities to support the creation of environmental collaborative monitoring networks. The chapter evaluates and compares two projects that explored citizen involvement within mobile environmental monitoring: 'PEOPLE' and 'Senses@Watch' projects.

Patrick Laube et al. in Chapter 14 argue that Geographical Information Science can contribute to discovering knowledge about the patterns made in space-time by individuals and groups within large volumes of motion data. The chapter introduces an approach to analysing the tracks of moving point objects using a methodological approach called Geographic Knowledge Discovery (GKD). Chapter 14 demonstrates that the integration of knowledge discovery methods within Geographical Information Science is an appropriate and powerful way to move beyond the snapshot with respect to motion analysis and provides a means to investigate motion processes captured in tracking data. The methods proposed in the chapter are illustrated using case studies from biology, sports scene analysis and political science.

Chapter 11

Spatiotemporal Accuracy in Mobile Phone Location: Assessing the New Cellular Geography

Pablo Mateos [1] and Peter F. Fisher [2]

[1] Centre for Advanced Spatial Analysis, University College London, England
[2] Department of Information Science, City University, England

11.1 Introduction

Mobile or cellular phones form part of the everyday life experiences of 80% of the adult population in developed countries and their use is growing (see Figure 11.1, which reports 2003 data). They have quickly become ubiquitous devices that go wherever their users go, surpassing their original purpose of an individual communication system to become a 'wearable computing' device.

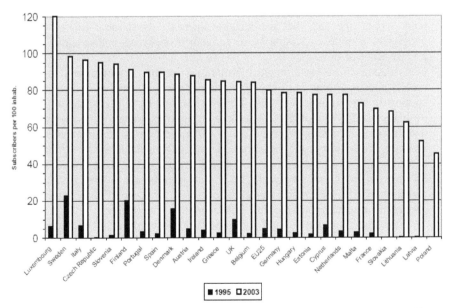

Figure 11.1. European mobile phone penetration. Number of subscribers per 100 inhabitants 1995 and 2003 (EU-25). Source: EUROSTAT (2005).

Dynamic and Mobile GIS: Investigating Changes in Space and Time. Edited by Jane Drummond, Roland Billen, Elsa João and David Forrest. © 2006 Taylor & Francis

Mobile phone *location*, together with the individual identification of its user, can provide a new methodology to understand population mobility in contemporary societies (Miller, 2004). However, from a geographic point of view, most research published on mobile phone location has primarily focused on the spatial information requirements to support Location Based Services (LBS), or its visualisations, at the *individual* user level (Mountain and Raper, 2001; Dykes and Mountain, 2003, 2002) rather than at a *society* level. A recent exception of this is the work of Shoval and Isaacson (2006).

The contribution presented in this chapter is based on the premise that mobile phone location (or its technological successors) ought to become a new spatial reference system, drawing a parallel with the development of the postcode to become the 'New Geography' a decade ago (Raper et al., 1992). The 'New Geography' of the turn of the millennium has been also defined as a 'Mobile Geography' (Amin, 2002; Fisher and Dobson, 2003) where society is no longer seen to be tied to spaces of fixity but rather to move in spaces of flows. The authors of this chapter believe that mobile phone location methodology allows the measurement of the mobility patterns of large groups of people, through the analysis of the 'spatiotemporal signature' of their mobile phones. However, such measurement is limited by constraints of the spatiotemporal *accuracy* imposed by the technology and thus configuring what is here defined as the 'New Cellular Geography' — a geography of cells through which people can be seen moving. In other words, the fact that the location accuracy is so poor makes the mobile geography a cellular geography of movement between coarse cells, and not a precise space of flows.

Assessing those technological limitations in the spatiotemporal accuracy of mobile phone location is of significant importance to social scientists interested in starting to understand the 'New Cellular Geography'. This chapter presents an evaluation of the spatiotemporal accuracy of mobile phone location with the aim of determining its most appropriate scales of application as an automated method to measure and represent the mobility of individuals or large groups of the people in contemporary cities. Measuring and understanding the spatiotemporal accuracy of information about mobile objects is a crucial requirement to build reliable dynamic and mobile GIS, the major theme of this book, and essential in determining the geographical scale of mobility that this technology can measure.

The research carried out and presented here analysed the level of spatiotemporal accuracy of mobile phone location available in the UK in 2004, as an early example of the technology easily accessible to any researcher interested in mobility studies. Furthermore, it also provides a baseline against which future enhancements of mobile phone location methods (i.e. 3rd Generation mobile phones, or voice-over Wi-Fi technology) can be compared with.

The second section of this chapter presents a brief overview of contemporary concepts of cities as spaces of flows to justify the need for new tools to measure urban mobility. Section 11.3 reviews a series of important issues surrounding mobile phone location technology, including society and mobile phones, new location uses and requirements, mobile phone location accuracy and privacy issues, in particular focusing on its spatiotemporal accuracy. The fourth section presents the methodology of the research carried out to assess the spatiotemporal accuracy of mobile phone location available in the UK in 2004, while the Section 11.5 summarises the analysis of the results. Finally, the sixth and last section offers some conclusions and drafts out the future developments and opportunities for mobile phone location to become the 'new cellular geography'.

11.2 Measuring the mobile society

> *'In contemporary societies mobility has become the primary*
> *activity of existence.'*

> (Prato and Trivero, 1985, cited in Thrift, 1996, p. 286)

Contemporary conceptions of cities and urban life give mobility a primary role as the major structuring component, using such metaphors as 'the space of flows' (Castells, 1989), or 'a place of mobility, flow and everyday practices' (Amin and Thrift, 2002). Cities are no longer seen as a bounded space around a single centre, or as an independent organic structure with well-defined borders, nor as an integrated system following specific rules within an 'outside world'. Instead cities are today perceived as nodes in a space of flows (Castells, 1996). Cities are thus seen as the relative space of the 'multiplex city' vs. the old order of the 'uniplex city' resulting from a 'splintering urbanism' (Graham and Marvin, 2001) or as a place of mobility, flow and everyday practices (Amin and Thrift, 2002). These and other authors (e.g. Bauman, 2000; Urry, 2003; White, 1992), no longer see cities as spaces of fixity, where order should be sought, but instead as an amalgamation of changing flows, a station in networks of distant socio-economic relationships, a relative space of complexity.

Despite a general consensus in this major turn on the conception of contemporary cities, there are conflicting theories about how contemporary cities should be understood and represented. This chapter does not aim to participate in the current urban debate, but instead proposes and evaluates a new methodology to represent such new conceptions of contemporary cities. There is a need for finding new representations of cities and contemporary urban life but it is acknowledged that the right tools to do it have not been available. As Sudjic summarises it: 'it is true that in its new incarnation, the diffuse, sprawling, and endlessly mobile world metropolis is fundamentally different from the city as we have known it (…) But the equipment we have for making sense of what is happening to our cities has lagged far behind these changes' (Sudjic, 1992, p. 297).

This lack of appropriate 'measuring equipment' is especially obvious in the traditional methods of social science research that try to map and understand the spatiality of the 'mobile society', since they fail to adequately measure its rapidly changing spatiotemporal dynamics. These attempts to understand mobility from the standpoint of population geography (primarily based on population census), geodemographics, and transportation research (through travel surveys), usually fall short of reflecting the complex reality of each individual's life (Cole et al., 2002).

Therefore, mobile phone location, or in fact any other technology that allows a mobile personal computer device capable of transmitting its location (such as a PDA-GPS – personal data assistants with global positioning system, Wi-Fi laptop computers, RF-ID tags – Radio Frequency Identification tags, etc.), offer a very viable solution to track the mobility of large groups of population with extensive societal coverage already built in.

11.3 A review of mobile phone location

This section will look into the technological detail of mobile phone location to help understand the results of the evaluation of its spatiotemporal accuracy, which is presented in subsequent sections. The following five sub-sections will briefly cover a few of the important issues that shape mobile phone location technology; society and mobile phones, new location uses and requirements, mobile phone location accuracy and privacy issues. The focus is thus on the impact of each of these issues on spatiotemporal accuracy, as background against which to interpret the results of the research presented later in the chapter.

11.3.1 Society and mobile phones

During the last ten years, mobile phones have become an integral part of contemporary societies, not only in developed countries, but also in the so-called developing countries, where on many occasions it is the only type of telecommunication technology available to its citizens (Agar, 2003). The number of mobile phones worldwide in 2005 was 2 billion subscribers with an estimated figure of 3 billion subscribers by 2010 (Informa Telecoms and Media 2005). In most countries mobile phone penetration had widely surpassed the number of fixed telephone lines in 2005.

This phenomenal growth is signalling the inherent essence of the mobile phone versus the fixed telephone; the mobile phone identifies and communicates the person who uses it, while the fixed telephone line connects the household or the company (Cairncross, 2001). It is this personal use and individual identification, together with the ubiquitous possibilities of nearly anytime/anywhere one-to-one communication that has made mobile phones so popular. Their diffusion has been much faster than any other mass technology in the past, to the point of starting to be considered today as an 'extension of the body' (Townsend, 2000).

These are the pivotal points that justify the use of mobile phones as the location technology for the purpose of tracking mobility in contemporary societies, as expressed in Section 11.2, and that can be summarised in the following three major advantages for social scientists:

- ❑ High penetration across most groups in society
- ❑ Accepted status as an 'always-on wearable device'
- ❑ The individual identification of its user.

This chapter concentrates in the spatiotemporal accuracy in mobile phone location and therefore other societal aspects of the mobile phone boom will not be covered here (for a review, see Lacohée et al., 2003), nor its technical evolution and future developments. Nevertheless, additional research is needed to identify the social groups that lie in the remaining proportion of society that is not using mobile phones, and establish the major factors in their decision, since they would not be covered by the methodology analysed here (e.g. see Osman et al., 2003).

11.3.2 Mobile phone location technology

Mobile phone operators need to know the geographic location of each mobile phone device in the network in order to be able to route calls to and from them, and to seamlessly transfer a phone conversation from one base station to a closer one as the user is moving while talking. This technical need was transformed into a commercial opportunity to increase the *Average Revenue Per User* (ARPU; Adams, et al., 2003), through what are now known as 'Location Based Services' (LBS).

LBS are all those services that use the location information of a mobile device to provide a user with location-aware applications (Fisher and Dobson, 2003). Such location information can be provided by the network operator, the mobile phone device, or a combination of both. The type of LBS applications initially proposed were very broad and creative and raised many expectations in the general public (Schofield, 2004). For example, one was offered the possibility to make requests of the type of 'where is the nearest...?' (ATM, petrol station, pharmacy, etc.), identify friends that walk near by, ask for navigation instructions if we are lost, know where is another person, or receive a promotion from a familiar store as we walk past it (Location based 'spam', see Chapter 3). Nevertheless LBS failed to deliver its promises at the turn of the century, and its huge forecasted market potential did not come to reality basically because the users have yet to find any true value in the few service options available (Zetie, 2004). This is partly because early services have been very restricted due to the poor location accuracy available, and the limited capabilities of both the hand-held hardware (screen size and quality, processing power and storage capacity) and the network data transfer speeds and bandwidth (Mountain and Raper, 2001). These issues are beyond the concern of this accuracy-focused chapter, however, and hereinafter only the location aspect of the much broader 'LBS field' will be considered in this analysis (the 'L' in LBS).

Despite the initial commercial failure of LBS, new legislation recently introduced by the US and the European Union (EU) requiring mobile phone operators to provide an accurate location for calls to emergency services, have acted as a catalyst for increased commercial support of LBS since 2001 (Chen, 2004). Worldwide revenue from LBS is now expected to increase to more than USD $3.6 billion by 2010, from $500 million in 2004 (Chen, 2004). Those legal requirements and their implications for location accuracy demands will be discussed in Section 11.3.3. The second characteristic of this LBS 'revival' is that the most successful applications are not those that offer location-aware contents to the mobile device user, but instead, those that provide the user's geographical location to a third party, together with some other value-added services, especially tracking of children, travelling employees or vehicle fleets (Zetie, 2004). These services have raised new privacy challenges that will be briefly exposed in Section 11.3.5. But all are based upon the ability of the system to report reliable and accurate locations of users, providing the rationale of the work reported in this chapter.

11.3.3 New location uses and requirements

The automatic location of persons or mobile objects is a broad research field and commercial market, being LBS for mobile phones just a subset of it. As computers become ever more ubiquitous and multifunctional, mobile computing devices 'need to know' their geographical location in order to perform certain functions (Costa-Requena et al., 2001). This broad field of ubiquitous computer devices is generally referred to as 'wireless computing', and the devices that reveal the location of the person using them (e.g. GPS, mobile phones, RF-ID tags, etc.) have been termed '*Personal Location Devices*' (PLD; Fisher and Dobson, 2003). Therefore, the discussion presented here should be viewed within the much wider subject of wireless computer device location, but has been reduced in this chapter to mobile phone location for simplicity due to the enormous popularity of that device.

A major distinction must be made between applications where the location information is only known by the *personal location device* (PLD) itself, its 'end-user', or the network operator, and those in which this information is passed onto a third party. A third party refers here to an organisation or person who requests the location information of a PLD, being a different entity from either the user of the PLD (1st party) or the network or technical operator of the PLD (2nd party) (Fisher and Dobson, 2003). The current debate and new thrust of mobile phone location applications in the last few years have been particularly centred on these third-party applications.

The situations in which society is ready to relinquish some individual privacy for a greater benefit are typically those where life is at risk, or justice is at issue. The privacy issues of disclosing such information are briefly exposed in Section 11.3.5. In this section the main two new uses of such 'potentially allowed disclosures' will be summarised, since they are shaping the new location requirements and are providing new scenarios for future research in this area: emergency services, and terrorism and crime prevention.

Emergency Services. In situations where life is at risk, the most important strand of new location applications fall within the emergency services arena (see also Chapter 12 on the use of mobile GIServices applications in disaster management). The dispatchers of emergency-response organisations recognised the growing problem of not being able to locate calls from mobile phones, which in many countries account for over 50% of total calls to emergency services (Salmon, 2003). The caller is usually not able to provide his/her location accurately to send a rapid response, especially under a panic situation or when outside the area of his/her daily wanderings (Hunt, 2004).

In the US a set of legislation known as 'e-911' (for 'enhanced 911', the federal emergency number) was approved by the Federal Communications Commission

(FCC) as early as 1996, requiring all the wireless communication operators to provide the *automatic location information* (ALI) of callers to 911 emergency services (Federal Communications Commission, 2004). The initiative was to be rolled out in two phases at the end of which wireless operators were required to provide tight location accuracy depending on one of two possible methods to provide ALI; 'network-based solutions', where the network calculates the location of the caller, or 'handset-based solutions', where the location is provided by the actual handset (requiring GPS-enabled phones). Those accuracies were (Salmon, 2003):

❑ For network-based solutions - 100 metres for 67% of calls, and 300 metres for 95% of calls.
❑ For handset-based solutions - 50 metres for 67% of calls, and 150 metres for 95% of calls.

The FCC has already fined several operators for millions of dollars for failing to meet those requirements (Weaver, 2003).

The importance of the e-911 initiative lies in the fact that, based on the legal pressure faced by mobile phone operators, the wireless location market has been exhausting all the technical possibilities that were financially feasible to provide an accurate ALI, speeding up dramatically the implementation of true LBS as a side effect (Branscombe, 2003). Similar but later efforts have been introduced in the European Union (EU) under the e-112 directive (European Commission, 2001), proposed in 2001 and finally approved in July 2003 (Branscombe, 2003), whose benefits are just starting to be realised by the emergency services sector in some EU countries (Hunt, 2004).

Terrorism and crime prevention. Other examples of the new uses of mobile phone location information by third parties that have already started to be accepted by society are related to the numerous recent measures to tackle terrorism or crime. Mobile phone location information is already being used by the police to track offenders, either in chases or as evidence for trials. Summers (2003) reports six different trials where mobile phone location evidence proved decisive in the conviction of murderers between 2002 and 2003, depicting this technology as 'the new fingerprint'. The value of this technology for police has been definitely proven during the Madrid and London train terrorist bombings, on 11th March 2004 and 7th and 21st July 2005, in which mobile phones played a major role in activating the bombs or capturing the terrorists (El Pais, 2004; BBC News 2005a).

However, the way that mobile phone location technology worked in 2004 could only *partially* help the police authorities, not only because of the location accuracy problems already mentioned, but more importantly due to the ephemeral nature of the location data that is not systematically stored. In the UK, the Home Office proposed a requirement for mobile phone operators to keep location information for 12 months and SMS for six months (Mathieson, 2003), a proposal which has been extended to the whole of Europe in 2005 (BBC News, 2005b). These proposals present an important challenge to individual privacy (BBC News, 2005c), as well as require a high level of commitment and investment from the operators, sometimes saturated by police requests (The Economist, 2005). On the other hand such datasets pose an enormous potential for the type of geographical analyses this chapter proposes (assuming that some degree of access to aggregated data is allowed), as well as a critical challenge to current ontologies of spatiotemporal representation, GIS data models and database storage and processing capacity.

11.3.4 Mobile phone location accuracy

There are several approaches to finding the PLD user's location employing various technologies and resulting in several geographical accuracies. Until recently, a distinction in the positioning technology of a PLD would clearly differentiate the type of device and its market sector (e.g. mobile phone, GPS or RF-ID tag). As these technologies have been miniaturised they are being combined in hybrid solutions that use more than one of these location methods to calculate its position (such as GPS-enabled phones). Nevertheless, in 2004 the majority of PLDs used were mobile phones (in a traditional sense) and still had a 'stand-alone' basic network positioning method. Therefore, since the interest of the research presented in this chapter was to measure the commonly available mobile phone location technology in the UK in 2004, only network-based location methods will be discussed in this chapter.

Mobile phone location methods (until GPS-enabled phones appeared in 2004) rely on the way operators structure the cellular network of transmitters for finding phones in their service territory and routing calls to them. The basic type of phone positioning is called *Cell-ID* location (Cell Identification), a method that requires little investment and provides poor accuracy since cells vary greatly in size, especially outside urban areas (Spinney, 2003). Accuracy of Cell-ID location depends on the size of the cell where the user is located (the greater the cell size, the less accurate the location estimate), cell size being dependent on several factors: the density of base stations that an operator has in an area, the power of their transmitters, the height over the ground of the transmitters and the obstacles around the base station (e.g. buildings, trees, local topography). As a result, the accuracy can vary from 500 metres to over 5 km (see Section 11.5.1 for the actual results and discussion).

A variation of the Cell-ID methodology is called Cell-ID++, which combines basic Cell-ID positioning, with Timing Advance (TA), and Network Measurement Results (NMR). TA corresponds to a distance estimate from the base station to the handset based on timings, while NMR measures the power of the signal received by the mobile phone from the adjacent base stations (Faggion and Trocheris, 2004). Cell-ID++ just estimates the radius around a base station where the mobile phone is likely to be located.

In order to improve location accuracy beyond Cell-ID solutions, many different methods have been proposed. The 'Angle of Arrival' (AOA) method requires a minimum of two base stations to determine the angle of arrival of the mobile phone signal, and the network can then work out the handset location by bilateration.

'Enhanced Observed Time Difference' (E-OTD), also called 'Time of Arrival' (TOA) (Zhao, 2002), measures the time difference between signals from three or more base stations to the mobile phone to calculate the distance to each station, and then obtains its position by the trilateration of those distances. All elements in E-OTD must be constantly synchronised. A different method that does not require time synchronisation is 'Time Difference of Arrival' (TDOA) in which the mobile phone position is determined by the network's servers based on trilateration but using time difference measurements rather than absolute time measurements.

All of these methods are very difficult to implement across the network because expensive equipment such as directional antennas, precise clocks or additional location-processing servers at base stations are required (Zhao, 2002). Additionally some of them require special software to run on the handsets, which would mean that it would only work in the most advanced ones. According to Mountain and Raper (2001) in 2001 neither did most of the UK networks provide these conditions, nor were the handsets ready, and even when all factors converge the accuracy was at best 200 metres in urban areas.

In the US the e-911 initiative has caused mobile phone operators to abandon network-based location solutions, due to the poor spatial accuracies of the network solutions described above, as well as the high infrastructure costs required to implement them across an extensive territory. Alternatively, they had rapidly started switching to GPS-enabled phones (Federal Communications Commission, 2004) that can give an accuracy of 20 metres and are much cheaper to subsidise at the handset level rather than invest in the network that would have been soon replaced by 3rd generation technology. New hybrid solutions include methods that combine both network and handset-based location calculation, of which Assisted-GPS (A-GPS) is the most promising and accurate one (Shoval and Isaacson, 2006), but had still not made its way to mainstream mobile telephony by 2005 and therefore will not be covered in the chapter.

11.3.5 Privacy issues

> 'What counts is not the barrier but the computer that tracks each person's position – licit or illicit – and effects a universal modulation.'
>
> (Deleuze, 1992, p. 5-7, cited in Thrift, 1996, p. 291)

Mobile phone location has attracted a wide interest in the critical analysis of its challenges to individual privacy, since it allows a person's location to be known by other people. This aspect, termed location privacy, has already been covered in Chapter 3 of this book, and only a few additional issues particular to mobile phone location will be briefly mentioned here.

The right to privacy in the handling of personal data is regulated in most of the countries by personal data protection legislation. These ensure that personal data should only be collected where necessary, should only be used for the purposes for which they were collected, should not to be disclosed to another group or agency without some sort of consent, and should be securely stored (Curry, 1999). Of interest here are the situations of disclosure of such personal information to another group or agency (named in this chapter as disclosure to a third party), and the requirement of specific user consent.

As mentioned in Section 11.3.3, society has already deemed it necessary to relinquish some individual privacy for a greater benefit from mobile phone location, typically in situations where life is at risk (emergency services), or crime has been committed. Fisher and Dobson (2003) propose a typology of seven scenarios for the use of personal location information, and establish whether disclosure of such information to third parties will be beneficial, and therefore should be legally permitted, and those in which it should not and thus should be prosecuted.

The aim of this chapter is to propose the use of mobile phone location technology as a proxy to understand mobility at societal level, and therefore this would require the consent of mass groups of society in order to be truly representative. The proposed approach should then be to ask the potential participants for their consent, ensuring the information is only used for the purpose of understanding urban mobility in aggregated ways. This is a similar scenario as with the information requested in a population census, although there all citizens are obliged to provide personal information on condition that it only be disclosed in a spatially aggregated form (Dale et al., 2000). Nevertheless, as Chapter 3 has already insisted, new legislation should regulate a 'new right to privacy in a geocoded world' (Curry, 1999), so that these beneficial uses are permitted and potential abuses prosecuted, before the technology is made 'blind' altogether, and its potential urban research benefits are lost.

11.4 Methodology for assessing spatiotemporal accuracy in mobile phone location

This section presents the methodology undertaken to measure the spatiotemporal accuracy of mobile phone location in the UK in 2004. The main objective of the research methodology selected is to illustrate, through meaningful examples, what are the characteristics, limitations and potential uses of different datasets containing current mobile phone location information for a group of users in a metropolitan area.

The main concern of potential users of this technology is the level of spatiotemporal accuracy of the location information that was commercially available in the main European markets at the time this research was carried out (2004). Therefore, the research presented in this chapter is an empirical exploration of the technology available in 2004 through a series of mobile phone location tests that surveyed the movements of a small sample of participants in the UK.

In the UK, the four mobile phone operators with a GSM licence in 2004 were: *O2*, *Vodafone*, *T-mobile* and *Orange*. A fifth operator called '*3*' held a 3rd Generation licence only (UMTS). Other 'mobile brands' in the market that could seem independent in fact operate using one of the four operators' networks listed above (e.g. *Virgin*). In this chapter the term 'operators' are used to describe the four GSM operators in 2004. In addition to the operators, mobile phone location services can be provided by any third party, under a contract with the operators to sell value-added services using location information. In the UK, at the time of writing (2005), there were around 12 different companies of this type, which will be referred to in this chapter as service providers.

Due to the difficulties of accessing mobile phone location data directly from the operators, it was decided to subscribe to several of those service providers and purchase from them the location service of the participant's mobile phones. Two different services were selected based on price, ease of access to and automation of the location information collection; 'FollowUs' (www.followus.co.uk) and 'Childlocate' (www.childlocate.co.uk).

Figure 11.2. Mobile phones and GPS used in Test 2 (photograph: Brenda Valdes).

A series of three different tests of mobile phone location were carried out in order to generate a minimum dataset that would enable the desired analysis to be performed. The mechanics of each of these tests and their objectives will be individually explained below in this section, and the overall results in Section 11.5. All of these tests basically consisted of locating several mobile phones through two of the service providers that offered a commercial option to track mobile phones with the consent of each user. The differences between the tests reside in the aim of each test, their geographical coverage, and the stages in which they were developed, providing different settings for the subsequent analysis of the data.

As a result of each test, a series of 'time-location stamps' (TLS) was collected. The concept of time-location stamps will be used in this chapter to refer to the spatiotemporal information of a single event in which both the geographical location of a mobile device (in a pair of coordinates) together with the date and time at which such a location was measured is stored in the project's database. The aims and methods of each of the three tests carried out can be summarised as follows:

❑   **Test 1** - Aimed to measure the intra-urban mobility patterns of a group of nine postgraduate students in the city centre of Leicester, UK. The test was carried out by approaching the potential participants through friends, since the sensitivity of information requested (personal location) demanded a minimum level of trust about the researcher that could not be expected from anonymous random participants. An account was opened with the service provider *Childlocate*, where the nine students' mobile phones were initially registered through SMS. The actual location tests were carried out during three consecutive weekdays in June 2004, at various temporal intervals each day and always within non-sleeping hours (typically 9:00 to 21:00h), in order to compare the outcome of several temporal resolutions in the data. The total number of location requests was 159 at an average temporal interval of 2.5 hours. The location requests were manually placed at *Childlocate* Website, one participant at a time. Only seven requests out of the 159 were not successful, therefore yielding 152 valid locations (availability rate of 95.7%).

❑   **Test 2** - Specifically addressed the need to measure the actual spatial and temporal accuracy of the mobile phone locations provided by the different phone operators, and analyse the differences between them, and throughout the different areas of the city centre of Leicester, in order to determine the factors for variations in accuracy. The test located three mobile phones from different operators (O2, Vodafone and T-mobile), travelling together with a GPS unit (Garmin GPS 12XL) through several areas of the city of Leicester, by car and walking. Figure 11.2 shows a photograph of the simple equipment used for this test. The different location estimates were compared between the different operators and also with the 'true' position determined by the GPS. The chosen service provider for this test was *FollowUs*, because it offered a facility to access historical location queries (Childlocate only offers a location in 'real time'), so that all locations could be batch-processed later. Moreover, the information about the locations could be accessed as a pair of British National Grid coordinates at 1 metre accuracy, substantially facilitating the transfer of the data to the GIS. The test consisted of 87 location requests placed from different points scattered around the study area, retrieving 83 valid TLS (95.4% availability rate). The temporal resolution of the mobile phone location requests in this test was considered irrelevant, because the importance of the test was to cover a dense number of different locations within the study area despite their time intervals. GPS data was collected at 30-second intervals, and later matched to TLS data by their closer time stamp through a temporal query, and then the spatial difference between GPS and TLS measurements was calculated. Figure 11.3 shows the scatter plot of those spatial differences by operator, while Figure 11.5 shows a map with the visualisations of the itinerary taken and the TLS locations by operator. Further analysis of these results is presented in Section 11.5.

❑ **Test 3** - Measured inter-urban movements, by tracking a single mobile phone during several car trips around the UK. The aim of this test was to assess the advantages of mobile phone location in measuring *inter-urban* mobility patterns as opposed to the *intra-urban* ones measured in Test 1. Test 3 was envisaged as an additional experiment to find the best scale of application for mobile phone location technology, in order to point out directions for further research. The test consisted in measuring both mobility patterns through the location of a mobile phone (a single operator in this case; O2) while moving by car between cities in the UK at distances between 100 km and 300 km. In order to perform this test, the service provider chosen was again *FollowUs* due to their ability to program automatic location requests, that for this test were set at every hour from 6:00 to 21:00h (the maximum temporal resolution allowed). The TLS data was retrieved in a different way for this test, since the automatic location requests to FollowUs send back an email with the mobile phone's position. The relevant information (mobile phone number, location coordinates, date-time stamp, postcode, etc.) was retrieved from the email and imported into the research database.

These three tests provided a total number of 309 different TLS with their corresponding estimated accuracy. These data were processed in a spatiotemporal database and analysed using ArcGIS and its *Tracking Analyst* extension. An overview of the general characteristics of the three tests is summarised in Table 11.1.

Table 11.1. Summary of location tests performed.

Test No.	Test to measure	No. phones located	No. time location stamps (TLS)	Service provider	Study area	Length of test	Temporal res. (daytime)
1	Intra-urban mobility tracking	9	152	ChildLocate	Leicester	3 days	2.3 hours
2	Location accuracy of operators	3	83	FollowUs	Leicester	2 days	N/A
3	Inter-urban mobility tracking	1	74	FollowUs	England	1 week	1 hour
	TOTAL		**309**				

11.5 The results: real accuracy of the 'new cellular geography'

The results of the methodology described in the previous section throw new light on the understanding of the 'new cellular geography', that would not have been possible to assess with the scarce secondary data available from the industry. These findings can be summarised around the topics of 1) location accuracy and configuration of the cellular geography, 2) temporal accuracy and 3) aggregation and visualisation.

11.5.1 Location accuracy and the structure of the cellular geography

The tests described in Section 11.4 confirmed that the spatial disposition of mobile phones' geography is determined by the location method used by all the GSM operators in the UK in 2004, the basic *Cell-ID* method, without even a slight enhancement such as *Cell-ID++* (Faggion and Trocheris, 2004). The results of Tests 1 and 2 revealed that the mobile phone location estimates provided by the operators as a pair of coordinates always coincide with a few common points in space that will be called here 'milestones' for simplicity. For example, the 152 TLS obtained in Test 1 were all located at 24 fixed 'milestones', and the location accuracy estimate at each 'milestone' was always the same. This means that the location reported is that of the base station or transmitter, and thus the reported location accuracy is determined by the cell size of each base station.

The consequence of the predominance of this location method is a very coarse spatial resolution, which determines the geographical scale of studies that could use mobile phone location as a proxy for mobility. The tests have demonstrated that the actual location accuracy provided by operators is substantially worse than commercially advertised, and worse than previously reported research in Japan (e.g. Hato and Asakura, 2001). The average estimated accuracy of actual locations in the tests performed is 3615 metres, with a minimum of 500 metres and a maximum of 5000 metres, while the values commercially advertised vary between 150-400 metres (Childlocate, 2004). At the time of writing this chapter, a year after the tests were performed, those commercial claims have changed their estimated accuracy to 150–900 metres (Childlocate, 2005).

Operators provide accuracy estimates for all locations given, but the tests showed that they always provide the worst possible accuracy estimate (cell radius), when in most cases the real accuracy is much better (somewhere between the cell radius and the centre of the cell). In fact the true accuracy, that is, the actual difference between the true position of the mobile phone (measured with a GPS) and the estimated position provided by the operator is generally significantly smaller (average 800 metres) than that reported by the operator (average 3615 metres). Figure 11.3 shows the scatter plot of those two measurements of accuracy for each position in Test 2, where the reported accuracy is plotted in the Y-axis and the true accuracy (GPS) in the X-axis. As it can be seen, except for five TLS out of 83, the reported accuracy is worse than the real one. In addition, all of those five TLS where the operators report a better accuracy than the real one are under the 2000 metres threshold, with differences of 100-500 metres above measured accuracy. This could also mean that at these shorter distances GPS error induced by buildings was also altering the measurement of the 'true' position.

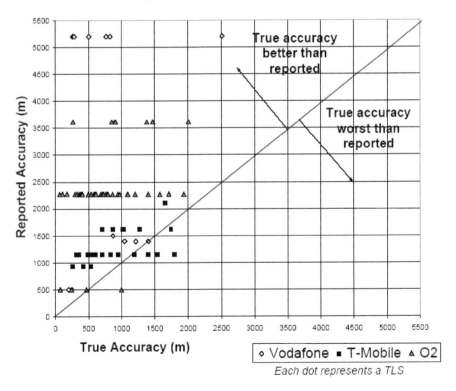

Figure 11.3. True vs. reported accuracy scatterplot (Test 2).

Another limitation of the 'cellular geography' resulting from this technology is the differences between the operators' location estimates provided for the same actual position (differences averaging 1800 metres between operators). This is due to different network layout (location of cell centroids), types of transmitter technology (sizes of cell radius), and height and topography, which influence reception coverage (as explained in Section 11.3.4). There is an additional complication of overlapping cells for the same operator: the mobile phone tends to 'stay' connected to the same transmitter even when there is a closer one, following an apparent rule of 'cell loyalty'.

Therefore, this technology is only appropriate for mobility studies that require locations that are less specific than the cell, i.e. they require a geographical resolution greater than about 3000 metres radius. Since a spatial resolution of that size would not be very meaningful to study intra-urban mobility for most cities, the most-indicated example of application would be in inter-urban mobility analysis. Test 3 proved the usefulness of such analysis and a visualisation of the type of inter-urban flows measured can be seen in Figure 11.6.

11.5.2 Temporal resolution

The time dimension is as important as the spatial one in dynamic and mobile GIS, especially in phone location, and many of the implications of Hägerstrand's 'Time Geography' (Hägerstrand, 1970) can be re-applied to this technology. The datasets will have a specific temporal resolution that will drive the type of applications in which it will be used. The minimum temporal resolution will be determined by the 'amount of mobility' that is to be monitored (i.e. high speed will require high temporal resolution).

The research identified the issue of assessing temporal accuracy, since not all the TLS were provided in real time. That is, there is a difference between the time when the location request was placed into the network and the time at which the mobile phone actually was at the reported location. This time difference will be called here 'temporal lag'. Figure 11.4 shows the frequency distribution of the temporal lag for the three tests' TLS (X-axis) for different time intervals (Y-axis). The figure shows that a significant proportion of TLS happened to be 'old locations', in other words, the location provided was showing where the phone was 'in the past', not at the time of the location request.

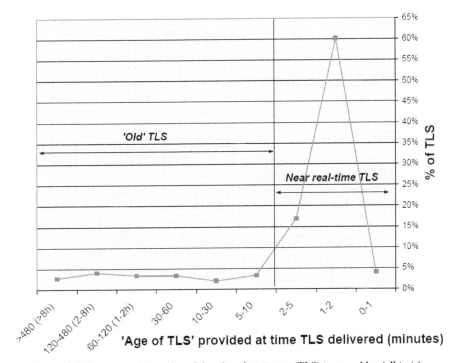

Figure 11.4 Frequency distribution of time-location stamps (TLS) temporal lag (all tests).

Out of the total TLS collected in the three different tests only 80% of them have a temporal accuracy of less than five minutes (which could be regarded as 'near real time' for most purposes), while 10% of them are between five and 60 minutes 'old', and the remainder 10% over one hour 'old' (with outliers more than eight hours later). Since all necessary devices were switched on at the time of testing and at the time of requesting TLS, the observed temporal lags can only be a consequence of the operators not always processing a new location request to the network but actually relying on stored information about where the mobile phone was the last time it was spotted in the network. Therefore, they assume the phone has not moved. This assumption is a major issue, which seems to have been overlooked by Location Based Services literature, and can have serious consequences for certain services such as emergency response applications.

11.5.3 Aggregating and visualising mobility

The visualisation of mobility flows places an enormous pressure on the creativity of aggregation and visualisation techniques to make sense of such data. Through the tests performed in this research it has been clear that the characteristics and amount of these data require new data models, aggregation methods and visualisation techniques. Figure 11.5 tries to offer a visualisation of Test 3 results, which only represented one person travelling for two days with three mobile phones and a GPS, a few data flows that obviously challenges traditional cartographic representations of movement.

The problem of aggregating the individual itineraries into meaningful common flows in order to search for general patterns is directly linked to the problem of previously selecting the right data model behind the representation of the phenomena being measured. One type of a spatiotemporal GIS data model to represent mobile phone location is a 'cell time-space' (Forer, 1998). For example, the datasets can be visualised showing cell-to-cell movements, amount of change of cell, differential cell intensity, etc. Another approach is to build density surfaces based on the number of occasions that a location is visited by the moving individual, and the time length of the 'stay' (Dykes and Mountain, 2002).

Further research is required into the different optimum data models and visualisations to understand the complexity of the 'mobile society', which increases with the number of subjects tracked, their spatiotemporal extensibility (Miller, 2004), and the temporal coverage and resolution of the datasets.

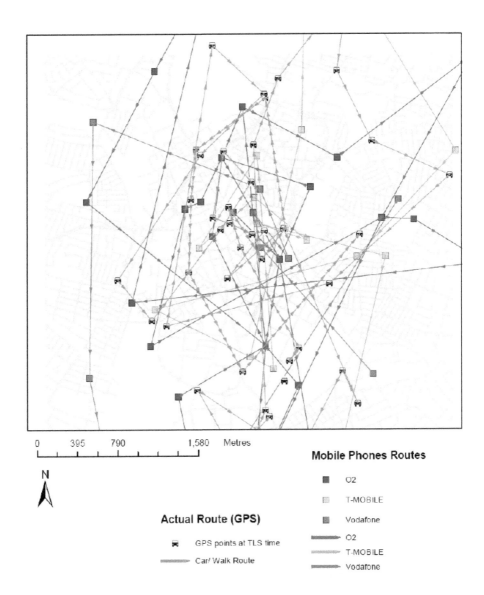

Figure 11.5. Visualisation of Test 2, GPS vs. TLS tracks through Leicester, UK. (See colour version following page 132.)

© Crown Copyright/database right 2004. An Ordnance Survey/Edina Digimap supplied service.

11.6 Conclusion and further developments

This chapter has presented the benefits of mobile phone location as a very efficient methodology to capture the mobility of large groups of the population, as well as some of its restrictions and challenges to become the spatial referencing system of the 'new cellular geography'.

Amongst its benefits, this technology presents significant advantages over GPS to track mobility, such as its low setup costs as the existing technology already covers nearly 80% of the adult population, its accepted ubiquitous presence in all aspects of everyday life, and its better urban coverage inside and between buildings. Amongst its restrictions, its current limited spatiotemporal accuracy makes it only suitable to measure inter-urban mobility, as the research presented here has indicated, and the reasonable need for user consent to disclose their location (Fisher and Dobson, 2003) limits the size of population sample than can be surveyed.

The results of the evaluation of spatiotemporal accuracy of mobile phone location carried out in the research presented here can be summarised as follows:

❑ Current spatial accuracy range: 1500–3000 metres
❑ Operators-reported spatial accuracy is substantially worse than the real one (measured by GPS)
❑ Temporal accuracy: five minutes to one hour
❑ Differences in accuracy are due to operator technology or base station placement, but not to third party service provider
❑ Recommended scales of application:
  ❑ Measurement of inter-urban or international mobility (as shown in Figure 11.6)
  ❑ Spatial resolution > 3000 metres
  ❑ Temporal resolution > 5 minutes

As technological developments in the area of personal mobile computing devices rapidly replace one another, the geographical accuracy and coverage of what has been traditionally known as 'mobile phones' is continuously improving. Assisted-GPS technology introduced in mobile phones, in mobile computers, and in other mobile objects, will allow more accurate spatiotemporal measurement no later than by the year 2010. This will yield urban researchers an innovative tool to measure intra-urban mobility at much finer geographical scale (below 20 metres) and with nearly total population coverage.

It is believed that once this point is reached (spatiotemporal resolution of less than 20 metres and ten seconds), mobile phone location will indeed be a new spatial reference system, drawing a parallelism with the evolution of the postcode to become the 'New Geography' a decade ago (Raper et al., 1992). This parallelism is based on the fact that if postcodes enabled the linking of many different datasets about the population to a unique spatial reference from which to undertake cross-sectional geographical analysis, mobile phones (or their technological successors) will soon allow the linking of many different datasets about individuals to their spatiotemporal flows, linking datasets through their different *timespaces*, facilitating the longitudinal analysis of the network society linking a space of 'cellular geographies'.

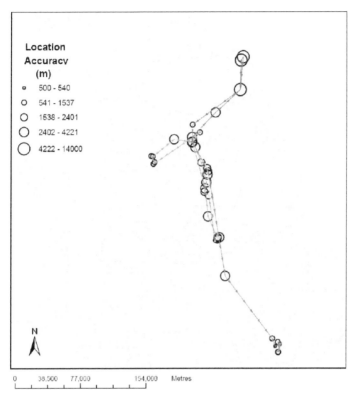

Figure 11.6. An example of inter-urban mobility application – Visualisation of Test 3 TLS through England.

© Crown Copyright/database right 2004. An Ordnance Survey/Edina Digimap supplied service.

Once mobile devices become the new spatial reference system to analyse population, as the postcode did in the 1990s, it is believed that from a social science perspective, legislation should be introduced to develop a new central government statistical survey effort that samples the personal location of the population (through mobile-phone-like devices) on certain survey days without the need for prior individual consent. This initiative should be based on similar guidelines as the national census of population to safeguard anonymity and require coverage of a large part of the population. This information should then be published and visualised in aggregated ways to preserve individual privacy, but that would allow access to much more accurate and frequent population mobility data for urban researchers and many other parties interested in the mapping of the 'new cellular geography'.

References

Adams, P., Ashwell, G. and Baxter, R. (2003) 'Location-based services - an overview of the standards', *BT Technology Journal*, 21, 1, pp. 34–43.

Agar, J. (2003) *Constant Touch: A Global History of the Mobile Phone,* Cambridge: Icon.

Amin, A. (2002) 'Spatialities of globalisation', *Environment and Planning A*, 34, pp. 385–399.

Amin, A. and Thrift, N. (2002) *Cities: Reimagining the Urban,* Oxford: Polity Press.

Bauman, Z. (2000) *Liquid Modernity,* Cambridge: Polity Press.

BBC News (2005a) *Tracking a suspect by mobile phone*, 3rd August, last accessed 8th September, http://news.bbc.co.uk/1/hi/technology/4738219.stm.

BBC News (2005b) *UK urges new phone record rules*, 7th September, last accessed 8th September, http://news.bbc.co.uk/1/hi/uk_politics/4221364.stm.

BBC News (2005c) *EU phone record plan in trouble*, last accessed 8th September, http://news.bbc.co.uk/1/hi/4225188.stm.

Branscombe, M. (2003) 'I'm on my way', *The Guardian,* 31 July, p. 15.

Cairncross, F. (2001) *The Death of Distance: 2.0 : How the Communications Revolution Will Change Our Lives,* (2nd Edition), London: Texere.

Castells, M. (1989) *The Informational City,* Oxford: Blackwell.

Castells, M. (1996) *The Rise of the Network Society,* Oxford: Blackwell.

Chen, A. (2004) 'After slow start, location-based services are on the map', *EWeek,* 12 July. Last accessed 05 Aug 2004, from: http://www.eweek.com/article2/0,1759,1621409,00.asp.

Childlocate (2004) *How accurate is the service?* http://www.childlocate.co.uk/faq.html; first accessed 23rd July 2004

Childlocate (2005) *How accurate is the service?* http://www.childlocate.co.uk/faq.html; last accessed 2nd September 2005

Cole, K., Frost, M. and Thomas, F. (2002) 'Workplace Data from the Census', in Rees, P., Martin, D. and Williamson, P. (eds.) *The Census Data System,* pp. 269 – 280,Chichester: Wiley.

Costa-Requena, J., Tang, H. and Del Pozo, I. (2001) 'SIP dealing with location based information', *Journal of Communications and Networks*, 3 (4), pp. 351–360.

Curry, M. (1999) 'Rethinking privacy in a geocoded world', in Longley, P., Goodchild, M., Maguire, D. and Rhind, D. (eds.) *Geographical Information Systems,* 2nd Edition, Vol. 2, pp. 757–766, New York: Wiley.

Dale, A., Fieldhouse, E. and Holdsworth, C. (2000) *Analyzing Census Microdata,* London: Arnold.

Deleuze, G. (1992) Postscript on the societies of control, *October*, 59, pp. 3–7 (cited in Thrift, 1996, p. 291).

Dykes, J. A. and Mountain, D.M. (2002) 'What I did on my vacation: spatiotemporal log analysis with interactive graphics and morphometric surface derivatives', in Wise, S., Brindley, P., Kim, Y.H. and Openshaw, C. (eds.) *GISRUK 2002 Conference Proceedings*, Sheffield, April 2002.

Dykes, J. A. and Mountain, D.M. (2003) 'Seeking structure in records of spatio-temporal behaviour: visualisation issues, efforts and applications', *Computational Statistics & Data Analysis,* 43, pp. 581–603.

The Economist (2005) 'Italian phone-tapping', *The Economist*, March 3rd, UK edition.

El Pais (2004) 'The telephonic trail' (*in Spanish*), *El Pais,* 06 May, p. 16.

European Commission (2001) 'Commission calls for study on caller location in mobile networks', *Official Journal of the European Communities,* S 105, 44. Last accessed 2/6/04, http://dbs.cordis.lu/cgibin/srchidadb?CALLER=NHP_EN_NEWS&ACTION=D&SESSION=&RCN=EN_RCN_ID:16877

Eurostat (2005) 'Around 80 mobile subscriptions per 100 inhabitants in the EU25 in 2003', *News Release- Eurostat Press Office,* 20/2005 - 7 February 2005.

Faggion, N. and Trocheris, A. (2004) 'Location-Based Services strengthen the strategic position of mobile operators', *Alcatel Telecommunications Review*, 4th Quarter 2003/1st Quarter 2004.

Federal Communications Commission (2004) *Enhanced 911*, [Online], Available: http://www.fcc.gov/911/enhanced/ [17 July 2004].

Fisher, P. and Dobson, J. (2003) 'Who knows where you are, and who should in the era of mobile geography?', *Geography*, 88, 4, pp. 331–337.

Forer, P. (1998) 'Geometric Approaches to the Nexus of Time, Space and Microprocess: Implementing a Practical Model for Mundane Socio-Spatial Systems', in Egenhofer, M. J. and Golledge, R. G. (eds.) *Spatial and Temporal Reasoning in Geographical Information Systems*, pp. 171–190, New York: Oxford University Press.

Graham, S. and Marvin, S. (2001) *Splintering Urbanism: Networked Infrastructure, Technological Mobilities and the Urban Condition*, London: Routledge.

Hägerstrand, T. (1970) 'What about people in regional science', *Papers of the Regional Science Association*, 24, pp. 7–21.

Hato, E. and Asakura, Y. (2001) 'Utilization of travel locus chart using mobile communication systems for GIS-T. 20–22 June. Paper presented at the *ASIA GIS 2001 Conference*, Tokyo.

Hunt, J. (2004) 'Route to rescue', *The Guardian*, [Online] 6 May, p. 19.

Informa Telecoms and Media (2005) *Global Mobile Forecasts to 2010*, 5th Edition, Informa Telecoms and Media, [Online], Available: http://www.informatm.com/marlin/20001001561/ARTICLEVIEW/mp_articleid/20017308046/mp_pubcode/IMG_press/media/true [06 September 2005].

Lacohée, H., Wakeford, N. and Pearson, I. (2003) 'A social history of the mobile telephone', *BT Technology Journal*, 21, 1, pp. 203–211.

Mathieson, S. (2003) 'Keeping 1984 in the past', *The Guardian*, 19 June, p. 24.

Miller, H. J. (2004) 'Activities in space and time', in Stopher, P., Button, K., Haynes, K. and Hensher, D. (eds.) *Handbook of Transport 5: Transport Geography and Spatial Systems*, St Louis, MD: Pergamon-Elsevier Science.

Mountain, D. M. and Raper, J. (2001) 'Positioning techniques for location-based services (LBS) characteristics and limitations of proposed solutions', *Aslib Proceedings*, 53(10), pp. 404–412.

Osman, Z., Maguire, M. and Tarkiainen, M. (2003) 'Older users' requirements for location based services and mobile phones', *Human-Computer Interaction with Mobile Devices and Services*, 2795, pp. 352–357.

Prato, P. and Trivero, G. (1985) 'The spectacle of travel', *Australian Journal of Cultural Studies*, 5, pp. 25–43 (cited in Thrift, 1996, p. 286).

Raper, J. F., Rhind, D. and Shepherd, J. (1992) *Postcodes: The new geography*, Harlow, UK: Longman.

Salmon, P. (2003) 'Locating calls to emergency services', *BT Technology Journal*, 21(1), pp. 28–33.

Schofield, J. (2004) 'Smart places', *The Guardian*, [Online] 25 March, p. 23.

Shoval, N. and Isaacson, M. (2006) 'The application of tracking technologies to the study of pedestrian spatial behaviour', *The Professional Geographer*, 58, (1), in press.

Spinney, J. (2003) 'Mobile positioning and LBS applications', *Geography*, 88, pp. 256–265.

Sudjic, D. (1992) *The 100 Mile City*, San Diego: Harcourt Brace.

Summers, C. (2003) 'Mobile phones - the new fingerprints', *BBC News*, 18 December [Online], Available: http://news.bbc.co.uk/1/hi/magazine/3307471.stm [3/08/04].

Thrift, N. (1996) *Spatial Formations*, London: Sage.

Townsend, A. (2000) 'Life in the realtime city: Mobile telephones and urban metabolism', *Journal of Urban Technology*, 7, pp. 85–104.

Urry, J. (2003). *Global Complexity*, Cambridge: Polity.

Weaver, H. F. (2003) 'T-Mobile agrees to $1.1 million fine for missing E911 deadlines', *RCR Wireless News*, 18 July, [Online], Available: http://rcrnews.com/cgi-bin/news.pl?newsId=14352. [4 August 2004].

White, H. (1992). *Identity and Control: A Structural Theory of Social Action,* Princeton, NJ: Princeton University.

Zetie, C. (2004) 'Location services find their way to the enterprise', *Information Week,* 02 August, [Online], Available: http://www.informationweek.com/story/showArticle.jhtml?articleID=26100784 [05 August 2004].

Zhao, Y. (2002) 'Standardization of mobile phone positioning for 3G systems', *IEEE Communications Magazine,* 40 (7), pp. 108–116.

Chapter 12

Mobile GIServices Applied to Disaster Management

Ming-Hsiang Tsou [1] and Chih-Hong Sun [2]

[1] Department of Geography, San Diego State University
[2] Department of Geography, National Taiwan University

12.1 Introduction

Disaster management (or emergency management) is unique among GIS applications because it deals directly with loss of human life and property damage. In September 2005, the tragic event of Hurricane Katrina in the U.S. demonstrated how important disaster management is. While the comprehensive implementation of disaster management systems can save thousands of people's lives, poorly implemented disaster *management* can of itself cause significant casualties, property damage and economic loss when the disaster happens.

On December 26, 2004, another example of poor disaster management was recognised after a massive 9.0 earthquake in the Indian Ocean. A horrifying tsunami destroyed coastline areas of 11 countries and caused an unbelievable number of deaths (over 150,000). People from around the world began to realise the power of Nature and how devastating hazards and loss can occur by underestimating her power. Some news reporters from the National Public Radio (NPR) in the U.S. commented that if these countries around the Indian Ocean had had a tsunami early warning system (such as the Pacific Tsunami Warning System used by the U.S. and Japan) hundreds of thousands of people would have been saved from the tsunami. However, the authors of this chapter disagree with this statement because a single tsunami warning system is not sufficient for the establishment of comprehensive disaster management. This chapter argues that what is really needed is an integrated mobile and distributed GIService, combined with the early warning systems, to support disaster management, response, prevention and recovery.

To create a comprehensive disaster management system, our society needs to rely on advanced geo-spatial technologies and services. Mobile GIS is one of the most vital technologies for the future development of disaster management systems. Mobile GIS and mobile Geographic Information Services (mobile GIServices) extend the capability of traditional GIS to a higher level of portability, usability and flexibility. Mobile GIS are integrated software and hardware frameworks for the access of geo-spatial data and services through mobile devices via wireline or wireless networks (Tsou, 2004). The unique feature of mobile GIS is the ability to incorporate Global Positioning Systems (GPS) and ground truth measurement within GIS applications.

This chapter introduces a new term, 'mobile GIServices', which describes a framework to utilise Mobile GIS devices to access network-based geo-spatial information services (GIServices). Mobile GIServices can be adopted in various GIS applications and scenarios, including car navigation systems, utility management, environmental monitoring and habitat protection tasks. Disaster management and emergency response are one of the most popular domains in the recent development of mobile GIServices.

For example, mobile GIServices can combine GPS and satellite images to assist the local government and emergency response teams in identifying potential threat areas, so critical 'hot zones' can be immediately created. Near real-time spatial analysis models supported by GIS could be used to rapidly generate the most effective evacuation routes and emergency plans during natural hazard events, including wildfires, floods and tsunamis. Wireless Internet-based GIS could also assist public policy officials, firefighters and other first responders with identifying areas to which their forces and resources should be dispatched. To accomplish these goals, it is important to introduce these new mobile GIServices technologies to emergency management personnel and related organisations. Also, emergency managers and first responders need to realise both the advantages and the limitation of GIS technologies in disaster management.

In the U.S., the percentage of agencies that used computers as tools for emergency operations (such as 911 or emergency calls, ambulance dispatch, evacuation procedures or rescue services) was 54.2% in 2001. The percentage using emergency management software (such as GIS or Management Information Systems – MIS) was 26.6% (Green, 2001). Although software usage has increased in the last few years, some emergency managers and staff are still reluctant to adopt computers and GIS for their main tasks (based on the authors' own experiences). One of the major obstacles is the concern for system portability and reliability. Traditional GIS are not considered portable by first responders (such as local police officers, fire fighters and emergency medical personnel who can arrive first and take actions to rescue people and protect property). Emergency managers are also worried that loss of electrical power during a disaster might cause the whole computer system to break down.

The recent development of mobile GIS and mobile GIServices might solve these problems, as proposed in this chapter, by providing their own independent power supply systems (batteries and Uninterruptible Power Supply – UPS) and having a great portability (cellular phones, Pocket PCs, etc.). In addition, this chapter discusses how the new wireless communication technologies, such as 4th Generation (4G) cellular phone systems, Wi-Fi and Wi-MAX techniques, might further improve the capability of mobile GIServices and support comprehensive information services for disaster management.

This chapter will first introduce the disaster management framework for mobile GIServices (Section 12.2) and then recent advances in mobile GIService technology

(Section 12.3). The discussion will focus on disaster management in three categories: emergency preparedness, emergency response and disaster recovery. Next, the Taiwan Advanced Disaster Management Decision Support System (TADMDS) will be introduced as a showcase of the integration of mobile GIServices with Web-based GIServices (Section 12.4). Finally, this chapter will conclude in Section 12.5 by highlighting the current limitations and possible future directions of Mobile GIServices technology.

12.2 The framework of disaster management

The term 'disaster' has various meanings and interpretations. To paraphrase Drabek and Hoetmer (1991) a disaster is defined as a large-scale event that can cause very significant loss and damage to people, property and communities. One important notion is that disasters are an outcome of risk and hazard (Cutter, 2003). Traditionally, there are two types of hazard: natural hazards (floods, typhoons, tornadoes, earthquakes, etc.) and technological hazards (chemical explosion, nuclear power plant meltdowns, terrorist attacks, etc.). In disaster management, we need to consider both potential hazards and potential community vulnerability (White, 1945). If an area or a local community has high-income level residents and very strong government support, the vulnerability of such a community will be low and will be more resistant to large-scale hazards. If a local community is poor and vulnerable, a small-scale earthquake or flood might cause significant human casualties and property damage. Human factors always play an important role in disaster management.

The development of GIS has occurred over three decades and there are many GIS applications focusing on disaster management and emergency response (Coppock, 1995). GIS can be used in real time for monitoring natural disasters (Alexander, 1991) and remote sensing imagery can be applied to emergency management (Bruzewicz, 2003). GIS modeling helped the Chernobyl nuclear disaster relief (Battista, 1994), the management of wildfires (Chou, 1992) and the assessment of community vulnerability (Chakraborty and Armstrong, 1996; Rashed and Weeks, 2003). There have also been many research efforts combining GIS with natural hazard risk modeling and risk management decision support systems (Zerger and Smith, 2003).

On the other hand, mobile GIServices are a very new research domain and their focus is different from traditional GISystems. With the progress of wireless technology and GPS, mobile GIServices for disaster management are likely to become very popular in the next few years. This chapter provides an overview of mobile GIS applications in disaster management. The chapter follows a conceptual framework developed by Drabek and Hoetmer in 1991, called 'comprehensive emergency management' to highlight the potential of mobile GIS applications. This framework has four temporal phases of disaster management: mitigation, preparedness, response and recovery (see Box 12.1 and Figure 12.1).

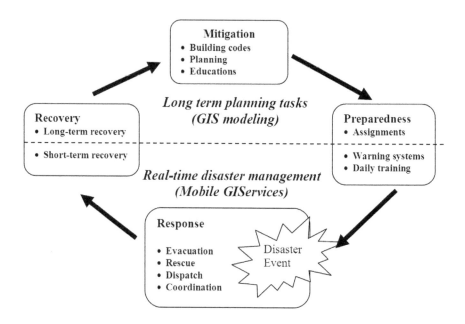

Figure 12.1. The role of mobile GIServices in disaster management (modified from Cova, 1999 and Godschalk, 1991).

Box 12.1 – The four temporal phases of disaster management (Federal Emergency Management Agency [FEMA], 2004).

(1) **Mitigation** – Mitigation actions involve lasting, often permanent, reduction of exposure to, probability of, or potential loss from hazard events. They tend to focus on where and how to build. Examples include: zoning and building code requirements for rebuilding in high-hazard areas; flood plain buyouts; and analyses of flood plain and other hazard-related data to determine where it is safe to build in normal times, to open shelters in emergencies or to locate temporary housing in the aftermath of a disaster. Mitigation can also involve educating businesses and the public on simple measures they can take to reduce loss or injury, such as fastening bookshelves, water heaters and filing cabinets to walls to keep them from falling during earthquakes.

(2) **Preparedness** – While mitigation can make communities safer, it does not eliminate risk and vulnerability for all hazards. Therefore, jurisdictions must be ready to face emergency threats that have not been mitigated away. Since emergencies often evolve rapidly and become too complex for effective improvisation, a government can successfully discharge its emergency management responsibilities only by taking certain actions beforehand. This is preparedness. Preparedness involves establishing authorities and responsibilities for emergency actions and garnering the resources to support them. A jurisdiction must assign or recruit staff for emergency management duties and designate or procure facilities, equipment and other resources for carrying out assigned duties.

(3) **Response** – Response is the third phase of emergency management and covers the period during and immediately following a disaster. During this phase, public officials provide emergency assistance to victims and try to reduce the likelihood of further damage. Local fire department, police department, rescue squads and emergency medical service (EMS) units are primary responders.

(4) **Recovery** – Recovery is the fourth and final phase of the emergency management cycle. It continues until all systems return to normal or near-normal operation. Short-term recovery restores vital life-support systems to minimum operating conditions. Long-term recovery may go on for months – even years – until the entire disaster area returns to its previous condition or undergoes improvement with new features that are less disaster-prone. For example, a town can relocate portions of its flood-prone community and turn the area into open space or parkland. This illustrates how recovery can provide opportunities to mitigate future disasters.

In term of mobile GIServices' tasks for each phase, there is a significant difference between real-time disaster management *needs* versus long-term disaster planning and mitigation *processes*. Mobile GIServices will be critical for real-time disaster management tasks rather than long-term planning processes. Many tasks in emergency response, recovery and preparedness will need critical geo-spatial information in real time and updated geo-data from the field personnel (fire fighters or police officers). Traditional GIS modelling and spatial analysis functions will be used mainly for long-term planning tasks in the mitigation phase and some tasks in long-term recovery and emergency preparedness, such as estimating recovery cost and assigning responsible zones (see Figure 12.1).

Real-time disaster management has very specific requirements that are significantly different from long-term mitigation planning for disaster management. The differences between real-time systems and the long-term planning process are summarised in Table 12.1. Most real-time disaster management will need to access the information immediately for warning systems, evacuation or responder dispatch efforts. Therefore many emergency managers will only tolerate 5-10 seconds response time from sending a GIS function request to getting an answer from the system. On the other hand, long-term recovery and mitigation tasks are less time-sensitive and the GIS analysis runtime can be more flexible ranging from ten minutes to several hours.

Table 12.1. Differences between real-time systems and long-term planning processes.

	Long-term planning tasks (GIS modelling)	Real-time disaster management (Mobile GIServices)
Response time	Flexible (1-10 days) (less sensitive to tasks)	Immediately (1-10 seconds) (sensitive to task operations)
Map types	Thematic maps (land use, census data, administration boundary, soil, etc.)	Pragmatic maps (roads and traffic updates, event locations, evacuation maps, GPS integration)
Numbers of users	Small number (1-10 people in administration level)	Large number (over hundreds – rescue teams, the first responders, and the general public)

Regarding map types, real-time emergency tasks will need 'pragmatic maps' with GPS functions for navigation, evacuation routes and traffic updates. Transportation is the key theme in many related emergency response tasks. On the other hand, a long-term mitigation plan will focus on thematic maps by using advanced GIS modeling and spatial analysis. There might be many layers used in this area, including land use, census data, administration boundary, soil, terrain, vegetation, etc. Regarding the number of users, in real-time emergency responses these will be much larger than in the long-term mitigation plan. Depending on the level of the disaster, the size of in-field agents in real-time emergency tasks could range from a dozen of people to hundreds of responders and staffs. In contrast, long-term mitigation tasks use GIS modeling for the planning processes that might only involve a few decision-makers or GIS professionals (ranging from one to ten people).

Therefore, to implement a comprehensive mobile GIService, we need to consider the technological challenges of large numbers of users, the response time and the nature of the GIS functions. Real-time emergency response tasks will need robust and user-friendly mobile devices. The devices must be robust (they may drop to the ground), easy to read (even under direct sunlight or in snow conditions), and have a long battery life and runtime.

The next section will start to focus on the mobile GIServices framework and how to adopt mobile GIServices into three disaster management phases: preparedness, response and recovery. Since most tasks at the mitigation phase are more related to GIS modelling and spatial analysis (see Figure 12.1), this chapter will not focus on the disaster mitigation tasks.

12.3 Mobile GIServices framework

The architecture of mobile GIServices utilises a client/server computing framework. Client-side mobile GIServices components are the end-user hardware devices, which display maps or provide the analytical results of GIS operations. Server-side components provide comprehensive geo-spatial data and perform GIS operations based on a request from the client-side components. Between client and server, there are various types of communication networks (such as hard-wired network connections or wireless communications) to facilitate the exchanges of geo-data and services. Figure 12.2 illustrates the six basic components of mobile GIServices: 1) positioning systems; 2) mobile GIS receivers; 3) mobile GIS software; 4) data synchronisation and wireless communication component; 5) geo-spatial data; and 6) GIS content servers (Tsou, 2004).

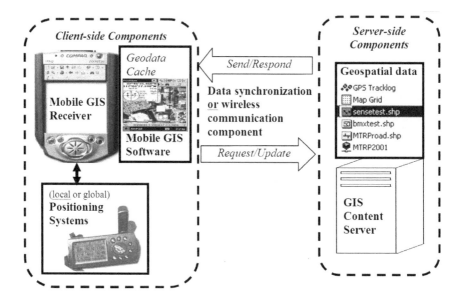

Figure 12.2. The architecture of mobile GIServices (Tsou, 2004). (First published in *Cartography and Geographic Information Systems*, **vol. 31 (3), p. 156.)**

Positioning Systems refer to the components that provide geo-referenced coordinate information (X, Y and Z-elevation) to mobile GIS receivers. There are two major types of systems, *local positioning systems* and *global positioning systems* (GPS). Local positioning systems rely on mechanical distance measurement or triangulation of the radio signals (or cellular phone signals) from multiple base stations in order to calculate the position of a device. GPS uses satellite signals to calculate the position of GPS units. Sometime, mobile GIS applications may require both types of positioning systems within urban areas to generate satisfactory results.

Mobile GIS receivers are small-sized computers or terminals that can display maps and locational information to end-users. The hardware components of mobile GIS receivers include CPU, memory, storage devices, input/output connections and display (screen) hardware. Pocket PCs, smart phones, tablet PCs and PDAs are the most popular mobile GIS receivers. Occasionally, notebook computers can be used as a mobile GIS receiver if connected to GPS and other mobile GIS components. However, most mobile GIS receivers require a very small-size hardware device to achieve their portability. The major differences between small mobile GIS receivers and traditional personal computers are smaller screen resolutions (240 x 300), limited storage space and slower CPU speed (Wintges, 2003).

Mobile GIS software refers to the specialised GIS software applications employed by mobile GIS systems. Because of the limitations of mobile GIS receivers (smaller display units, limited storage, etc.), the design of mobile GIS software needs to focus on specific GIS operations (geocoding, address matching, spatial search, routing services, map display, etc.) rather than encompassing comprehensive GIS functions. For example, the functional design of LBS software is quite different from the functions provided in field-based GIS packages. Most mobile GIS software packages are lightweight, customisable and are designed to function with positional systems (such as GPS tracking).

Geo-spatial data are specifically designed GIS layers or remotely sensed imagery used for mobile GIS applications. Because of the limited storage space in mobile GIS receivers, most GIS data needs to be compressed or subset from their original extents. Usually, the mobile GIS receivers store geo-spatial data in a *geo-data cache*[2], located in a temporary GIS storage space or a flash memory card. Often customised datasets are downloaded and synchronised from GIS content servers. One alternative approach is to utilise wireless communications to access needed portions of large-sized GIS layers and/or remotely sensed imagery from the content server directly.

Data synchronisation/wireless communication components support the linkages between mobile GIS receivers and GIS content servers. These linkages could be real-time wireless communications (via Wi-Fi or cellular phone signals) or cable-based data synchronisation communications (via USB or serial ports). Both mechanisms provide two-way communications. For cable-based connections, the GIS content servers *send* geo-data to the receivers (stored in geo-data cache) and the receiver uploads *updated* geo-data back to the content server. For wireless communication, the mobile GIS receivers *request* a specific service or map from the GIS content server, and the server *responds* to the request by sending the new map to the receiver. To facilitate two-way communications, several middleware or data synchronisation software packages (such as Microsoft ActiveSync or Web Services) are required for mobile GIS applications. If both mobile receivers and GIS content servers have networking capabilities, Internet-based protocol, such as TCP/IP and HTTP, can provide very effective communication channels for mobile GIS applications.

GIS content servers are stand-alone GIS workstations or Web-based servers providing geo-spatial data or map services to mobile GIS receivers. Most cable-based mobile GIS receivers use stand-alone GIS workstations as content servers. Wireless-based mobile GIS receivers may require advanced Web servers or wireless Internet map servers for accessing geo-spatial data. In some instances, one mobile GIS receiver may be used to access multiple Web-based servers at the same time in order to integrate multiple GIS layers. A single GIS content server can also provide data and services to multiple mobile GIS receivers simultaneously.

Mobile GIServices can provide geo-spatial information and GPS coordinates for field-based personnel conducting remote field (*in situ*) GIS tasks. For example,

---

[2] Geo-data cache is the temporary local memory storage for saving geo-spatial data downloaded from the GIS content server. Therefore, if the connection between the client and content server is not available, mobile GIS units can still use the local geo-data cache to perform parts of GIS tasks.

landscape architects may use mobile GIS devices to display a remotely sensed image as a background in a remote field location and then draw a preliminary design for a tree line based on GPS locations. To enable comprehensive mobile GIServices, wireless communication is essential for connecting mobile GIS devices and GIS content servers. Box 12.2 introduces recent progress in broadband wireless technology, Wi-Fi and WiMAX, which can provide comprehensive communication channels for mobile GIServices.

Box 12.2 – Recent progress in broadband wireless technology, Wi-Fi and WiMAX, which can provide comprehensive communication channels for mobile GIServices.

Wi-Fi/WiMAX data network systems are a promising category for broadband wireless mobile GIS communication. Both Wi-Fi and WiMAX are wireless network standards defined by the IEEE 802 LAN/MAN Standards Committee (IEEE 802 LAN/MAN Standards Committee, 2004). The IEEE 802 committee forms multiple working groups in developing Local Area Network (LAN) standards and Metropolitan Area Network (MAN) standards, such as 802.3 (Ethernet), 802.11 (Wireless LAN), 802.15 (Wireless Personal Area Network – WPAN), and 802.16 (Broadband Wireless Access – WiMAX). Currently, the most common wireless LAN infrastructure is the IEEE 802.11 (or Wi-Fi) technology. IEEE 802.11 specifies the physical and Media Access Control (MAC) layers for operation of Wireless Local Area Networks (WLAN). The 802.11 standard provides for data rates from 11Mb/s to 54Mb/s (Pandya, 2000). The term, Wi-Fi (wireless fidelity), is the global brand name across all markets for any 802.11-based wireless LAN products. Many computers, PDAs, printers, etc. have begun to adopt Wi-Fi or IEEE 802.11 as their major communication channels.

There are four extensions in the 802.11 technology as follows:

- 802.11.**a** provides up to a 54Mbps transfer rate in the 5GHz band (referred to as Wi-Fi5).
- 802.11.**b** is the most popular extension and can provide up to 11Mbps data transfer rate in the 2.4GHz band. Because of the different radio frequency, 802.11b devices are not compatible (accessible) to 802.11a signals.
- 802.11.**g** provides up to a 20+ Mbps data transfer rate in the 2.4GHz band. Since the 802.11g and 802.11b standards are using the same radio frequency, 802.11g devices are backward compatible to 802.11b signals.
- 802.11.**n** is a new technology (available in late 2005) to upgrade the 802.11a and 802.11g. 802.11.n adopts MIMO (multiple input multiple output) technology to provide faster communication speed up to 200 Mbps. The 802.11n can be integrated with 2.4 GHz 802.11g or 5 GHz 802.11.a together (Janowski, 2005).

WiMAX is an emerging IEEE 802.16 standard for broadband wireless wide area network (WWAN) or metropolitan area network (MAN) applications. WiMAX can provide a larger coverage of service area than Wi-Fi. Its communication

Box 12.2 (cont.)

signals can cover a 4-6 mile range (or up to 20 miles for the long distance setting). With such range and high throughput, WiMAX is capable of delivering backhaul[3] for carrier infrastructure, enterprise campuses and Wi-Fi hotspots (Intel, 2004).

WiMAX can provide broadband to areas that do not have cable or DSL services. Presently, WiMAX includes two steps of the IEEE 802.16 technology. The IEEE 802.16d is the first step, which will be used to specify large area wireless communication via outdoor antennae in a fixed location. A fixed WiMAX service can provide up to 75 Mbps speed with Sub 11Ghz radio frequency. The IEEE 802.16e will be the next step (under development), used to specify portable wireless hardware for mobile WiMAX services. The new mobile WiMAX can provide roaming capability and enable more persistent connectivity within a service area (Intel, 2004). Mobile WiMAX use 2-6 GHz frequency with 30 Mbps communication speed.

The progress of wireless communication, mobile devices and GPS devices has significant impact for the development of mobile GIServices. Recently, more and more mobile GIS applications have been encountered in data gathering, vehicle navigation and emergency response situations. The next sections highlight the potential of mobile GIServices for various tasks in disaster management. Section 12.3.1 describes how mobile GIServices can be used for emergency preparedness, Section 12.3.2 evaluates how mobile GIServices can be used for emergency response, and Section 12.3.3 discusses the use of mobile GIServices for disaster recovery.

12.3.1 Mobile GIServices for emergency preparedness

As discussed in Section 12.1, many essential tasks in emergency preparedness will benefit from real-time mobile GIServices, such as emergency warning systems and daily personnel training activities. Mobile GIS can be used to create various early warning systems by combining wireless remote sensors with GPS and mobile devices. For example, in 2002 the Department of Transportation in California (Caltran) collaborated with the University of California at San Diego to install GPS-enabled sensors in all major bridges in California. In preparedness for earthquakes, flooding events or any physical damage occurring to bridges, the remote sensors automatically send out warning signals to the control and command centers giving the new GPS positions via wireless networks. By comparing the old GPS position with the new one, transportation managers can immediately detect any potential

---

[3] This means that the WiMAX can become a network backbone (with huge bandwidth) for mobile communication needs (data transmission from hundreds of mobile clients to a centralized access point). Backhaul is similar to the meaning of backbone, but 'backhaul' is used more often for mobile communication channels, c.f. 'backbone' used for wired networks, like Ethernets or the Internet.

damage on the bridges and make quick responses (see High Performance Wireless Research and Education Network [HPWREN], 2005). Mobile GIServices combining wireless remote sensors with GPS can also be applied in other disaster warning systems, such as earthquake prediction, tsunami warning and wildfire management.

Another example of emergency preparedness is the development of comprehensive intelligent vehicle transportation systems at the California Institute of Telecommunication and Information Technology, or Cal(IT)2 (see CAL(IT)2, 2005). Cal(IT)2 collaborated with U.S. National Science Foundation and Caltra to develop pervasive computing and communications systems for intelligent transportation, called 'Autonet'. This project utilises mobile networks among cars with GPS to estimate and predict travel conditions and optimise traffic management. For example, a traffic light can automatically adjust its change frequency based on the numbers of vehicles nearby the traffic light. When an emergency vehicle approaches the traffic light, it will change its lights to allow the emergency vehicle to pass first. Therefore, all vehicles become cooperating partners in the traffic management system. By tapping into the information storage and processing power available within each vehicle, the traffic management authority can approach system-optimal control.

Mobile GIS equipment can also be adopted in the daily training activities of firefighters, police officers and emergency responders (such as the '911' or '999' services). At the preparedness phase, there are great potentials and many future applications for mobile GIServices. However, there are still some challenges for mobile GIServices installation, including how to update GIS databases and road conditions, how to provide secure wireless communication channels, and how to balance performance and portability. The most challenging part is to improve the usability of the software and hardware and to create a quick learning curve for the first responders. Current user interface design in mobile GIS software and hardware is still too complicated and difficult to use (Tsou, 2004). For example, one popular Mobile GIS software package, ESRI ArcPAD, only has tiny fixed-size buttons (Figure 12.3) for general map viewing functions (Zoom-In, Zoom-out or Pan).

12.3.2 Mobile GIServices for emergency response

Emergency response is the most critical phase in disaster management. Mobile GIServices can play a very important role in evacuation, dispatch and vehicle tracking. To activate an evacuation plan, emergency managers have to gather the most updated geo-spatial information from the field as quickly as possible. By combining mobile GIS software, wireless communications and GPS, *in situ* agents (firefighters, police officers) can report the ground truth immediately via Wi-MAX or cellular networks. For example, if police officers found a possible terrorist attack target, they could immediately submit the hot zone and publish the information to every police officer in the nearby area. The mobile GIServices will update the critical information more effectively and efficiently than traditional radio signal conversions and report mechanisms (Figure 12.3).

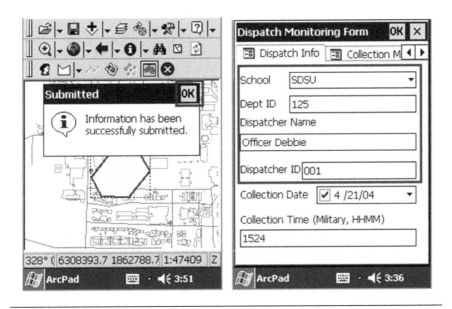

Figure 12.3. A police officer can submit a hot zone by using mobile GIServices.

Mobile GIServices can also be applied for the first responder dispatch monitoring systems (Figure 12.4). A Web-based mapping interface can gather everyone's field report and updates together and display these changes immediately for emergency managers. The dispatch monitoring system will be used to integrate the field-based information for facilitating better emergency responses.

In addition to real-time GIS updates and dispatches, another important task in emergency response is to use GPS tracking of all in-field agents. Figure 12.5 shows a Web-based GPS tracking map browser, which illustrates a simulated security officer carrying a Pocket PC with GPS functionality across the campus of San Diego State University. In the Web browser, the dot can dynamically move according to the GPS signals received from the security officer's Pocket PC. This prototype testing was created by using ESRI's Tracking Server (beta-version) with customised ArcPAD GPS functions. Due to the lack of WiMAX wireless channels on the campus, the system relied on a GPS simulator to create the real-time GPS signals feeding to the Internet Map Server via TCP/IP.

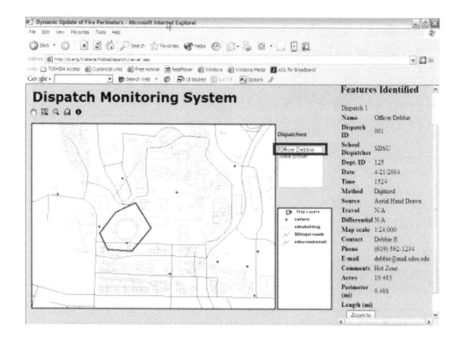

Figure 12.4. A Web-based emergency dispatch monitoring system.

These examples illustrate the potential of mobile GIServices in emergency response. There are other possible response tasks for mobile GIServices, such as rescue team coordination, choosing appropriate shelter locations for disaster victims, creating the best evacuation routes for the residents, etc. Some tasks might need to combine both spatial analysis functions and the real-time mobile GIServices.

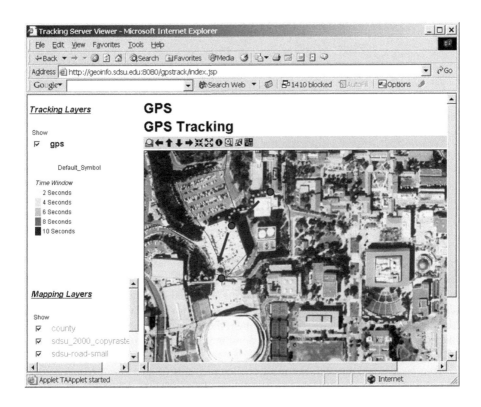

Figure 12.5. Web-based real-time GPS tracking services. The moving (black/red) dot indicates a person with the mobile GIS device sending GPS location back to an Internet map server in real time. See colour insert following page 132.

12.3.3 Mobile GIServices for disaster recovery

Mobile GIServices can be applied in both long-term recovery tasks and short-term recovery tasks. Long-term recovery tasks, such as re-building damaged houses, ecological conservation and restoration and community reconstruction, require years of efforts to recover the damages caused by disasters. Mobile GIServices might be able to help in some tasks, such as the detection of land use change in environmental remediation. Figure 12.6 shows an example of using remote sensing imagery and GPS to compare the land cover changes for different vegetation types after a wildfire event.

Figure 12.6 shows different colours indicating different types of land cover change (Tsou, 2004). For example, the green colour indicates areas of increasing leaf cover within the study area. Test participants used the GPS to locate their positions on the colour-coded land use map during assessments of land cover changes around the event area.

Figure 12.6. Land cover changes detection by mobile GIServices (Tsou, 2004). The different coloured polygons indicate the different types of landcover changes. (First published in *Cartography and Geographic Information Systems,* **vol. 31 (3), p. 164.) See colour insert following page 132.**

For short-term recovery tasks, the major focus is to restore vital life-support systems to minimum operating conditions. Repairing damaged roads and bridges, providing clean water and electricity are the major tasks involved with these short-term recovery tasks. Mobile GIServices can help the emergency managers to accomplish these tasks by using utility mapping combined with GPS to identify the scale of damages in specific areas.

In a large-scale disaster, the most challenging part of short-term recovery is the coordination between different rescue teams. For example, hundreds of different rescue teams participated in the 2004 Tsunami disaster recovery and rescue work. Their experiences indicate that it is extremely difficult to coordinate these rescue teams in order to cover all damaged areas. Mobile GIS and GPS mapping will be an excellent tool to help the coordination of such tasks in the future. Another example is the rescue effort after the September 11, 2001 terrorist attacks in New York, US. The Fire Department of New York (FDNY) joined city, state and federal agencies to recover the Ground Zero site. Fire fighters used a ruggedised GPS receiver [4], manufactured by a company called 'LinksPoint and Symbol', with hand-held data terminals to collect geo-spatial information from the debris of the buildings (see Forbes, 2002). The FDNY was tasked with the responsibility of documenting the items recovered from the rubble of the disaster zone and recording information regarding location, time and type of item found – information critical to both the ongoing investigation and analysis of the event. This information was also being used by other agencies involved in the investigation. This example demonstrates a

---

[4] When equipment or a device has been modified or its shield and reliability enhanced to become weather-resistant, and shock-resistant, it is called 'ruggedised', i.e. a 'tough' design for devices for any types of environment or usages.

great potential for adopting mobile GIServices for post-disaster investigation and recovery works.

The following section will now use a case-study from Taiwan to demonstrate the actual implementation of disaster management systems by combining mobile GIServices and Web-based GIS.

12.4 Case-Study: Taiwan advanced disaster management decision support system

The National Science and Technology Program for Hazards Mitigation (NAPHM) is an integrated and inter-disciplinary program, which is sponsored by the National Science Council (NSC) of Taiwan and is operated by the National Taiwan University (NTU), for hazards mitigation-related research and technology development. The overall goal of this program, which started in 1997, is to implement hazard mitigation research to effectively reduce the risk of and loss to the general public and society arising from natural hazards. The main objectives of this program are as follows (Yen et al., 1997):

❑   To provide a comprehensive technological framework for practical hazards mitigation efforts.

❑   To consolidate the efforts of the government agencies and disaster management communities involved in order to promote, systematically, hazard mitigation research.

❑   To integrate hazard-related research results and transfer them to feasible procedures so that they can be implemented effectively.

❑   To develop appropriate methodologies for the potential analysis, risk assessment and scenario simulation of natural hazards. The methodologies will be fine-tuned before they can be used to develop a hazard mitigation plan within the jurisdictions of various government agencies.

One unique feature in the design of NAPHM is to adopt mobile GIServices within the disaster management system. Disaster managers and other decision makers can utilise an integrated disaster information network (IDIN), which adopts a distributed architecture via wireless mobile networks and mobile GIS devices (Figure 12.7).

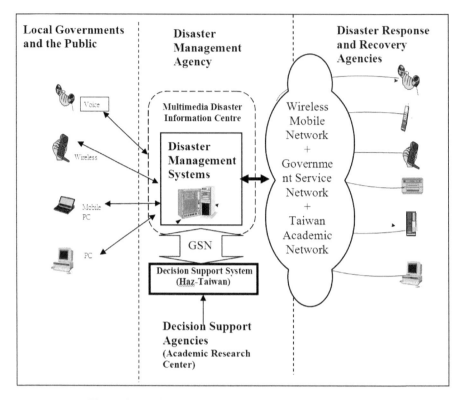

Figure 12.7. An integrated disaster information network (IDIN).

There are some considerations to be made in the designing of a comprehensive IDIN. The first issue is the network connectivity. A robust and reliable network connectivity lies at the heart of any successful disaster management application. However, in reality there is a wide range of network types with different bandwidths for transmitting data and information. Successful prototypes should have enough network bandwidth to disseminate information across various connection types.

The second issue is to create a wireless mobile communication environment. Wireless computing (e.g. notebook computers, slim-notebooks, Pocket PCs, PDA's and ever-increasing hand-held mobile phones, pagers and other electronic devices) is essential for the communication among first responders, emergency managers and decision makers. Mobile computing and wireless communication will become the key technology for disaster management because the mobile computing platform can be widely used in real-time emergency response, ground-truth measurements and disaster management networks.

Figure 12.8 illustrates a typhoon information display system. The system can display current and predicted typhoon routes (the top box on the left side), satellite image of typhoon (the middle box) and radar information (the bottom box on the left side) together. Users can choose one of the maps to switch to full-size display (the right-side window) and some animation functions. Therefore, one can track the past, current and future predicted status of typhoon in this system (Figure 12.8).

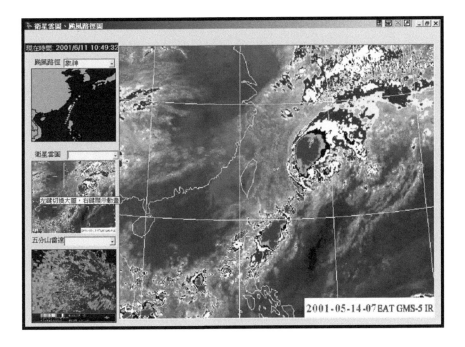

Figure 12.8. Typhoon information display system for Taiwan. (Top box on the left side shows the predicted path of typhoons, middle box shows the satellite image, and bottom box shows local radar information. Users can click on the left-mouse button for a full-size display or click on the right button to display animation.) See colour insert following page 132.

In order to efficiently manage the response to a disaster, the damage events report templates were designed to be used by the first responders (which are also the mobile GIS users) to submit their findings into a GIS (Figure 12.9). The location or zone of each damage event can be submitted by mobile GIS devices and then displayed on the system with information on the classification, magnitude and handling of the situation. The interface of the system can also display videos or photos to aid emergency responses.

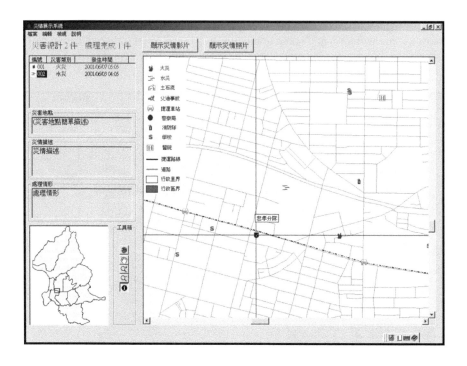

Figure 12.9. Damage condition display system in Taiwan that can track multiple disaster recovery and rescue tasks. (Top table on the left side shows the ID of each disaster and their symbols – red icon for fire, blue icon for flood. Second table describes the disaster location and street names, third table describes the disaster situation and the damage level, fourth table shows the rescue and recovery actions taken by local government. Additional functions provide video and photos of the disaster taken at different locations. This screen shot shows that there are two major disaster events: one processed and one still awaiting action.) See colour insert following page 132.

In general, the Taiwan disaster management system illustrates the need for geo-spatial information during various disaster events, such as typhoons, earthquakes and floods. The typhoon information display system has been successfully implemented by the Central Weather Bureau in Taiwan. The system has been used for five years (since 2001) to help the general public and the media tracking the path of typhoons and their impacts.

This section introduced an integrated disaster information network (IDIN) for supporting multiple decision support systems in disaster management by using the Internet, mobile GIServices, and wireless communications. The integrated spatial decision support systems can provide decision makers with critical and real-time information for hazards mitigation and emergency response, including potential hazard areas, hazard loss estimation and scenario simulation of various hazard mitigation options. The next section discusses the future development and limitation of mobile GIServices.

12.5 Conclusion and future developments

Disaster management is a complex domain of human activity involving multiple agencies and stakeholders. A collaborative approach utilising state-of-the-art mobile GIServices can facilitate a comprehensive and functional disaster management plan. This chapter introduced the basic components of mobile GIServices and their potential disaster management role in three main phases: preparedness, response and recovery. The GIS industry has started focusing on mobile GIS applications and the development of mobile hardware/software (Peng and Tsou, 2003), such as ESRI's ArcPAD, Mapinfo's MapXtend, and mobile Google map (Google Inc., 2005). However, there are still some major impediments in the development of mobile GIServices.

The first impediment is the lack of comprehensive user interface designed specifically for mobile GIServices. Most current mobile GIS software still follows the legacy concepts of desktop GIS interfaces. The tiny, sensitive stylus pen and the small on-screen keyboard input method are not the right choice for mobile GIServices in the emergency context. Direct voice commands and an easy, touchable screen simply used by human fingers (that may be wearing gloves) are more appropriate for emergency responders and in-field workers (see Figure 12.10).

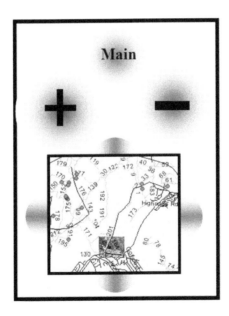

Figure 12.10. A simplified user interface design for displaying maps in a touch-screen mobile GIS device.

The second limitation of current mobile GIServices is the lack of real-time data collection and distribution mechanisms. It was difficult to verify the accuracy of submitted geo-spatial data from fieldwork. Currently, a GIS professional has to manually convert the data submitted from field workers to the Web-based GIService framework. Some predicted advances in Web services technologies and improvement in distributed database functions might solve these technical problems in the future. However, it is always dangerous to rely on automatic data conversion without verifying the data accuracy and data quality.

The third impediment is the integration of spatial analysis and GIS modeling into mobile GIServices. Many emergency tasks and disaster management works will need advanced GIS analysis functions that require significant computing power and computer memory. Most mobile GIS devices are tiny and only have very limited computing capability. The pre-processing and post-processing time for spatial analysis and remote sensing images might prevent the adoption of mobile GIServices for real-time response tasks due to the hardware limitations. One possible solution is to send the complicated GIS model and spatial functions via the Internet to remote GIS engine services. Then, the analysis results will be sent back to the mobile GIS devices via the network.

The final issue is the lack of alternative display methods for mobile GIServices. Since most mobile GIS devices are small and fragile, emergency responders and managers might be reluctant to use small screens on Pocket PC or cellular phones to share their maps with others. One possible alternative is to print out paper maps directly from mobile GIS devices since paper maps are easy to carry and there will then be no need for batteries in the field. It would be useful if users could print paper maps directly from their mobile GIS devices via wirelessly portable printers or from a built-in printer inside a Pocket PC or a notebook computer.

In summary, this chapter introduced an integrated mobile GIServices framework that can provide comprehensive services for disaster management tasks. The chapter has argued that mobile GIServices is a very promising field with very high demands from both field-based workers and the GIS vendors. With the progress of new wireless communication technology and GPS techniques, mobile GIServices can help to monitor the dynamic changes in the real world and provide vital information to prepare and prevent natural hazards or human-made disasters. Hopefully, with the efforts from GIS professionals and GIS developers, the advancement of dynamic and mobile GIS research will protect people from various hazards in the future and improve their quality of life.

Acknowledgments

This paper forms a portion of the NASA REASoN-0118-0209 project at San Diego State University. Funding by the NASA REASoN Program and matching funds from San Diego State University Research Foundation are acknowledged and greatly appreciated.

References

Alexander, D. (1991) 'Information technology in real-time for monitoring and managing natural disasters', *Progress in Physical Geography,* vol. 15, pp. 238–260.

Battista, C. (1994) 'Chernobyl: GIS model aids nuclear disaster relief', *GIS World,* vol. 7, no. 3, pp. 32–35.

Bruzewicz, A. J. (2003) 'Remote sensing imagery for emergency management', in Cutter, S. L., Richardson, D. B. and Wilbanks, T. J. (eds.) *Geographical Dimensions of Terrorism,* pp. 87–97, New York: Routledge.

CAL(IT)2 (2005) *The Development of Comprehensive Intelligent Vehicle Transportation Systems at the California Institute of Telecommunication and Information Technology, or Cal(IT)2,* [Online], Available: http://www.calit2.net/briefingPapers/transportation.html [23 July 2005].

Chakraborty, J. and Armstrong, M. P. (1996) 'Using geographic plume analysis to assess community vulnerability to hazardous accidents', *Computers, Environments and Urban Systems,* vol. 19, no. 5/6, pp. 341–356.

Chou, Y. H. (1992) 'Management of wildfires with a geographical information system', *International Journal of Geographical Information Systems,* vol. 6, pp. 123–140.

Coppock, J. T. (1995) 'GIS and natural hazards: an overview from a GIS perspective', in Carrara, A. and Guzzetti, F. (eds.) *Geographical Information Systems in Assessing Natural Hazards,* vol. 6, pp. 21–34, Amsterdam: Kluwer Academic.

Cova, T. (1999) 'GIS in emergency management', in Longley, P. A., Goodchild, M. F., Maguire, D. J. and Rhind, D. W. (eds.) *Geographical Information Systems,* Second Edition, pp. 845–858, New York: John Wiley & Sons, Inc.

Cutter, S. L. (2003) 'GI science, disaster, and emergency management', *Transactions in GIS,* vol. 7, no. 4, pp. 439–445.

Drabek, T. E. and Hoetmer, G. J. (eds.). (1991). *Emergency Management: Principles and Practice for Local Government,* Washington, DC: International City Management Association.

Federal Emergency Management Agency (FEMA) (2004). *Independent Study Course Material Download: IS-1 Emergency Manager: An Orientation to the Position,* [Online], Available: http://training.fema.gov/EMIWeb/IS/is1.asp [20 July 2005].

Forbes, M. (2002) 'Fire Department of New York mobile recovery database developed at Ground Zero using GPS', Online weekly news at *The Wireless Developer Network* [Online], Available: http://www.wirelessdevnet.com/articles/linkspoint/ [21 October 2005].

Godschalk, D. R. (1991) 'Disaster mitigation and hazard management', in Drabek, T. E. and Hoetmer, J. G. (eds.) *Emergency Management: Principles and Practice for Local Government,* pp. 131–160, Washington, DC: International City Management Association.

Google Inc. (2005). *Mobile Google map,* [Online], Available: http://mobile.google.com/loc_search.html [21 July, 2005].

Green, W. G. (2001) 'E-emergency management in the US: a preliminary survey of the operational state of the art', *International Journal of Emergency Management,* vol.1, no.1, pp. 70–81.

High Performance Wireless Research and Education Network (HPWREN) (2005). *HPWREN Project Website,* [Online], Available: http://hpwren.ucsd.edu/news/021115.html [23 July 2005].

IEEE 802 LAN/MAN Standards Committee. (2004) *IEEE 802 Detailed Overview,* [Online], Available: http://grouper.ieee.org/groups/802/ [1 June, 2005].

Intel (2004). *Understanding WiMAX and 3G for Portable/Mobile Broadband Wireless, Technical White Paper,* [Online], Available: http://www.intel.com/netcomms/technologies/downloads/305150.pdf [30 March 2005].

Janowski, D. D. (2005) 'New ways to go wireless', *PC Magazine,* March 22, 2005, 24(5), pp. 98–102.

Pandya, R. (2000) *Mobile and Personal Communication Systems and Services,* New York: IEEE Press.

Peng, Z. R. and Tsou, M. H. (2003) *Internet GIS: distributed geographic information services for the Internet and Wireless Networks,* New York: John Wiley & Sons, Inc.

Rashed, T. and Weeks, J. (2003) 'Assessing Social Vulnerability to Earthquake Hazards through Spatial Multicriteria Analysis of Urban Areas', *International Journal of Geographic Information Science*, vol. 17, no. 6, pp. 549–576.

Tsou, M. H. (2004) 'Integrated Mobile GIS and Wireless Internet Map Servers for Environmental Monitoring and Management', Special Issue on Mobile Mapping and Geographic Information Systems in *Cartography and Geographic Information Science*, vol. 31, no.3, pp. 153–165.

White, G. F. (1945) *Human Adjustment to Floods*, Chicago: The University of Chicago.

Wintges, T. (2003) 'Chapter 25: Geodata Communication on Personal Digital Assistants (PDA)', in: Peterson, M. P. (ed.) *Maps and the Internet*, pp. 397–402, Oxford, UK: Elsevier Science Ltd.

Yen, C. L., Tsai, Y. B. and Chen, L. C., (1997) *Planning Report for the National Science and Technology Program for Hazards Mitigation,* Taiwan, National Science Council (in Chinese).

Zerger, A. and Smith, D. I. (2003) 'Impediments to using GIS for real-time disaster decision support', *Computers, Environments and Urban Systems*, vol. 27, pp. 123–141.

Chapter 13

Citizens as Mobile Nodes of Environmental Collaborative Monitoring Networks

Cristina Gouveia [1], Alexandra Fonseca [1], Beatriz Condessa [2] and António Câmara [3]

[1] Centre for Exploration and Management of Geographic Information, Portuguese Geographical Institute, Portugal
[2] Department of Civil Engineering and Architecture, Instituto Superior Técnico, Technical University of Lisbon, Portugal
[3] Environmental Systems Analysis Group, Faculty of Sciences and Technology, New University of Lisbon, Portugal

13.1 Introduction

Monitoring systems have been used widely to increase knowledge of the state of the environment. They are responsible for collecting and registering the baseline data of environmental systems. Monitoring is more than taking measurements; it is about learning the current state of the system, the system dynamics, the impact of management actions and how the information collected can be used to reach management goals. According to Boyle (1998) and Vaughan et al. (2003), monitoring systems have evolved to link monitoring information to the decision-making process. However, according to Vaughan et al. (2003) the major limitation of environmental monitoring is the ability to provide timely identification and warning of emerging problems to the public, stakeholders, research personnel and managers. Additionally, monitoring systems have shown difficulties in providing information to raise awareness, educate and provide the basis for informed decisions.

Due to the temporal and spatial characteristics of environmental data, GIS has been used to support environmental monitoring activities (Larsen, 1999; Gao, 2002). Mobile GIS, in particular, has been explored to support fieldwork, facilitating data collection and management (Tsou, 2004; Chapter 12, Tsou and Sun, in this book). In general, mobile computing applications together with mobile communications create new opportunities for environmental monitoring. For example, mobile GIS together with mobile communications can provide location-aware monitoring data and facilitate the collection of real-time data (Tsou, 2004).

Dynamic and Mobile GIS: Investigating Changes in Space and Time. Edited by Jane Drummond, Roland Billen, Elsa João and David Forrest . © 2006 Taylor & Francis

The exploration of such technological developments may support the creation of non-traditional approaches within environmental monitoring.

Community participation within environmental monitoring systems has been one approach followed to increase public awareness and education on environmental problems and to provide timely information to citizens and decision makers (Vaughan et al., 2003; Cuthill, 2000). Moreover, public participation within environmental monitoring may contribute to increasing the knowledge on the state of the environment. Presently, the impact of volunteer monitoring initiatives is limited mainly due to a lack of data credibility and difficult data access and use. A collaborative framework is required to support volunteer tasks and increase the impact of volunteer initiatives (Gouveia et al., 2004). The creation of environmental collaborative monitoring networks (ECMN) is proposed in this chapter as a framework to promote citizen participation within environmental monitoring, while supporting the use of citizen-collected data.

In ECMN, citizens are the nodes of a monitoring network that uses collaboration among its partners to facilitate volunteer monitoring activities. ECMN are committed to increase the impact of volunteer monitoring initiatives, namely by supporting the use of citizen-collected data by other stakeholders. Additionally, they intend to contribute to increase the knowledge on the state of the environment and educate the public on environmental issues.

Mobile computing and communication, together with the evolution of sensing devices, have created new opportunities to support the creation of ECMN. These technological developments may support collaboration among citizens allowing to link isolated initiatives and promoting volunteer monitoring. It is possible to envision a future where the common citizen equipped with information appliances, ranging from data loggers to smart sensors, contribute with their local data to increase the knowledge on the state of the environment, overcoming spatial and temporal gaps of the traditional monitoring systems. Early warning systems to protect environmental quality may emerge and benefit from these equipped and motivated citizens avoiding larger damages on the environment.

The major goal of this chapter is to explore the use of mobile computing and communications together with sensing devices to support citizens within their monitoring activities. It evaluates the possibility of creating a mobile collaborative monitoring network where each node is a citizen with no predefined location and willing to participate within environmental monitoring. The chapter starts by presenting the spatial, temporal and social characteristics of environmental monitoring networks in Section 13.2. It goes on describing environmental collaborative monitoring networks as a way to overcome some drawbacks of traditional monitoring networks (Section 13.3). The opportunities created by mobile technologies to support citizen involvement within environmental monitoring are analysed in Section 13.4 and the building blocks of mobile collaborative networks are then proposed in Section 13.5. To illustrate the issues involved in the implementation of mobile environmental collaborative monitoring networks two examples are analysed: the PEOPLE project and Senses@Watch (Section 13.6).

Finally, conclusions and lessons learned from the analysis of the examples are presented and major research questions are identified (Section 13.7).

13.2 Environmental monitoring networks and their spatial, temporal and social characteristics

Environmental monitoring activities are strongly associated with the nature of environmental variables, which act in different temporal and spatial scales. The weather is a good example of the unpredictability of environmental variables across temporal and spatial scales. Some variables to be understood require long-term monitoring, such as the case of tributyltin (TBT) antifoulants that may cause, for example, shell deformity and larval mortality in some molluscs (Satillo et al., 2001), while others are event driven and require real-time measurements, such as the concentration levels of carbon monoxide. The design of environmental monitoring systems should consider the frequency, duration and peaks of variables as well as time of response of the sensor or measuring device. Data acquisition systems may be time-based, value-based or hybrid, depending on data characteristics and system goals.

The spatial scale may vary from local to regional and global. The issues of scaling within ecological monitoring and its implication for sensor deployment are addressed by Withey et al. (2002). Location is therefore one of the key attributes of environmental variables. The measuring devices used to monitor environmental variables can be fixed-location or portable (see Table 13.1). Fixed-location measuring devices are normally used as part of a continuous, on-line monitoring system. Continuous monitoring has the advantage of enabling immediate notification when there is an upset. Portable measuring devices can be used to analyse any point in the system, but have the disadvantage that they provide measurements only at one point in time.

Table 13.1. The spatial component of environmental variables and monitoring measuring devices (adapted from Markowsky et al., 2002).

	Fixed Measuring Devices	**Mobile Measuring Devices**
Fixed targets	The use of fixed sensors to monitor, for example, soil characteristics such as moisture, temperature and nutrient levels.	Monitor specific locations using portable measuring devices. The use of sensors and robotics that move to specific locations to monitor environmental variables.
Mobile targets	Fixed air quality monitoring stations.	Organism tracking: coupling electronic tags to migratory birds. The use of air quality diffusive samplers by citizens.

To maximise spatial coverage and the representation of monitoring activities while reducing the costs involved, environmental monitoring networks have been established. These networks have been designed for a variety of applications and goals, and are responsible for collecting and registering the baseline data of environmental systems. Table 13.2 presents examples of criteria to consider when designing a monitoring network.

Table 13.2. Examples of network design criteria.

Criteria	Observations
Spatial coverage	From local to global scales. Site selection process must account for spatial variability and distribution. Other data such as demographics, land use information are inputs for site selection.
Simplicity	Easy operation and maintenance. Criteria such as the possibility to perform straightforward data analysis are also considered.
Representation	It may involve criteria such as capture of local maximums or assurance of randomised site selection.
Minimise costs	Instruments are usually expensive. It may imply a combination of fixed and mobile stations.
Duration and frequency	Capture the temporal dynamics. Estimate both long and short-term trends. Applications such as early warning systems require real-time data while other may use average data.
Public acceptability that risk is monitored	Network design should consider social components such as the case of fears and perceptions.

Data collection is the main activity of environmental monitoring networks. Data collection procedures depend on the variable being measured, the spatial and temporal coverage and the equipment available. However, the data collected by environmental monitoring networks present spatial and temporal gaps, which restrict the usefulness of such systems. On the other hand, one of the major limitations of environmental monitoring is to provide timely information to the public, stakeholders, research personnel and managers (Vaughan et al., 2003), which constrains the public debate on the state of the environment. Additionally, monitoring systems in the past have shown difficulties in providing information to raise awareness, educate and provide the basis for informed decisions.

Non-governmental organisations (NGOs) and concerned citizens have made some voluntary efforts to collect data on the state of the environment, contributing to overcome some of the above-mentioned limitations of monitoring networks.

Examples can be found since the early 1900s in projects such as the National Audubon Society Christmas Bird Count. A review of the history of volunteer monitoring is presented by Lee (1994). Volunteer initiatives intend not only to inform the public about the state of the environment, but also to support citizens to take action and participate within environmental decision making. Additionally, volunteer monitoring data have been integrated with professional data and used by NGO, researchers and public agencies to overcome spatial and temporal gaps in official monitoring systems (Stokes et al., 1990; Root and Alpert, 1994; Au et al., 2000; Fortin, 2000; Lawson, 2000; Young-Morse, 2000). On the other hand, volunteer initiatives may intend to educate citizens about the environment and the methods to evaluate its quality. The GLOBE project, where primary and secondary students carry out scientifically valid measurements in the fields of atmosphere, hydrology, soils and land cover, is an example of an educational initiative.

However, the impact of volunteer-collected data is limited mainly due to a lack of data credibility. Additionally, the organisation and motivation of volunteer projects restrict the impact of such initiatives. Volunteer monitoring is usually organised around particular motivations or events (for example the above-mentioned National Audubon Society Christmas Bird Count). This scenario results in isolated data collection points, not ensuring spatial, temporal and thematic coverage and above all, not facilitating the integration of citizen-collected data with other initiatives. The challenge is to link citizens and their data collection activities creating a monitoring network that promotes data representativeness.

A collaborative framework is required to support volunteer tasks and increase the impact of volunteer initiatives. Networks are good organising tools due to their flexibility and adaptability (Castels, 2001). The creation of environmental collaborative monitoring networks, as proposed in this chapter, may be a way to promote citizen participation within environmental monitoring.

13.3. Environmental collaborative monitoring networks

Traditionally, in environmental monitoring networks, the nodes are sensors or measuring devices connected to data loggers. In the case of automated networks, these nodes are connected by telemetry and the data are transferred to a central node, which is usually the owner of the network. However, the creation of a network that takes advantage of volunteer monitoring initiatives implies a different approach, where public involvement and collaboration play a major role.

In ECMN, the nodes are citizens or groups of citizens willing to participate within environmental monitoring, while the links show relationships or flows between the nodes (see Box 13.1 that describes the characteristics of ECMN nodes and links). The characteristics of the nodes vary according to the tasks performed by each citizen or group of citizens. Each node may play different roles from data collection to data management or activism promotion.

ECMN should consider the diversity of volunteer initiatives, from individual complaints to formal data collection activities, and take advantage of citizen efforts. For the purposes of illustration, the sequence of steps involved in the creation of an

ECMN is presented in Figure 13.1. Citizen involvement begins with the acknowledgement of an environmental problem and the motivation to contribute for its resolution and understanding (Step 1). In an ECMN context, citizens use a back-end information infrastructure to make public their personal concerns and ask around what is known about the issue (Step 2). If the idea is to report an environmental problem, citizens may avoid this step and go ahead and collect data (Step 3). Step 2 allows identifying other citizens who may have similar problems. These new participants (Step a) may volunteer to collect data or share their knowledge on the topic. From this interaction, ECMN participants may establish and agree procedures, e.g. data collection protocols (Step b).

Box 13.1 - Characteristics of the nodes and links of environmental collaborative monitoring networks (ECMN).

Nodes: Citizens or groups of citizens equipped with environmental sensors or relying only on human senses. The tasks involved intend not only to support data collection and management but also to promote virtual collaboration among concerned citizens. Citizen motivations and characteristics vary.

Links: Links may represent data transfer, networking with other volunteers or scientists, or data access, namely, access to learning materials. Links may be tangible, when nodes are physically connected through information and communication technologies (ICT), or intangible when referring to relationships within communities of volunteers. Links may connect nodes of the same network or connect different networks, namely volunteer and official environmental monitoring networks. They may allow for one-way or two-way communication.

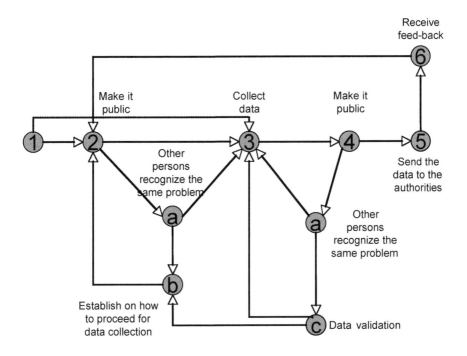

Figure 13.1. Steps involved in citizen participation within an ECMN.

Like in traditional environmental monitoring networks, data collection is the central activity within an ECMN. It is through data collection that citizens may contribute to increase the knowledge on the state of the environment. Each citizen or group of citizens may rely solely on human senses (e.g. smell, vision) to collect data or can be equipped using instruments with different levels of sophistication and accuracy. According to the data collection procedures, citizen initiatives may create fixed or mobile collaborative networks (see Table 13.1).

Data collection initiatives condition data characteristics, which may be quantitative or qualitative and may include factual data and opinions. In general, data collected by citizens are spatial, temporal and have strong multi-media characteristics. ECMNs encourage citizens to make public the data collected (Step 4, Figure 13.1), which may attract other citizens, scientists and environmental professionals. These new participants may review the project data, for example, by comparing the data to other sources of data (Step c). As they look at the issue from a different perspective, they may suggest improvements and even might join the project. These processes are useful to validate the data and increase data credibility.

The involvement of the administration and the authorities plays a major role in the development of ECMN (Steps 5 and 6, Figure 13.1). Although ECMN should be independent and owned by the citizens who participate in their activities, authorities should be part of the process as early as possible. Authorities' roles may range from funding to approving quality assessment/quality control (QA/QC) plans. According to Castells (2001) the ability of citizen networks to reach out to a broader user base is highly dependent on institutional support from an open-minded administration.

Technology itself provides a part of the answer for linking the nodes and supporting the multiple types of relationships or flows among them. Information and communication technologies such as the Internet and wireless communications may allow the improvement of coordination and management activities, particularly in networks beyond a certain size and complexity, which have difficulties in coordinating functions, focusing resources on specific goals and in accomplishing a given task.

Moreover, technological developments such as mobile communication and computing together with sensing devices are creating new opportunities for data collection. The emergence of sensor networks based on wireless communications for environmental monitoring is one example that illustrates the impact of such technological developments. However, volunteer environmental monitoring initiatives have, at different levels, difficulties in accessing these new technologies and in ensuring their correct use. Nevertheless, the increasing availability of personal gadgets, such as phone cameras and GPS, may support citizen data collection and even may favour the collection of non-traditional types of data with interest for environmental monitoring. For example, phone cameras may promote the collection of photos of environmental variables such as oil spills.

On the other hand, the use of technology should also address social behaviour and organisation to sustain volunteer monitoring activities. At least three kinds of social behaviour are necessary: 1) people must participate; 2) people must have access to technology, be able to use it and perform the needed maintenance; and 3) citizens must manage social dynamics, recruiting new nodes, promoting social interaction and rewarding desirable behaviours. Without these social issues, even sophisticated tools and infrastructure will not sustain the creation and maintenance of ECMN.

A review of the opportunities created by mobile computing and communications within collaborative environmental monitoring may enable understanding of how it can be used to support the developments of ECMN. This is described in the next section.

13.4 Mobile computing and communication opportunities for collaborative environmental monitoring

This section aims to analyse the opportunities brought by mobile computing and communication for collaborative environmental monitoring, through the presentation of the main technological developments and applications. Integrated networks are becoming a reality as a result of wireless communication developments providing fully distributed and ubiquitous mobile computing and communications. The increasing number of services for mobile users is changing the nature and scope of computing and communication (see also Chapter 11, Mateos and Fisher, in this book).

Mobile computing and communications supported by the pervasive use of the Internet (Rosen et al., 1998; Larsen, 1999; Hale et al., 2000; Vivoni et al., 2002) are having a major impact on all environmental monitoring activities, since they have created new forms of data collection, access, processing and communication. Additionally, sensors used within environmental monitoring to detect and measure a wide range of physical, chemical and biological variables are becoming smaller, cheaper and smarter. In fact ICT have supported the development of micro-sensors integrated onto a single chip with a processor – the so-called smart sensors. This new breed of sensors is creating new opportunities for *in situ* environmental monitoring (both by professionals and citizens) and was considered by Saffo (1997) as the next wave of innovation.

Given the spatial nature of environmental monitoring (see Section 13.2) mobile GIS and location based services (LBS) are examples of developments that might have a significant impact in environmental monitoring activities and particularly in the promotion of citizen involvement in those activities. Mobile GIS is one of the technological developments that have been explored for environmental monitoring. It has been used to provide integrated mobile geo-spatial information services that support and help optimise field-based management tasks. Data collection is one of the privileged areas of application of these mobile spatial systems (Peng and Tsou, 2003). Monitoring and change detection of natural habitat areas can be accomplished in real time by integrating GPS, wireless communication and Internet Mapping facilities. Tsou (2004) describes a mobile GIS prototype allowing natural habitat preserve managers and scientists to access Internet map servers via their mobile devices, such as pocket PCs, notebooks, or personal digital assistants (PDA) during their field trips. Users can conduct real-time spatial data updates and/or submit changes back to the Web server over the wireless local area network (WLAN).

These capabilities can be very helpful for collaborative environmental monitoring. Real time updating of environmental data or the access to baseline data for the monitored area can bring a value added to volunteer activities within their monitoring tasks. Nevertheless, mobile GIS is not an accessible resource for citizens. It represents a significant investment and requires know-how that is not compatible with the nature of volunteer involvement in monitoring activities. In Chapter 12 of this book, Tsou and Sun discuss this type of problem in relation to the use of mobile GI Services for emergency preparedness. Location based services can be considered more appropriate for monitoring activities. It corresponds to a more widely available technology not requiring special skills and running in more widespread equipment (e.g. mobile phones). Location based services may illustrate the utility that the integration of sensors, computers and wireless communication may bring to different monitoring activities such as data access, exploration and communication. They allow users to request information from several databases from their PDAs or mobile phone, filtering the information based on the location, time and user profile relevance.

Although within application areas other than environmental monitoring, location based services developed within projects such as WebPark (Dias et al., 2004a; Krug et al., 2003) and LiveAnywhere Traffic (Ydreams, 2002) may illustrate the usefulness of this technological integration. WebPark, a research and development project co-funded by the European Commission (EC), developed a platform that enables the deployment of location based services in natural areas (Dias et al., 2004a, 2004b). It provides information to the visitors of natural areas through the use of smart phones and GPS. A WebPark guide is a mobile Website that has dynamic content that changes with the visitors' location, time and interests. It is considered an environmental education tool and a way for promoting the park information. LiveAnywhere Traffic (Ydreams, 2002) is a full-featured mobile traffic information system that processes street-camera video feeds, road sensor data and sends real-time traffic information to users' mobile phones. It allows end-users to outsmart traffic jams and side-step delays, by making information available any time, anywhere.

Mobile environmental information systems (MEIS) is another research project in the field of location based services that aims to explore the possibilities for ambient aware mobile applications in the domain of environmental information systems (Antikainen et al., 2004). Several mobile environmental applications have been created and used within this project to demonstrate the possibilities of mobile applications in the collection, use and transmission of ecological data for both private and organisational users. It includes the experimental development of prototypes for MEIS such as one for visiting a university botanical garden or the application to support professional biologists or amateur nature observers in their field surveys.

Location based services translate the spatial context dependency allowing access to user location-dependent useful data, within the monitoring activities. They also allow inserting location-based information that can be shared with others, and receiving location based alerts associated to specific features or facts of interest to the monitoring work. The Municipal Master Plan Mobile Interactive Visualisation System — a research project developed in 2001 by the Portuguese National Centre for Geographic Information (CNIG) and the Environmental Systems Analysis Group (GASA) of the New University of Lisbon (Portugal) under the leadership of the authors of this chapter — is an example of a location based service that intends to facilitate access to visualisations and data on Municipal Master Plans by the public. The mobile application includes a visualisation system with automatic zooms and the use of anchors at the municipal level (main roads and railroads) and at the local level (public buildings) besides a new legend system.

The impact of such technological developments within environmental monitoring is illustrated by STEFS — Software Tools for Environmental Study (Vivoni et al., 2002) a project that also uses other types of sensors. STEFS is an integrated system for data collection on mobile computers using a GPS and a water-quality sensor to collect data, which are sent through a wireless network, to a database server. Mobile mapping software records and maps the exact locations where the environmental readings are taken. The level of integration of the different technologies in STEFS is still very weak (water-quality sensors and a GPS connected to laptop PC and a mobile phone). In spite of the weak integration, this project highlights the potential brought by the integration of these different technologies for environmental monitoring activities.

Besides the advantages, pointed out, of integrating the different technologies, Reingold (2003) refers to the integration of wireless communication, computers and sensors as the next wave of innovation, since it may allow the creation of communities of people that cooperate in ways never before possible. In fact, the present ability of integrating these technologies into small units such as mobile phones represents one of the more practical ways for equipping citizens for collecting and communicating environmental data. This possibility of integrating sensors, computers and wireless communications into such widespread equipment can bring new opportunities for the involvement of volunteers in environmental monitoring activities.

Mobile phones have gained a rich set of input modes besides acoustic input and output; some incorporate accelerometers to detect gestures, orientation, or photographic and video input. Soon all mobile phones will know where they are, which will bring a degree of location awareness into various personal computing devices (Estrin et al., 2002). Moreover, it may be possible through mobile phones, to access a broad range of services: from entertainment, such as viewing television shows and music videos, to personal assistance tools such as online multi-media organising and publishing. The availability of such devices and services is just a matter of the market.

Other opportunities for environmental monitoring can be foreseen from other recent developments such as 'moblogs' (Reingold, 2003). These mobile Weblogs consist of content posted to the Internet from a mobile or portable device, such as a cellular phone or PDA. According to Reingold, (2003), the use of a mobile phone or other mobile device to publish content to the World Wide Web (WWW), whether that content is text, images, media files or some combination of the above, provides the tools citizens need to publish independent reports of news events as they are happening. Moblogs can act as a new way for catalysing collective action on the Internet that can be applied within voluntary environmental monitoring activities.

The next section analyses how these opportunities might be used within a mobile ECMN context. It identifies the characteristics of a mobile ECMN, the requirements to build such a mobile network and the advantages for the involvement of volunteers in environmental monitoring.

13.5 The application of mobile technologies to environmental collaborative monitoring networks

The emergence of mobile computing and communication favours the creation of mobile networks, where node location is not predefined and varies. The idea is to take advantage of mobile technologies to support citizens to collect and communicate *in situ* environmental data. Although similar to any mobile monitoring network, where measurements are made using portable equipment, mobile collaborative networks have their own characteristics. In mobile collaborative networks each node is a citizen that may collect data from more than one site. Spatial coverage of mobile collaborative networks is variable since data collection points vary with time. Nevertheless, the use of mobile nodes allows the increase of the number of data collection points.

The spatial coverage of each node may be represented by a set of points or a line depending on whether measurements are discrete or continuous. Continuous measurements favour the collection of personal exposure data to a given variable. An example of such type of measurements is presented in PEOPLE project (described in Section 13.6), where volunteers use a diffusive sampler to measure their personal exposure to benzene within their daily activities. Discrete measurements may be event-driven (for example citizen complaints about illegal dumping) or may have a predefined frequency. The characteristics of the measurements are related to the equipment used.

The equipment used by mobile network nodes may vary but it needs to be portable and easily available. Due to their popularity and the integration of communication and computing technologies, mobile phones and hybrid PDA are the most convenient way to equip citizens to become nodes of collaborative monitoring networks. Such equipment supports citizen data collection activities without being too intrusive, facilitating the integration of monitoring activities within the daily activities of citizens. This integration promotes less formal data collection initiatives, such as citizen complaints on a specific problem, and favours the use of sensory data.

Mobile phones and hybrid PDA are particularly useful to register and communicate data. Moreover, some of this equipment integrates useful sensing devices, such as microphones and cameras that may also be used to detect and measure environmental variables. For example, mobile phones that incorporate a sound level analyser may be used to capture, register and communicate environmental noise data. On the other hand, nodes may add other sensing devices to the minimal configuration, such as GPS among others. Examples like the STEFS project (mentioned in Section 13.4) show the possibility to loosely couple a wide range of environmental quality sensors with mobile phones.

The building blocks of mobile environmental collaborative monitoring networks are the same as in any ECMN, although mobile networks have some specific requirements (see Table 13.3). Mobile networks require easy-to-use and portable sensors to facilitate the integration of monitoring activities within the citizen's daily life. The limited computation power of each node requires an Internet based system that facilitates data access and management. Such system should easily communicate with mobile phones; for example send and receive data through short messaging system (SMS) or multi-media messaging system (MMS).

Another specific requirement of mobile networks is the need to determine the node position. Location is important to geo-reference the data collected, but also to enable each node to receive context-aware data, such as maps or data from official networks, to facilitate data collection activities. The location system depends on the equipment used by each node and may rely on the mobile phone location system or may be based on the use of GPS. Among the systems available it is possible to identify Assisted GPS, Cell ID, Time of Arrival and Radio Frequency-Based Systems, which have different associated accuracy errors (see Chapter 11, Mateos and Fisher, in this book).

Mobile collaborative monitoring networks may allow collecting data on sites difficult to access and out-of-reach of fixed stations. The major advantages of mobile environmental collaborative monitoring networks are that:

❑ Mobile networks increase data collection points, although the data collection devices may be less sophisticated then the ones used in fixed monitoring networks. Mobile networks can thus contribute to overcome thematic, spatial and temporal monitoring gaps.
❑ They favour real-time data collection since each node is able to perform instantaneous communication.
❑ Location and temporal data may be automatically associated with the data, facilitating data collection and management.
❑ They offer a highly flexible network configuration and the possibility to activate network nodes based on their location in relation to specific events.
❑ They offer easy communication among nodes, promoting community building and coordination among nodes.

Mobile collaborative environmental monitoring networks take advantage of having citizens equipped with sensors (even if only the human senses) and tools to register and communicate environmental data. Due to their characteristics, mobile environmental monitoring networks are appropriate to support fast site surveys and early warning systems since they may allow the collection of data in a high number of points in a short period of time.

On the other hand, the use of equipment that integrates sensors and mobile computing and communication by volunteers is still in an early stage and several issues remain to be investigated. The possibility to automatically locate volunteers and the data they collect is an advantage of using such technologies. However it may create problems concerning citizen privacy (see also Chapter 3, in this book). Moreover, the diversity of equipment used may create data integration problems; network management and coordination may be hampered by this.

Table 13.3. Major features of ECMN building blocks and mobile requirements.

ECMN building blocks	Major features	Mobile requirements
Motivated citizens	Citizen motivation may range from an individual agenda (e.g. reputation, anticipated reciprocity, personal learning and enjoyment) to more altruistic motives such as the attachment or commitment to certain values or ideals (Kollock, 1999; Lakhani and Wolf, 2001). Problems associated with NIMBY (Petts, 1999) may affect citizens' involvement in ECMN, distorting in some cases the goals of the monitoring network.	Personal exposure data explore the link between health and environment. Motivations such as personal learning and enjoyment favour the participation within mobile networks, which promote the collection of outdoor variables and personal exposure.
Sensing devices	May range from sensors that register human sensory data to sensors that detect and register variables not detected by human senses (e.g. carbon monoxide). Criteria such as usability and affordability of a sensor are particularly important when choosing sensors for volunteer initiatives.	Sensors have to be portable and easy to carry.
Back-end information infrastructure	Includes communication services such as telephone, Internet and cellular connection, resources such as ancillary data, training materials and guidelines, and collaborative spatial tools such as annotation tools. The resources and collaborative spatial tools are incorporated in a collaborative spatial system.	It uses cellular networks to communicate. For example SMS or MMS are a powerful way to support community activism. The Internet is used to access to the common resources, such as data and collaborative tools. The collaborative spatial system should enable to receive and send data to mobile phones. Characteristics of the mobile phone may vary as well as the cellular networks: GSM, GPRS, UMTS. The back-end infrastructure should be used to determine each node location.

Notes: GPRS - General Packet Radio Service; GSM - Global System for Mobile Communications; MMS - multi-media messaging system; NIMBY – Not in my back yard; SMS - short messaging system; UMTS - Universal Mobile Telecommunications System.

It is important to evaluate the specificity of each building block to understand how mobile environmental collaborative monitoring networks may be implemented and identify the major research questions. An analysis of projects that illustrate different approaches to ECMN building blocks may contribute to understand the major issues involved in the creation of such types of framework. The experience from two of such projects is described in next section.

13.6 Examples of projects that explore ECMN building blocks

This section aims to analyse two examples of projects identifying their strengths and weaknesses for the creation of environmental collaborative monitoring networks. The experience from two projects — PEOPLE and Senses@Watch — aiming at promoting citizen involvement in planning and environmental monitoring is discussed. These projects, although with different scopes, and mobile computing and communication approaches, explore in different ways the building blocks of an environmental collaborative monitoring network (see Table 13.3). However they both intend to promote public participation and explore the use of volunteers in mobile monitoring.

The PEOPLE project, described in Section 13.6.1, presents some of the issues involved in equipping volunteers with mobile sensors to collect personal exposure data to specific pollutants on daily citizen trajectories. Senses@Watch, described in Section 13.6.2, illustrates a collaborative spatial system, which is a component of the back-end information infrastructure required to support citizens' activities of environmental data collection and identification of environmental complaints. The description presented focuses on the mobile component of the system.

13.6.1 PEOPLE Project: Assessing citizens' air pollution exposure in European cities

This example relates to the use of volunteers with mobile sensors to collect environmental data on personal exposure to environmental pollutants. The Population Exposure to Air Pollutants in Europe (PEOPLE) project is a research project promoted by the Institute for Environmental Sustainability (IES) from the Joint Research Centre (JRC) involving six European countries (see PEOPLE Project, 2002). In Portugal, project partners included universities, NGOs and central public administration bodies. Moreover, it expected with the support of other organisations, such as radio stations and local municipalities, to raise project awareness and recruit volunteers.

The PEOPLE project aims to assess outdoor, indoor and personal exposure levels to air pollutants in European cities, focusing on emissions from transport and smoking. It intends to support health impact assessment as the study focuses on carcinogenic pollutants presenting long-term effects on human health. One relevant goal of this project is to support local, national and European decision making, and to raise citizen awareness of air quality in general, and in particular the impact of personal lifestyles (mode of living, mode of transport, smoking habit). In parallel, the PEOPLE project aims to monitor city environments through the production of contour maps of the background city-wide pollution levels as well as monitoring the places we inhabit (e.g. domestic indoor) or visit (e.g. shops). The comparison of the mobile data with the fixed data helps to define if personal exposure is significantly different from environmental data, in particular data used to define compliance with air quality directives.

Campaigns in participating cities were organised with the involvement of citizens, scientists, decision makers and the media. In each city, diffusive samplers were used to monitor personal exposure and environmental pollution levels of benzene. Benzene was selected as the first pollutant to be measured, considering that it is carcinogenic and associated with the risk of the development of leukaemia. Each citizen selected to participate was provided with a simple measurement device, and requested to expose the sampler to ambient air for 12 hours on their body during a well-specified day of the working week. PEOPLE campaigns were completed in Brussels and Lisbon (22 October 2002), Bucharest and Ljubljana (27 May 2003), Madrid (3 December 2003) and Dublin (28 April 2004). No automatic positioning (e.g. GPS) was used.

The project allowed the collection of data that was only possible through the use of a network of volunteers equipped with diffuse tubes. The results obtained showed that citizens are sometimes exposed to levels of pollutants above the limits even when fixed site monitoring does not give evidence of those violations.

This project exemplifies the logistics required to involve volunteers using mobile sensors: from equipment purchase to distribution. On the other hand, the project institutional framework, which was designed to support organisational issues such as recruiting volunteers or supporting volunteers, is an interesting example of collaboration among different stakeholders: from researchers, to public administration and NGO. The support of the media and NGO was also an important contribution.

The project PEOPLE represents an isolated initiative, which reduces the impact of its effects not only in terms of the usefulness of the gathered information but also in terms of citizen engagement and awareness on environmental protection. Additionally, the technology used does not take advantage of the characteristics of mobile computing and communication, such as automatic registry of data collection and volunteer position. For example, the sensors used (diffusive samplers) do not provide results in real time, which hampered volunteer awareness and future involvement. It would be useful to explore mobile computing and communication technologies that provide real-time data collection to understand the costs and benefits involved.

13.6.2 Senses@Watch: The use of sensory data collected by concerned citizens

The project Senses@Watch illustrates some components of the 'back-end' information infrastructure required to support the collection and use of environmental data by concerned citizens. Senses@Watch is a research project developed, under the leadership of the main authors of this chapter, by the Portuguese National Centre for Geographic Information (CNIG) now integrated in the Portuguese Geographical Institute (IGP), the College of Sciences and Technology from the New University of Lisbon (FCT-UNL), and also the College of Psychology and Education (FPE) from the University of Lisbon.

The Senses@Watch project aims to define and evaluate strategies to promote the use of citizen-collected data through their senses while monitoring the state of the environment, including the information used in environmental complaints (see Senses@Watch Project, 2002). Senses@Watch project involves the use of information and communication technologies to support and promote a wider use of the data collected by citizens. A prototype of a Web-based collaborative site is being developed including an interface for mobile phones.

The Senses@Watch collaborative site (see Figures 13.2 and 13.3) intends to support citizens to collect and manage the data within isolated initiatives. The design of the system follows the well-known metaphor of postcards. The main idea is that citizens may use the Web or mobile phones to create their postcards with photos, sounds, graphics and text describing an environmental problem. Such e-cards can be published on the WWW and sent to the authorities in charge. Considering the major tasks performed by citizens when filing a complaint, the prototype provides the access to three types of tools:

❑ Tools to support data collection and processing—this group includes tools to geo-reference, annotate data and create metadata. To minimise the ambiguity of a reference to a place, citizens may use a gazetteer and maps to support the geo-referencing of a complaint. Data annotation tools give the user the possibility to underline specific issues within the images through the association of graphics, sounds and text to an image. Through the available metadata the user can make available information that supports the assessment of data fitness and reuse. For example, data quality indicators can be built based on the history of each individual contribution to the system.

❑ A case library intending to support the creation of multisensory messages in the context of environmental public participation—it includes prerecorded images annotated with graphics, icons, texts and non-spoken sounds, as well as short textual descriptions that translate sensory data into environmental quality information.

❑ Data access and visualisation tools, namely thematic, temporal and spatial searches. These tools are based on Web mapping services. They allow users to access the data collected by other citizens and overlay it to data from other sources such as orthophotos.

To illustrate some of the potential benefits of using mobile phones to support public participation within environmental monitoring one prototype is currently under development. The mobile application being developed within the project intends to explore research questions such as what type of tools should be available through mobile phones to support citizens within their complaint process, how should data be presented according to the size and quality of most mobile phones, and what type of interface is more appropriate (Fonseca and Gouveia, 2005).

Access to the full contents of Senses@Watch collaborative site is available through a WAP (Wireless Application Protocol) interface. An application that automatically geo-references the data collected by citizens using their cameraphones is still under development since it requires the operator's agreement to provide such service and the consent of the user to release personal information such as the user's location. Since geo-referencing using the positioning system of the operator may be inaccurate (especially within rural areas) the application being developed includes the possibility to use a gazetteer together with maps, to allow users to identify the location of the data collection and reduce the ambiguity of a reference to a place.

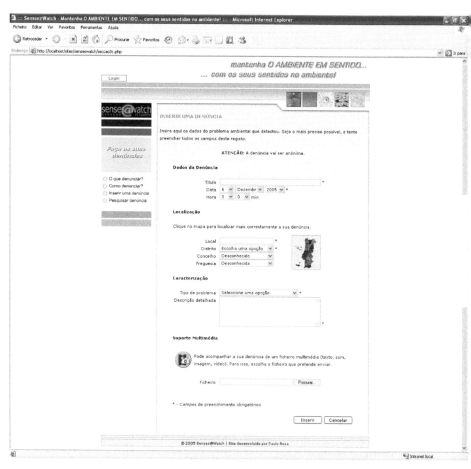

Figure 13.2. Senses@Watch collaborative site (December 2005 version): Insertion of data on an environmental complaint, including the possibility of data geo-referencing through the use of a Webmapping application.

The collaborative spatial system proposed within Senses@Watch is Web-based but has an interface to mobile phones. It allows the exploration of non-traditional types of environmental data such as images, sounds and videos in association with spatial information, which are predominant in non-formal data collection initiatives such as citizen complaints. The Senses@Watch project has found that the creation of tools and methodologies to facilitate data collection, access and validation may help to overcome some of the problems associated with data quality. It proposes a collaborative spatial system, which is one component of the 'back-end' information infrastructure required to develop ECMN. Furthermore, the collaborative spatial system intends to explore data and tools provided by spatial data infrastructures, such as gazetteers.

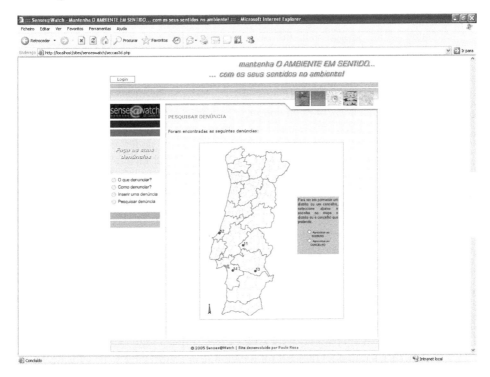

Figure 13.3. Senses@Watch collaborative site (December 2005 version): visualisation of existing specific data on complaints at the national level.

13.6.3 Comparative analysis and lessons learned from the two projects

Four major lessons can be obtained from the analysis of the two projects (see Table 13.4 that summarises the main characteristics of the two projects). First, involving concerned citizens in mobile environmental monitoring may allow the collection of non-traditional types of monitoring data: from personal exposure, in the case of PEOPLE, to sensory data, in the case of Senses@Watch. It allows an increase in the amount of available environmental data, both spatially and temporally. Nevertheless, data credibility is still a major issue in volunteer monitoring. The steps involved in the creation of ECMN must favour triangulation as, according to LeCompte and Goetz (1982) triangulation increases validity and reliability. In an ECMN context the integration of different perspectives, from citizens to experts, may contribute to data triangulation.

The second lesson is related to the need to have tools to support the tasks such as data collection, data access, data management and community building. The use of mobile technologies favours the collection of multi-media and sensory data, which requires the development of tools to facilitate data integration and management. For example, more research is needed to understand how to integrate and manage different types of data such as voice messages, SMS and MMS. These tools may take advantage of existing information infrastructures. For example, gazetteers available within spatial data infrastructures may be used to support users' geo-referencing needs. On the other hand, the Senses@Watch project is an example where tools that intend to facilitate data collection and management, such as data annotation tools, are integrated in a collaborative system that target the specific needs of volunteers.

Another lesson brought by the projects presented is linked to the institutional and organisational framework required to support the logistics involved in volunteer projects. For example, activities such as recruiting new nodes and maintaining the motivation of the existing ones require a framework that facilitates communication and collaboration among the people involved. Different organisation models can be found. The two projects presented have followed different approaches. The project PEOPLE uses a top-down approach while the Senses@Watch intends to take advantage of bottom-up initiatives. More research is required to analyse the organisational and institutional models that favour the creation of ECMN.

The fourth lesson concerns the role of ICT within ECMN. Mobile technologies are one of the recent achievements that can change the way people can collaborate. In fact the popularity of mobile cellular phones makes them an attractive device to support citizenship activities such as voluntary environmental monitoring. The major barriers are related to the large variety of mobile devices and the costs associated with the communication services. Related to the creation of systems to support public participation, issues like users' privacy and the need to have an attractive business model (at least for the mobile operator) have to be considered.

13.7 Conclusion and future developments

Monitoring systems have been widely used to increase knowledge on the state of the environment and have evolved to link monitoring information into the decision-making process. The usefulness of current environmental monitoring systems is however restricted as they present specific problems associated with the existence of spatial and temporal gaps, and reflect limitations on providing timely identification and warnings of emerging problems to the public, stakeholders, research personnel and managers. Additionally, they do not effectively raise awareness, do not contribute significantly to environmental education and do not adequately provide the basis for informed decisions (Vaughan et al., 2003).

Table 13.4. Characteristics of two projects containing building blocks of an ECMN.

	People Project	Senses@watch
Goals	Air pollutant personal exposure monitoring by volunteers. Awareness raising.	Promote collaborative environmental monitoring.
Motivations	Increase the knowledge on the environment especially health and environment issues.	Identification of environmental problems. Explore the data included in citizen complaints.
Technology	Diffuse samplers used by citizens within their daily trajectories.	Web site with a mobile interface. Mobile phones to collect and send data.
Spatial coverage	Urban areas.	Urban and natural environments.
Deliverables	Data on air pollutant personal exposure.	Environmental collaborative monitoring site.
Strengths	Collect personal exposure data that would be difficult to collect in a different context. Institutional framework, including from NGO to media and public administration.	Explores the use of ICT to promote collaboration.
Weaknesses	Isolated initiative. Technologies used for data collection and communication do not favour citizen awareness and lifestyle changes.	R&D project, which does not include a sustainable institutional framework. Lack of field testing.

Public participation in environmental monitoring might help to overcome some of these problems, but a collaborative framework is required to support volunteer tasks and increase the impact of volunteer initiatives. The creation of ECMN, as discussed in this chapter, may contribute to promote citizen participation within environmental monitoring, while supporting the use of citizen-collected data. In these networks the nodes are interested citizens who volunteer to participate from different motivations.

Mobile communication and computing may be used to link citizens and therefore create new opportunities to support the creation of ECMN. The use of such technologies together with portable sensors may enable citizens to collect data within their daily activities, for example in their trips from home to work. Mobile GIS is one of the technological developments that have been explored for environmental monitoring. Its capabilities can be very helpful for collaborative environmental monitoring but at present it is not a widely accessible resource for citizens as it has significant investment and know-how requirements. Location based services can be considered more adequate to volunteer monitoring activities as they do not require special skills and run in more widespread equipment (such as mobile phones). These factors are very important when dealing with teams of volunteers with different levels of expertise and involved in environmental monitoring tasks such as data collection and environmental advocacy.

The opportunities brought by mobile technologies have not been fully explored as very few projects intend to create networks of citizens that participate within environmental monitoring. However, there are some examples of projects that explore mobile collaborative networks building blocks. In this chapter the PEOPLE and Senses@Watch projects were presented as two attempts to explore citizen involvement within mobile environmental monitoring. Four main lessons were derived from the analysis of the two projects (see Section 13.6.3): 1) involving concerned citizens within mobile environmental monitoring may allow the collection of non-traditional types of monitoring data but data credibility is the key; 2) it is necessary to have tools to support the different tasks (e.g. data collection, data access, data management and community building); 3) institutional and organisational frameworks are crucial to support the logistics involved in volunteer projects; and 4) mobile technologies, such as mobile phones, can change the way people collaborate, but issues of privacy need to be taken into account.

Mobile applications for collaborative environmental monitoring present advantages such as automatic data geo-referencing, the possibility to instantaneously communicate and access data, and "anytime anywhere" accessibility. For example, mobile technology instantaneous communication favours the collection of real-time data and allows the activation of nodes according to their location. Such technologies are particularly useful in supporting the creation of early warning systems that operate based on data gathered locally. These systems can be used in situations such as forest fires or floods, involving volunteers in rural areas and communities, and may benefit from the communication improvements brought by mobile technology.

In summary, mobile technologies may support citizen involvement within environmental monitoring as they create new opportunities for data collection and facilitate communication among citizens. Mobile technologies may be available anytime anywhere enabling to increase temporally and spatially the data collection points. Additionally, they automatically associate time and location to the data collected. However, the use of mobile technologies to promote public involvement within environmental monitoring requires further investigation to address questions such as: how may technology be used to improve data credibility? How we integrate the different types of data collected using mobile phones? Which environmental variables are more suitable for collection using mobile computing and communication? How can mobile phones be equipped to make them environmental monitoring kits? Which sensors are available? Which can be integrated and which should be coupled? Which business models may support the use of mobile technologies by citizens involved within environmental monitoring? These research questions imply the exploration of technological and socio-economic variables. Moreover the development life cycle of mobile devices, such as mobile phones, constrains the research in this application area.

Acknowledgements

This research was partially funded by the POCTI/MGS/35651/99. The authors would like to thank all the members of the project team.

References

Antikainen, H., Bendas, D., Marjoniemi, K., Myllyaho, M., Oivo, M., Colpaert, A., Jaako, N., Kuvaja, P., Laine, K., Rusanen, J. Saari, E. and Similä, J. (2004) 'Mobile environmental information systems', *Cybernetics & Systems*, 35, 7–8, pp. 737–751.

Au, J., Bagchi, P., Chen, B., Martinez, R., Dudley, S. A. and Sorger, G. J. (2000) 'Methodology for public monitoring of total coliforms, Escherichia coli and toxicity in waterways by Canadian high school students', *Journal of Environmental Management*, vol. 58, no. 3, pp. 213–230.

Boyle, M. (1998) *An Adaptive Ecosystem Approach to Monitoring: Developing Policy Performance Indicators for Ontario Ministry of Natural Resources*, Masters Thesis, Department of Environment and Resource Studies, University of Waterloo, [Online], Available:, http://ersserver.uwaterloo.ca/jjkay/grad/mboyle/th_pdf.html [January 2005].

Castels, M. (2001) *The Internet Galaxy, Reflections on the Internet, Business, and Society*, Oxford: Oxford University Press.

Cuthill, M. (2000) 'An interpretive approach to developing volunteer-based coastal monitoring programmes', *Local Environment*, vol. 5, no. 2, pp. 127–137.

Dias, E., Beinat, E., Rhin, C. and Scholten, H. (2004a) 'Location aware ICT in addressing protected areas' goals', in Prastacos, P. and Murillo, M. (eds.) *Research on Computing Science*, vol. 11 (special edition on e-Environment), Mexico City: Centre for Computing Research at IPN, pp. 273–289.

Dias, E., Beinat, E., Rhin, C. Haller, R. and Scholten, H. (2004b) 'Adding value and improving processes using location-based services in protected areas', in Prastacos, P. and Murillo, M. (eds.) *Research on Computing Science*, vol. 11 (special edition on e-Environment), Mexico City: Centre for Computing Research at IPN, pp. 291–302.

Estrin, D., Culler, D., Pister, K. and Sukhatme, G. (2002) 'Connecting the physical world with pervasive networks', *IEEE Pervasive Computing*, vol. 1, no. 1, pp. 59–69.

Fonseca, A. and Gouveia, C. (2005) 'Spatial multimedia for environmental planning and management', in Campagna, M. (ed.) *GIS for Sustainable Development. Bringing Geographic Information Science into Practice Towards Sustainability*, Boca Raton, FL: Taylor and Francis.

Fortin, C. (2000) 'Minneapolis-St Paul area volunteer monitoring: A coordinated approach' in Sixth National Volunteer Monitoring Conference, Austin, Texas, retrieved from http://www.epa.gov/owow/volunteer/proceedings/sixth/session2.html#paul

Gao, J. (2002) 'Integration of GPS with remote sensing and GIS: Reality and prospect', Photogrammetric Engineering and Remote Sensing, vol. 68, no. 5, pp. 447–453.

Gouveia, C, Fonseca, A., Câmara, A and Ferreira, F. (2004) 'Promoting the use of environmental data collected by concerned citizens through information and communication technologies', Journal of Environmental Management, 71, pp. 135–154.

Hale, S., Bahner, L. and Paul, J. (2000) 'Finding common ground in managing data used for regional environmental assessments', Environmental Monitoring and Assessment, vol. 63, no. 1, pp. 143–157.

Kollock, P. (1999) 'The economies of online cooperation: Gifts and public goods in cyberspace' in Smith, M. and Kollock, P. (eds.), Communities in Cyberspace, London: Routledge, pp. 220–239.

Krug, K., Mountain, D. and Phan, D. (2003) 'WebPark – Location-based services for mobile users in protected areas', GeoInformatics, March, pp. 26–29.

Lakhani, R. and Wolf, R. (2001) 'Does free software mean free labor? Characteristics of participants in open source communities', Boston Consulting Group Survey Report, Boston, MA, [Online], Available: www.osdn.com/bcg/

Larsen, L. (1999) 'GIS in environmental modeling and assessment', in Longley, P., Goodchild, M., Maguire, D. and Rhind, D. (eds.) Geographic Information Systems, Volume 2, New York: John Wiley and Sons, pp. 999–1007.

Lawson, R. (2000) 'Coordinating Monitoring in the Lake Michigan Basin', in US EPA, Proceedings of the Six National Volunteer Monitoring Conference, Austin, Texas, [Online], Available: http://www.epa.gov/owow/volunteer/proceedings/sixth/session2.html#paul

LeCompte, M. and Goetz, J. P. (1982) 'Problems of reliability and validity in ethnographic research', Review of Educational Research, 52, pp. 31–60.

Lee, V. (1994) 'Volunteer monitoring: A brief history', The Volunteer Monitor, vol. 6, no. 1, [Online], Available: http://www.epa.gov/volunteer/spring94/ppresf16.htm.

Markowsky, G., Nagel, D., Mitchell, D., Thot-Thompson, J. and Williams, T. (eds.) (2002) 'Anywhere, Anytime, Any Size, Any Signal: Scalable, Remote Information Sensing and Communication Systems', Report from an NSF-sponsored workshop, January 14–15, 2002, ACCESS Center Arlington, VA.

Peng, Z.R. and Tsou, M.H. (2003) Internet GIS: Distributed Geographic Information for the Internet and Wireless Networks, New York: John Wiley.

PEOPLE Project (2002) PEOPLE Project site, http://ies.jrc.cec.eu.int/Units/eh/Projects/PEOPLE/, last accessed 30/09/05.

Petts, J. (1999) 'Public participation and environmental impact assessment' in Petts, J. (ed.) Handbook of Environmental Impact Assessment, 2 volumes, Malden, MA: Blackwell Science, pp. 145–177.

Reingold, H. (2003) Smart Mobs. The Next Social Revolution, Norwood, MA: Perseus Publishing.

Root, T. L. and Alpert, P. (1994) 'Volunteers and the NBS', Science, 263 (5151), p. 1205.

Rosen, E. C., Haining, T. R., Long, D. D. E., Mantey, P. E. and Wittenbrink C. M. (1998) 'REINAS: A real-time system for managing environmental data', International Journal of Software Engineering And Knowledge Engineering, vol. 8, no. 1, pp. 35–53.

Saffo, P. (1997) 'Sensors: The next wave of innovation', Communications of the ACM, vol. 40, no. 2, pp. 93–97.

Satillo, D., Johnston, P. and Langston, W. J. (2001) 'Tributyltin (TBT) antifoulants: a tale of ships, snails and imposex' in Harremoës, P., Gee, D., MacGarvin, M., Stirling, A., Keys, B., Wynne, S. Guedes Vaz (eds.). Late Lessons from Early Warnings: The Precautionary J. Principle 1896-2000, Luxembourg: Office for Official Publications of the European Communities pp. 135–143.

Senses@watch Project 2002, Senses@watch Project Site, http://panda.igeo.pt/senses/sp/english/index.asp, last accessed 30/09/05.

Stokes, P., Havas, M. and Brydges, T. (1990) 'Public participation and volunteer help in monitoring programs: An assessment', *Environmental Monitoring and Assessment*, vol. 15, no. 3, pp. 225–229.

Tsou, M. (2004) 'Integrated mobile GIS and wireless internet map servers for environmental monitoring and management', *Cartography and Geographic Information Science*, vol. 31, no. 3, pp. 153–165.

Vaughan, H., Whitelaw, G., Craig, B. and Stewart, C. (2003) 'Linking ecological science to decision-making: Delivering monitoring information as societal feedback', *Journal of Environmental Monitoring and Assessment*, vol. 88, no. 1, pp. 399–408.

Vivoni, E. R., Camilli, R., Rodriguez, M. A. A., Sheehan, D. D. and Entekhabi, D. (2002) 'Development of mobile computing applications for hydraulics and water quality field studies', *Hydraulic Engineering Software* IX, Montreal: WIT Press.

Withey, A., Michener, W. and Tooby, P. (eds.) (2002) 'Scalable Information Networks for the Environment (SINE)', *Report of an NSF-sponsored workshop* (San Diego Supercomputer Center, October 29–31, 2001).

Ydreams (2002) *LiveAnywhere Traffic. Let your phone be your pilot,* [Online], Available: http://www.ydreams.com/solutions.php?sec=2&s_sec=1&s_s_sec=.

Young-Morse, R. (2000) 'Real-time detection of phytoplankton' in USEPA (ed.) *Proceedings of the Sixth National Volunteer Monitoring Conference*, Austin, Texas, [Online], Available: http://www.epa.gov/owow/volunteer/proceedings/sixth/session2.html#paul.

Chapter 14

Analysing Point Motion with Geographic Knowledge Discovery Techniques

Patrick Laube [1], Ross S. Purves [2], Stephan Imfeld [2]
and Robert Weibel [2]

[1] School of Geography and Environmental Science, University of Auckland, New Zealand
[2] Department of Geography, University of Zurich, Switzerland

14.1 Introduction

Mobility is key to contemporary life. In a globalised world, people, goods, data and ideas move in increasing volumes at increasing speeds over increasing distances, and more and more leave a digital trail behind them. More and more such tracking data is automatically collected in large databases. Exploring the dynamic processes afforded by the study of such digital trails—in other words *motion*—is an emerging research area in Geographical Information Science.

This chapter argues that Geographical Information Science can centrally contribute to discovering knowledge about the patterns made in space-time by individuals and groups within large volumes of tracking data. Whereas the representation and visualisation of motion is quite widespread within the discipline, approaches to actually *quantitatively* analysing motion are rare. Hence, this chapter introduces an approach to analysing the tracks of moving point objects, which are considered as the most basic and commonly used conceptualisation in representing motion in geography.

The methodological approach adopted is Geographic Knowledge Discovery (GKD)—an interactive and iterative process integrating a collection of methods from geography, computer science, statistics and scientific visualisation (Miller and Han, 2001). Its goal is the extraction of high-level information from low-level data in the context of large geographic datasets (Fayyad et al., 1996). This chapter sets out to illustrate that the integration of knowledge discovery methods within Geographical Information Science provides a powerful means to investigate motion processes captured in tracking data.

The chapter is structured as follows. Section 14.2 provides a literature overview on analysing point motion, identifies some shortcomings and proposes a set of objectives that the remainder of the chapter attempts to address. In Section 14.3 the central tenets of the proposed motion analysis approach are introduced. The methods are illustrated in Section 14.4, using case studies from biology, sport's

Dynamic and Mobile GIS: Investigating Changes in Space and Time. Edited by Jane Drummond, Roland Billen, Elsa João and David Forrest. © 2006 Taylor & Francis

scene analysis and spatialisation of political science data. Section 14.5 critically discusses this methodological approach to the mining of motion data. The chapter concludes by identifying the key steps made in integrating knowledge discovery techniques in Geographical Information Science for analysing motion and gives an outlook as to possible future work.

14.2 Motion analysis in Geographical Information Science

This section discusses the role of motion analysis in the field of Geographical Information Science and associated disciplines. The potential and limitations of recent work are discussed, and a set of objectives underpinning the work presented in this chapter are formulated.

The analysis approach proposed in this chapter focuses on the motion of points. Although all three fundamental abstractions of spatial entities, points, lines and polygons, may move in space and time, the most common representation of moving objects is points. Be it for tracked animals, taxi cabs or carriers of location-aware devices, the simplest way to track motion is to specify location at any time t by either a record of (x,y,t) coordinates or by a record of (x,y,z,t) coordinates. Thus, the prime object of interest of this chapter is the *moving point object*, irrespective of its real-world counterpart.

The most basic conceptualisation of the path of a moving point object is the so called 'geo-spatial lifeline' (Hornsby and Egenhofer, 2002; Mark, 1998). Mark (1998, p. 12) defines a geo-spatial lifeline as a 'continuous set of positions occupied in space over some time period'. Geo-spatial lifeline data usually consists of discrete fixes, describing an individual's location in geographic space with regular or irregular temporal intervals.

14.2.1 Visual exploration of motion data

The simplest way to visualise the motion of a moving point object is to map its complete trajectory on a Cartesian plane. Labelling of intermediate positions can add temporal information to the track in order to visualise the object's past locations. The symbology and the colour of the trajectory can also code motion speed, acceleration or motion azimuth (Dykes and Mountain, 2003).

Adding time as a third dimension allows the visual representation of trajectories in 3-D. Thus, increasing computational power in recent decades has given rise to a diverse set of applications adopting the space-time aquarium data model suggested by Hägerstrand's time geography (Hägerstrand, 1970). Most prominent is the work by Forer's group on visualising (and analysing) student lifestyles and tourism flows in New Zealand (Forer, 1998; Huisman and Forer, 1998; Forer et al., 2004).

Most static visualisations of motion can be animated by browsing through the temporal dimension. Andrienko et al. (2000) propose the 'dynamic interval view' in a case study of migrating storks. The interval view shows trajectory fragments during the current interval. In their prototype application for transport demand modelling, Frihida et al. (2004) provide an animated 2-D map view to dynamically visualise individual space-time paths. Tools for the animated visualisation of motion have recently found their way into commercial GIS. For example, ESRI offers the ArcGIS Tracking Analyst extension to visualise tracking data. It features various symbology options and a sophisticated playback manager. However, its power lies almost exclusively in the functionality provided to define events and to visualise where and when they occur.

Exploratory data analysis (EDA) of motion data aims to find potentially explicable motion patterns. 'Modern EDA methods emphasise the interaction between human cognition and computation in the form of dynamic statistical graphics that allow the user to directly manipulate various 'views' of the data. Examples of such views are devices such as histograms, box plots, q-q plots, dot plots, and scatter plots' (Anselin, 1998, p. 78). Kwan (2000) proposes a set of 3-D techniques to explore disaggregate activity-travel behaviour from travel diary data. Kraak and Koussoulakou (2004) present an exploratory environment featuring alternative views, animation and query functions for motion data.

As an excellent example of the exploratory analysis of motion data Brillinger et al. (2004) present a set of techniques applied to a huge collection of VHF telemetry tracked elk and deer. *Parallel boxplots* of the square roots of objects' speed by hour of the day are used to analyse circadian rhythms. Collapsing all available data for one time of day creates 'temporal transects' well suited to descriptive statistics. Decomposing the object's velocity to cardinal directions using a separate 'X-component velocity plot' and a 'Y-component velocity plot' provides insights on the directional bias in the joint motion of a group. Finally, 'vector fields' address the issue of the spatial distribution of motion properties and provide a sophisticated overview of the motion of a group moving in a distinct area over a distinct time period.

However, most exploratory approaches stop at representation and delegate the analytical process to user interpretation. Furthermore, many visualisation approaches focus on position, ignoring inherent motion properties such as speed, acceleration, motion azimuth and sinuosity. However sophisticated the exploratory tools may be, the human capability to recognise complex visual patterns decreases rapidly with an increasing number of investigated trajectories and larger numbers of moving objects as shown in Figure 14.1. Kwan (2000, p. 197) states that 'although the aquarium is a valuable representation device, interpretation of patterns becomes difficult as the number of paths increases...'. Thus, the exploratory power of 'flying through the space-time aquarium' is, in general, limited to a small number of moving point objects.

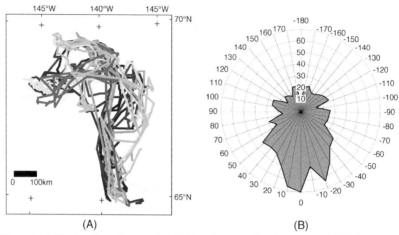

(A)   (B)

Figure 14.1. Exploration of geo-spatial lifelines. (A) Mapping the geo-spatial lifelines of moving point objects in a static map ignores completely the temporal aspect of motion and leads to confusing representations, as illustrated here with the tracks of only a dozen caribou migrating by the Beaufort Sea during two seasons. (B) The turning angle distribution of the same group of caribou illustrates the directional persistence in their motion (0° for straight on). See colour insert following page 132.

14.2.2 Descriptive statistics of motion data

Individual lifelines or aggregations of many lifelines and lifeline segments can be statistically described with respect to motion quantifiers such as travel distances, speed, acceleration, motion azimuth and sinuosity. The appropriate statistical description of motion is an important precondition for simulating motion processes, for example, in the field of behavioural ecology.

For many ecological questions, for instance animal metapopulation dynamics, knowledge about the dispersal capability of animals is necessary and acquired through extensive empirical and theoretical research (Berger et al., 1999). Berger et al. identify three frequently used linear mobility measures to describe an individual's motion in ecological field studies: mean daily movement, maximal distance between two fixes and the mean activity radius (that is the average distance between the capture point and all consecutive fixes).

In behavioural ecology, frequency distributions of 'step length' and 'turning angle' are investigated to gain an overall impression of the motion of the animals under study (e.g. Hill and Häder, 1997; Ramos-Fernandez et al., 2004). Directional persistence is often a key issue investigating turning angle distributions (see Figure 14.1B). Trajectories are normally characterised using frequency distributions of discrete classes between –180° and 180° (e.g. Schmitt and Seuront, 2001; Ramos-Fernandez et al., 2004). When describing the motion direction, a motion azimuth (absolute direction with respect to North) distribution is sometimes preferred over the turning angle. Radar plots visualise the turning angle distributions around the compass card in a very illustrative way.

Mean values and frequency distributions may give an appropriate overview of the way that certain moving point objects move in space and time. However, summarising the complex motion phenomena found, for instance, in the geo-spatial lifelines of seasonally migrating caribou in just a few holistic statistical descriptors removes all dynamic aspects of the motion process. The authors argue therefore that descriptive statistics are not well suited to acquiring more insights into individual motion patterns or inter-object relations in the motion process.

14.2.3 Knowledge discovery and data mining in motion data

Tracking motion processes very rapidly generates very large datasets. The Database Management Systems (DBMS) community, especially researchers interested in Spatiotemporal Database Management Systems (STDBMS), has introduced various approaches to querying databases covering moving objects (e.g. Sistla et al., 1998; Güting et al., 2003; Grumbach et al., 2003). However, querying a database means retrieval of stored objects, collections of objects or their observations from a database. Aronoff (1989) and Golledge (2002) argue that motion *analysis*, in contrast, must go beyond mere querying and requires the production of new information and knowledge that is not directly observed in the stored data. Thus, the aim of motion analysis must be to derive value-added knowledge about motion events.

In recent years various techniques developed especially for large volume and multi-source data, such as Knowledge Discovery in Databases and its component data mining, have entered the field of Geographical Information Science. Fayyad et al. (1996, p. 40) define Knowledge Discovery in Databases (KDD) as the 'nontrivial process of identifying valid, novel, potentially useful, and ultimately understandable patterns in data'. Data mining is just one central component of the overall knowledge discovery process denoting the application of specific algorithms for extracting patterns from data.

Miller and Han (2001) identified unique needs and challenges for integrating KDD into Geographical Information Science because of the special properties of geographic data. Hence, they propose the development of specific Geographic Knowledge Discovery (GKD) and geographic data mining approaches. The latter 'involves the application of computational tools to reveal interesting patterns in objects and events distributed in geographic space and across time. These patterns may involve the spatial properties of individual objects and events (such as shape, extent) and spatiotemporal relationships among objects and events in addition to non-spatial attributes of interest in traditional data mining' (Miller and Han, 2001, p. 16).

Although the ideas of geographic knowledge discovery match very closely the requirements for analysing motion, very few approaches actually mining motion data are found in the literature. Frihida et al. (2004) propose a knowledge discovery approach in the field of transport demand modelling. Their approach is designed to extract useful information from an origin–destination survey, i.e. to build individual space-time paths in the space-time aquarium. In a similar context Smyth (2001) presents a knowledge discovery approach to mine mobile trajectories. The overall goal of this research is to gain knowledge from mobile trajectories in order to design better, more scalable and less expensive location based services. The data mining algorithms describe chunks of trajectories using many measurable parameters (such as speed, heading, acceleration), then identify the behaviour of each chunk, and finally store these characteristics in a behaviour warehouse, that is to say, assign found motion patterns to archetypical behaviours ready to allocate to new data. For example, a car driver using an in-car mobile navigation system may benefit from guidance to petrol stations, automatically allocated to the stored behaviour 'driving on the motorway'.

Knowledge discovery is a promising approach to the problems of analysing motion. In contrast to analytical approaches emerging from a cartographic or GIS tradition that adopt a static view comparing snapshots, knowledge discovery adopts a process view, where events and processes are analysed rather than their instantaneous stamping in static space. Thus, the integration of knowledge discovery in GIS may help the discipline to move 'beyond the snapshot' (Chrisman, 1998, p. 85).

14.2.4 Key objectives for the analysis of point motion

From the background of this section a set of key objectives underpinning the research presented in this chapter were developed.

❑ Distinct motion events can be considered as detectable patterns in motion data. Thus, this chapter shall explore the development of tools integrating knowledge discovery techniques and Geographical Information Science for analysing motion data.
❑ Assuming that mining motion data is a reasonable approach to analysing motion data, this chapter shall introduce data mining techniques that allow the automatic detection of motion patterns.
❑ Given that data mining may find irrelevant or useless patterns, methods will be developed that help to define and discriminate between relevant motion patterns and meaningless patterns.

14.3 Mining motion patterns – a geographic knowledge discovery approach

This section reports on conceptual Geographical Information Science developing an integrated geographic knowledge discovery approach for analysing geo-spatial lifelines of groups of moving point objects. The overall goal is to conceptualise and implement a flexible framework to find user-defined motion patterns in the trajectories of groups of moving point objects. Section 14.3.1 introduces a family of basic motion patterns and a way to formalise and detect these patterns. Section 14.3.2 extents the motion pattern family including flocking and convergence processes. Finally, Section 14.3.3 provides a means to evaluate the relevance of the found patterns.

14.3.1 Characterising, formalising and detecting motion patterns

The key concept of the proposed geographic knowledge discovery approach is to compare the motion parameters of moving point objects over space and over time (Laube and Imfeld, 2002). Suitable geo-spatial lifeline data consist of a set of point objects each featuring a list of fixes, tuples of (x,y,z,t).

The approach focuses on the basic knowledge discovery steps 'data reduction and projection' and 'data mining', respectively (Fayyad et al., 1996). The first step consists of a transformation of the geo-spatial lifeline data into an analysis matrix featuring a time axis, an object axis and motion attributes (i.e. speed, change of speed and motion azimuth). It is assumed that specific motion behaviour and interrelations among the moving point objects are manifested as patterns in the analysis matrix. Thus, as a second step, formalised motion patterns are matched on the analysis matrix. In contrast to most exploratory approaches, motion pattern detection does not rely on pure visual exploration but offers the user to automatically search the data for patterns that appear to be reasonable given the issue under investigation.

The knowledge discovery approach follows the principle of syntactic pattern detection where simple patterns serve as primitives for the construction of more complex patterns (Jain et al., 2000). A set of generic patterns form the starting point for the composition of arbitrarily complex patterns. The following introduces the three example motion patterns, illustrated in Figure 14.2D.

❑ Constancy: One object expresses constant motion properties for a certain time interval.
❑ Concurrence: A set of objects express the same motion behaviour at a certain time.
❑ Trend-setter: One object (the trend-setter) anticipates the motion behaviour of a set of other objects.

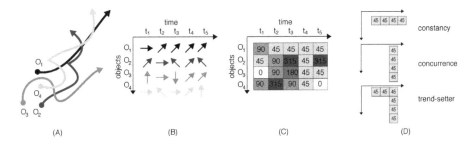

Figure 14.2. Mining motion patterns. The geo-spatial lifelines of four moving point objects (A) are used to derive at regular intervals the motion azimuth (B). In the analysis matrix consisting of classified motion attribute values (C) generic motion patterns such as 'constancy', 'concurrence' or 'trend-setter' are matched (D).

A formal language for describing motion patterns was needed, allowing the user to formalise the motion patterns of interest for the issue under study. Consequently, in Laube et al. (2005), a pattern description formalism adopting elements of the commonly used regular expression formalism (regex) as well as of basic mathematical logic was proposed. Whereas regex is used to search and manipulate strings, the proposed pattern description formalism is used to search motion patterns in tracking data.

A few examples will serve to illustrate some basic motion patterns as well as their formal description. A single deer heading north-east for a sequence S of four consecutive time steps is formalised as constancy pattern $P = S([45]\{4\})$. In contrast, the incident I of four deer all heading north-east at the same time is formalised as concurrence pattern $P = I([45]\{4\})$. Investigating group dynamics in a herd of deer one might search for an individual initiating travel in a north-east direction before all other members of the herd. Such a trend-setter pattern P is shown in Figure 14.2D. Deer O_1 anticipates at time t_2 two time steps in advance the motion of all other deer at t_4.

$$
P = \begin{cases} S([45]\{3\}) : t_{e-2}, ..., t_e \\ I([45]\{4\}) : t_e \end{cases} \tag{14.1}
$$

Decomposing the analysis matrix into its rows (motion attribute arrays) and columns (time-slices) allows use of derivatives of classical string pattern matching algorithms.

14.3.2 Spatially constrained motion patterns

The motion patterns introduced so far have focused purely on properties describing the motion of moving point objects, explicitly excluding their absolute positions. Excluding absolute positions is a valid approach to reducing the complexity of the motion process. However, moving point objects do not manifest complex interrelations solely in their motion properties but also in changes of their arrangement in absolute space. A set of spatially constrained patterns extends the family of motion patterns, incorporating the (dynamic) arrangement of the moving point objects in absolute (geographic) space. The proposed spatially constrained patterns can describe, for example, flocking behaviour as well as convergence and divergence processes (Laube et al., 2004).

Proximity measures known from the field of spatial data handling are used to express proximity relations between moving point objects. For instance, a 'flock' pattern is built of a concurrence pattern by adding a spatial constraint (see Figure 14.3). The spatial constraint can be an enclosing circle, a bounding box or an ellipse. In other words, a flock moves in the same direction, at the same time *and* place.

To understand aggregation patterns both the relative and absolute positions of moving point objects must be considered. Consider as an example the motion process performed by a set of thirsty antelopes converging from all directions to a water hole in the savannah. All of a sudden the antelopes perceive some hungry crocodiles in the shallow waters and flee from the water hole in all possible directions. This episode in the lifelines of the antelopes clearly expresses the dynamic aggregation pattern of objects converging, but at the same time the involved antelopes never expressed a static spatial cluster during that episode and perhaps never will. Thus, the idea of a 'convergence' pattern is not to make a forecast for a subsequent cluster, but it is a motion pattern in its own right, an intrinsically spatiotemporal one. Conversely, moving point objects moving around in a cluster may never converge. For example, cars circling the *Arc de Triomphe* in Paris form a cluster, but while they are on the 'roundabout' they are not converging. Even though convergence and clustering are often spatially and or temporally related, there need not be a detectable relation in an individual data frame under investigation. Wildlife biologists may be interested in several aspects of such an aggregation pattern: Which individuals are converging? Which are not? When and where does the process start, when and where does it stop?

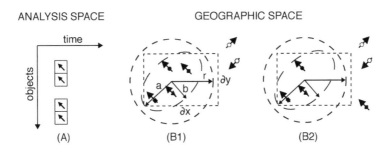

Figure 14.3. The spatially constrained motion pattern *flock*. The figure illustrates the constraints of the pattern flock in ANALYSIS SPACE (A, the analysis matrix) and in the GEOGRAPHIC SPACE (B). Fixes matched in the analysis space are represented as solid dots, fixes not matched as empty dots. Spatial constraints are represented as ranges with dashed lines. Whereas in situation (B1) the spatial constraint for the absolute positions of the fixes is fulfilled, it is not in situation (B2): The fourth object lies outside the range.

From an algorithmic perspective the convergence pattern identifies areas where many moving point objects appear to be converging, as estimated by extrapolated motion vectors (see Figure 14.4). A convergence pattern is found if the extrapolated motion azimuth vectors of a set of m moving point objects intersect within a range of radius r within a given temporal interval i. This pattern is intrinsically dynamic and exists uniquely neither in space nor in time, but only in a dynamic view of the world.

14.3.3 Evaluation of data mining approach

It has been recognised in the knowledge discovery literature that discovery systems can find a glut of patterns, many of which are of no interest to the user (Silberschatz and Tuzhilin, 1996; Padmanabhan, 2004). In the knowledge discovery approach introduced so far, the user has no means by which to estimate the 'interestingness' of the extracted patterns. Thus, as a first attempt to assess the interestingness of motion patterns, Laube and Purves (2005) propose comparing pattern occurrence in synthetic data based on random walk trajectories with pattern occurrence in observation data.

Silberschatz and Tuzhilin (1996) propose unexpectedness as a measure of interestingness of patterns. They argue that patterns are interesting because they contradict our expectations, given by our system of beliefs. The approach proposed in this chapter to capturing such beliefs is to generate synthetic lifelines using Monte Carlo simulations of random walks.

The concept of 'constrained random walk' (CRW) is used to simulate lifelines that have similar statistical properties to the observed data (Wentz et al., 2003). The constraints are given by frequency distributions of step length and turning angle derived from observation data (see Figure 14.1B). In a second step the number of patterns found in the synthetic data is compared with the number found in the observational data. The underlying assumption is that those patterns which appear to be outliers from the stochastic properties of the simulations are those which one can attach some initial interestingness to, prior to further investigation by the user (see Figure 14.5).

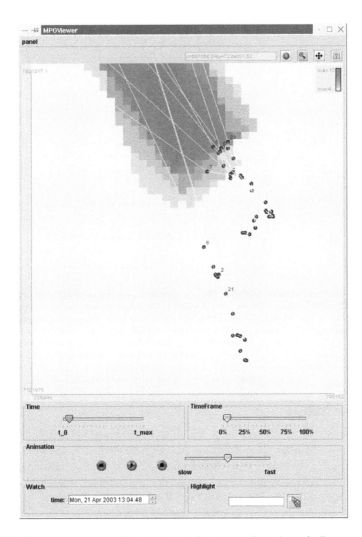

Figure 14.4. Convergence pattern. In the prototype implementation a dynamically computed grid highlights convergence areas (dark) where many extrapolated motion vectors (light rays) of migrating caribou intersect.

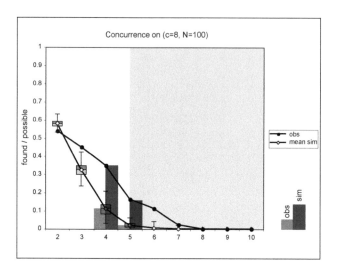

Figure 14.5. Pattern interestingness. Whisker plots help to assess the interestingness of found patterns. The plots compare the number of found patterns in the observation data (obs) with the number found in the simulated (sim). The *x* axis represents the extent of the pattern, in this case the number of objects building a concurrence pattern within migrating caribou (as described in the next section). The ratio on the *y* axis represents the number of patterns found compared to the number of patterns possible. When the ratio of the number of patterns observed is an order of magnitude different from the simulated data some qualitative notion of interestingness can be extracted. In this example this holds true for patterns with more than five moving point objects.

14.4 Case studies

A key test of the usefulness of a knowledge discovery system is its ability to identify known patterns. Therefore case studies from diverse fields were used to test and improve the concept. These case studies included animal tracking data, soccer scene analysis and a spatialisation application in political science (Laube et al., 2005; Laube and Purves, 2005). However, the following section only illustrates the knowledge discovery process in wildlife biology, using an example of investigating the migration patterns of a caribou herd.

The Porcupine Caribou Herd Satellite Collar Project is a cooperative project that uses satellite radio collars to document the seasonal range and migration patterns of the Porcupine Caribou Herd in northern Yukon, Alaska and Northwest Territories (Fancy et al., 1989; Fancy and Whitten, 1991; Griffith et al., 2002). The example given here focuses on a subgroup of the herd, consisting of ten individuals simultaneously tracked over almost two years, starting from March 2003 (Figure 14.6). The task is to check whether the known migration behaviour is expressed in motion patterns introduced earlier in this chapter.

analysis matrix

result frame

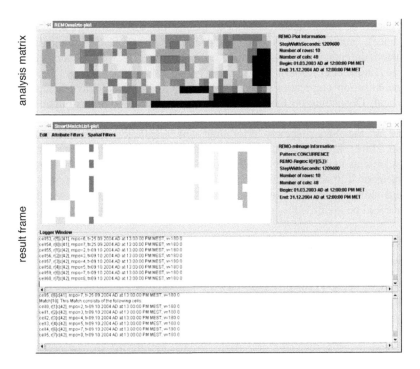

Figure 14.6. Porcupine Caribou case study. (Top) Analysis matrix for the motion azimuth of the caribou from Figure 14.1. Rows represent caribou individuals, columns represent time steps with an interval of two weeks. Eight azimuth classes are coded in greyscale, black refers to no data. (Bottom) Mined concurrence patterns of at least 5 individuals moving in the same direction. The found patterns correspond to northward spring migration and the southward autumn migration. The found motion patterns are highlighted in a results matrix and logged in the text frames below.

As shown in Figure 14.6 the knowledge discovery process can identify relatively distinct migration events in the motion azimuth, represented by the concurrence patterns of at least five individuals. The bars represented in different shades of grey represent instances of coordinated northward spring migration and southward autumn migration. The pattern extent of five caribou was chosen using the pattern evaluation experiments (see Figure 14.4), assigning this concurrence width a certain interestingness.

The spring migration in 2003 furthermore expressed a convergence pattern as illustrated in Figure 14.4. In April 2003 a large number of extrapolated motion vectors converge around a spot located around 68°N and 140°W. This area is the calving area of the Porcupine caribou herd near the Beaufort Sea and is seasonally visited, as can be seen in Figure 14.1A. Having mined this distinct convergence pattern allows the wildlife biologists, for instance, to indicate when this migration process starts.

Sport's scene analysis is another promising application field for spatiotemporal analysis (Iwase and Saito, 2003). In tracking data covering a short episode of a soccer game between Japanese student teams, an off-side trap situation (i.e. concurrence), straight strikes (i.e. constancy) and a player anticipating the coordinated defending backward move of the team (i.e. trend-setter) could be identified (Laube et al., 2005).

A spatialisation application illustrates changes in the ideological landscape of Switzerland. Depending on their voting behaviour, the 185 administrative districts of Switzerland are repeatedly located in an ideological space mapping value conflicts in the Swiss society (Hermann and Leuthold, 2003; Hermann and Leuthold, 2001). Irrespective of their political and social meaning, the districts can be considered as moving point objects, expressing a highly coordinated motion in a two-dimensional space. Concurrence detection could reproduce the well-known rightward shift of certain districts in the 1990s. Interestingly, some trend-setting districts could be identified as anticipating this process (Laube et al., 2005).

14.5 Mining motion patterns – a promising approach to analysing motion?

The following section critically examines the approach of mining motion patterns in order to analyse motion. It reports on a discrepancy between expectations, which are to some extent technology driven and dependent, and the actual availability of motion data. Then it discusses the issues in pattern detection, the relevance of mined patterns and investigates some granularity issues. Finally the section proposes the strict separation of static arrangements and dynamic aggregation processes.

14.5.1 Motion data – the crux with real data

The data mining approach introduced in this chapter has been applied to a variety of motion data. Although data mining is designed to make use of the nearly unlimited computing power of today's computers, and is thus especially suited for large numbers of moving point objects and very long tracking periods, the case studies illustrated in the previous section all consist of rather small numbers of objects, ranging from a few individuals to roughly 200 objects. This illustrates that the actual availability of individual tracking data lags behind what might be expected from the technological advances and, to some extent, the type of location-aware devices. Thus, it is still hard to find case study data showing these properties.

There are several reasons for the lack of tracking data for large groups of individuals. First, tracking large animals, such as caribou, over a long time period is expensive and laborious. Therefore, many animal tracking studies are limited to a relatively small number of tracked individuals. Furthermore, much available data about moving point object is very constrained by the underlying data model. Tracks of mobile phones, for example, do not disclose x,y coordinate observations, but only cell information (see also Chapter 11 on spatiotemporal accuracy in mobile phone location). Event-delimited data originating from moving object database applications (e.g. Sistla et al., 1998), such as a taxi cab fleet management systems, feature long static periods and rare updates and are thus not suited for repeatedly (inter-) relating the motion properties of moving point objects. Still another reservation must be made with objects that move on a network, for instance cars moving on a street network. Investigating their motion in space-time potentially may perhaps reveal more about the structure of the traffic network and little about the behaviour of the car drivers.

Furthermore, geoprivacy issues may limit data availability. The main concern with respect to spatiotemporal GIS is its potential for a rapid integration of spatial information and personal information (Dobson and Fisher, 2003). Ethical concerns and objections may additionally constrain the availability of geo-spatial lifeline data in the years ahead. Hence, data mining approaches for motion data could share for some time the problems of the space-time aquarium, remaining an elegant and promising concept, yet suffering from a lack of true applications.

However, the popularity of the object-oriented paradigm and the related proliferation of agent-based simulation approaches in Geographical Information Science increase the availability of artificial motion data (Brown et al., 2005; Pfoser and Theodoridis, 2003). As has been seen in the evaluation section, the great advantage of artificial data is its total controllability. Every dimension of artificial data can be produced at arbitrary granularities. Artificial life forms are always visible, healthy, don't die, don't get shot, don't lose their GPS receiver, don't need privacy, and are willing to report their location at any desired time. Thus, techniques to simulate moving point behaviour efficiently from small numbers of tracked individuals are one potential approach to reducing some of the challenges listed above.

14.5.2 Motion patterns – the danger of simplicity

The motion pattern mining introduced in this chapter follows the hierarchical approach of syntactic pattern recognition, providing a set of simple primitives to compose complex patterns. This strategy allows the composition of arbitrary motion patterns satisfying the yet unforeseen needs of potential users. The sequence of defining patterns, their subsequent detection in an analysis matrix and the final quantification of the data mining results is applicable to a wide range of motion phenomena.

The key advantage of having such a simple and user-friendly approach may be at the same time its most dangerous weakness. Motion patterns such as *concurrence* or *trend-setter* are descriptive, based on well-known everyday motion events, and thus easy to learn and apply for potential users – and sometimes easy to find. It appears very obvious that during the migration period in the caribou example a concurrence pattern can be found. However, in common with every other knowledge discovery approach, motion patterns can be mined that appear significant but, in fact, are not (Silberschatz and Tuzhilin, 1996; Laube and Purves, 2005). Thus, with its user-friendly simplicity such an approach may encourage careless use. To overcome this dilemma, one must first include expert knowledge to define and mine reasonable patterns and, second, have a strategy to assess the meaning and relevance of the mined patterns.

Mining predefined motion patterns is an example of the data-mining task of 'retrieval by content'. Thus, a user has a pattern of interest in mind and wishes to find similar patterns in the data (Hand et al., 2001). Potential users, scientific experts in fields such as biology, geography or sociology, must compose motion patterns that potentially lurk in geo-spatial lifelines. In other words, they must have an idea of what they expect to find. Investigating the Porcupine caribou data, biologists would, for instance, expect to find patterns of seasonal migration, a well-known phenomenon with Porcupine caribou. In most scientific projects such knowledge is available.

Nonetheless, the dependency on expert knowledge may make analysis subjectively dependent on the experts' skills. Hence, a careful examination of the statistical background of the process under study may help to establish a more objective means to estimate the meaning of mined motion patterns. 'Data mining is a legitimate activity as long as one understands how to do it correctly, data mining carried out poorly (without regard to the statistical aspects of the problem) is to be avoided' (Fayyad et al., 1996, p. 40). Our experience has been that seemingly significant patterns may in fact be statistically unsurprising.

This chapter reported on a first attempt to estimate the *interestingness* of mined patterns through their 'unexpectedness'. The use of Monte Carlo simulated constrained random walks to learn more about expected pattern occurrence is a straightforward yet appropriate choice. Including step length and direction change angle distributions (see Figure 14.1B) this approach is based on two features describing trajectories that are considered crucial by behavioural ecologists (Turchin, 1998). The method can furthermore be used to examine useful configurations of pattern matching sessions (i.e. attribute granularities, pattern lengths), which may not be obvious. Therefore it gives the data mining a more objective dimension, since the composition of the searched patterns is not solely dependent on the (potentially biased) knowledge of the expert user.

However, constrained random walks do not include 'in path auto-correlation' that may be a characteristic for certain motion processes. Certain animals, for example, may always turn east after heading north, or may express whole sequences with only very short or only particularly long steps. Such patterns could be found in the tracks of animals performing seasonal migrations, showing trajectories expressing in path auto-correlation with respect to migratory versus sedentary intervals (Bergman et al., 2000). As an alternative to the constrained random walk so-called 'transition matrices' of 'Markov Chains' could be used, that explicitly allow consideration of in-path auto-correlation in the form of direction change matrices (Jonsen et al., 2003; Jones and Smith, 2001).

14.5.3 Granularity issues – the burden of discretisation

As Erwig et al. (1999, p. 281) state very concisely, 'abstract models are simple, but only discrete models can be implemented.' Abstract models allow one to make definitions of spatiotemporal entities in terms of infinite sets, without worrying whether finite representations of these exist and what the implications of such representations are. It is thus very simple and straightforward to view a moving point as a continuous curve in space-time. But when it comes to implementation only finite or reasonably small sets are usually stored and manipulated in computers. Hence, the proposed conceptualisation of motion patterns is based on a discretisation of the geo-spatial lifelines; that is a set of fixes, connected by straight segments. However fine the temporal granularity may be on the temporal axis of the analysis matrix (Figure 14.2D), discretised time lines give rise to a number of granularity issues which need to be carefully addressed.

The discretisation of the temporal axis can be considered as analogous to the modifiable areal unit problem (MAUP) (Openshaw, 1984). Whereas the classical MAUP involves spatial aggregations, discretising lifelines different temporal aggregations are considered. Arbitrary aggregations may emerge from the mapping of potentially irregularly sampled fixes on to the regular analysis matrix, determined by an analysis granularity and a starting time. If the temporal units were differently specified, one might observe very different patterns and relationships. With respect to the analysis of discretised motion patterns on discretised geo-spatial lifelines it might make sense to talk about a 'modifiable temporal unit problem' or MTUP.

A further critical issue is the interplay of sampling and analysis granularity. Undersampling a lifeline causes information loss, oversampling may drown out the track's signal and may even feign auto-correlation between successive moves (Laube and Purves, 2005). In order to avoid semantic mismatches the knowledge discovery process must be performed at an adequate granularity. In other words, the phenomenon under investigation should further suggest suitable analysis granularities. Searching for seasonal migration patterns in the Porcupine caribou case study one should not choose an analysis granularity of hours, introducing noise caused by daily movement patterns irrelevant to the research question. The fix-sampling rate should define the finest possible analysis granularity. Nevertheless we should carefully examine the data as oversampling might have already occurred at the data collection process.

The numerical experiments performed in Laube and Purves (2005) showed that different aspects of granularity influence the results of the pattern detection process. Firstly, the granularity of the attribute classification influences the number of found patterns. So for example, the finer the attribute classification (e.g. eight classes with $45°$ motion azimuth intervals instead of four classes with $90°$ azimuth intervals), the less patterns are mined. But if a pattern can be found within a fine granularity, it is often the case that this pattern is statistically unusual, and thus interesting. Secondly, the pattern extent p (pattern length with patterns over time, pattern width for patterns across objects) influences the number of found patterns. So for example, the smaller the extent p of a pattern, the more likely it is to appear by random chance.

14.5.4 Integrating dynamic patterns in space and time

Traditional geographic analysis investigates heterogeneity in static space alone, trying to discover associations, aggregations and topological relationships between spatial entities (Golledge, 2002; Miller and Wentz, 2003; O'Sullivan and Unwin, 2003). On the other hand, time series or trend analysis methods adopted, for instance, in the field of ecology, are developed to find patterns in time alone and thus neglect the spatial dimension (e.g. Bjørnstad and Grenfell, 2001). However, when it comes to understanding motion phenomena, both dimensions must be integrated (Massey, 1999).

Whereas patterns such as constancy, concurrence and trend-setter involve only the motion properties of moving point objects relative to each other, the flock pattern clearly illustrates that some motion patterns must also involve the absolute locations of the moving point objects. A group of caribou only forms a herd (flock) if they move in spatial proximity to each other.

Furthermore aggregation processes such as convergence are intrinsically dynamic, that is spatiotemporal in nature, and cannot exist in either space or time alone. It is here therefore proposed to strictly separate the process of convergence from the final static cluster as its optional outcome (Laube et al., 2004). Methods designed to identify the static outcome of a convergence (such as cluster detection algorithms) are not suited to identify the dynamic process of convergence. In contrast, concepts such as the proposed aggregation pattern convergence, built from a set of motion extrapolation vectors, truly integrate space and time and are better suited for motion analysis. Thus, knowledge about the process of convergence is inherently dynamic, and cannot be extracted from static point distributions or static lifeline maps such as Figure 14.1A.

14.6 Conclusion and future developments

The concluding section first collects a set of insights gained from applying knowledge discovery and data mining techniques to the analysis of geo-spatial lifelines before anticipating possible directions for future research in the analysis of point motion data.

14.6.1 Conclusions

In recent years Geographical Information Science has moved from a data-poor and computation-poor period to a data-rich and computation-rich period (Miller and Han, 2001). Technological advances can be expected to increase this production of individualised trajectory data by orders of magnitude. Telecommunication services or customer loyalty card systems already automatically produce amounts of data that push the analysts' capabilities beyond their limits. Compared to rather visually oriented analysis techniques, such as exploratory spatial data analysis, mining motion patterns, as a form of geographic knowledge discovery, exhibits a high potential to cope with the emergent large volumes of tracking data. Nonetheless, privacy issues must be addressed before large-scale analysis of such data are possible, in order to overcome the well-grounded fears addressed by Dobson and Fisher (2003).

This chapter introduced a geographic knowledge discovery approach to mining motion patterns of geo-spatial lifelines. Its implementation and successful application in various case studies illustrated the potential of mining motion patterns. The following list illustrates how the objectives framed earlier in this chapter have been achieved:

❑ Mining motion patterns allows the integration of space and time, and hence analysis of the dynamic aspects of motion. Integration of knowledge discovery techniques with Geographical Information Science is an appropriate and powerful means to move beyond the snapshot with respect to motion analysis.

❑ Adopting a syntactic pattern detection approach and providing a pattern description formalism in principle allows potential users to compose arbitrarily complex patterns from simple motion pattern primitives. Such data mining is applicable to many types of motion data, as was illustrated in this chapter by three case studies, and is extensible.

❑ Expert knowledge of the process under investigation is needed to mine reasonable patterns, but mined motion patterns must be further evaluated through investigations of the statistical background of the process under study.

❑ Where real observation motion data are lacking or suffer from poor quality, carefully synthesised artificial motion data offer a feasible alternative to studying some processes and in particular to experimenting with data-mining approaches.

However, only the widespread application of geographic knowledge discovery *in practice* can prove its usability. Only as more applications are developed which address the special challenges laid down by the *dynamic* nature of spatiotemporal motion data will we be able to say that the integration of Geographical Information Science and data mining shows why *spatial* (and of course spatiotemporal) is *special* with respect to knowledge discovery.

14.6.2 Future developments

As shown in Section 14.2 several research groups have started to tackle in recent years the problem of movement analysis to broaden spatiotemporal Geographical Information Science. Despite recognition of the problem, major obstacles still prevent universal solutions. The concepts presented above have the potential to be extended in many ways.

Tracking data are in many cases not perfect. Lifeline data, which emerge from biological field research, suffer especially from uncertain or incomplete trajectories due to tracking system failures. Some work has already been done to handle uncertain (Moreira et al., 1999; Pfoser and Jensen, 1999; Trajcevski et al., 2004) and incomplete (Wentz et al., 2003) tracking data. However, the influence of imperfect tracking data on the results of geographic knowledge discovery and geographic data mining remains an open research issue.

The motion patterns introduced in this chapter are based on crisp concepts: they require a crisp attribute classification and the computation of derived motion properties happens at crisp times. One could argue that such 'deterministic' patterns mismatch with the rather smooth and continuous motion processes seen in many application fields. Consider for instance the following sequence of motion azimuth values:

$$P = S\ (45,45,45,45,60,45,45)$$

Should one now consider this rather strong persistence for one motion azimuth as being a constancy pattern? The influence of fuzzy patterns on the process of geographic knowledge discovery has not yet been investigated in detail by the authors.

Furthermore, the patterns introduced have focused on the purely geometric aspects of geo-spatial lifelines. This approach explicitly excluded the semantics of the investigated phenomenon; that is it did not consider any attribute information about the moving entities and the circumstances and environment they were moving in. It is argued in this chapter that not considering the semantics was a valid approach for investigating motion with a geometric focus on agents moving through space. However, real-life motion phenomena are usually linked to the host geography and to activity states of the individual, and do not disclose these complex interrelations in the geometry of their lifelines alone. Understanding and potentially predicting the motion of people, animals or other agents requires the integration of the geometric properties of their motion with semantic information describing the moving entities as well as the environment harbouring the motion. For example, a social scientist analysing peoples' motion will likely see value in investigating their cultural background, their socio-economic status, their purpose of travel or their means of transport. Wildlife biologists may want to incorporate sex, age or the physiology of moving animals. Furthermore, any assumption of agents moving in a featureless, homogeneous space does not hold for the complex motion of intelligent agents. Much greater insight can be gained by working with (x, y, t, a) data (Forer, 2002), where a stands for the attributes of the objects involved. Future research must thus develop conceptual approaches and analytical tools that explore the geometric *and* the semantic properties of motion.

Motion patterns do not need to involve only motion properties at a single time, either speed, change of speed, motion azimuth or sinuosity. In contrast, some motion processes may express characteristics in more than just one motion dimension. A foraging behaviour may, for instance, be characterised by a low speed and a high sinuosity at the same time. Similarly, investigating seasonal migration one might expect high-speed values only in two directions linking the summer and the winter habitat of a species. Multidimensional motion patterns are another promising direction future research could take.

The motion patterns introduced in this chapter are mined in a featureless and homogeneous space, not constraining or affecting the motion of the objects. However, for many application fields, space is heterogeneous, strongly influencing the motion process. Think, for example, of the insurmountably steep walls of a valley, strongly influencing the motion azimuth of migrating caribou. Or consider as another example individual-based models in behavioural ecology investigating habitat preferences. With the 'radial distance functions' Imfeld (2000) proposed an exploratory analysis approach to analyse motion with respect to its environment. One obvious option for future research is to move on in this direction and to identify and characterise a set of motion patterns emerging from motion in heterogeneous space.

Furthermore it would be interesting to investigate the influence of interactions between moving point objects. This may relate to interactions between individuals in a group as well as interactions between groups of objects. An example of the former is the influence of the hierarchical order in a group on motion-relevant behaviour (e.g. trend-setter behaviour). Competitive pressure between animal groups of different species may serve as an example for the latter case and may be expected to have an influence on motion patterns. The 'geographic ecological modelling systems' (GEMS) introduced by Westervelt and Hopkins (1999), allowing individuals to engage in interactions such as predator-prey, mating and symbiosis, may give some hints on future research direction with respect to interaction.

Finally, the proposed motion patterns assume that the moving point objects move without constraints in a featureless space. In many cases this is an oversimplification of reality, knowing full well that for instance animals actually frequently follow corridors or migrate in valleys rather than on ridges. However, such motion of objects through a landscape requires different forms of analysis to be explored.

Acknowledgements

The authors wish to thank the Porcupine Caribou Technical Committee, the Porcupine Caribou Management Board and the Wildlife Management Advisory Council (North Slope) for providing the excellent caribou data.

References

Andrienko, N.V., Andrienko, G.L. and Gatalsky, P. (2000) 'Towards exploratory visualization of spatio-temporal data', *3rd AGILE Conference on Geographic Information Science*, Helsinki Espoo, pp. 137–142.

Anselin, L. (1998) 'Exploratory spatial data analysis in a geocomputational environment', in Longley, P. A., Brooks, S. M., McDonnell, R. and Macmillan, B. (eds.) *Geocomputation: a Primer*, pp. 77–94, New York: John Wiley & Sons.

Aronoff, S. (1989) *Geographic Information Systems: A Management Perspective*. Ottawa: WDL Publications.

Berger, U., Wagner, G. and Wolff, W.F. (1999) 'Virtual biologists observe virtual grasshoppers: an assessment of different mobility parameters for the analysis of movement patterns', *Ecological Modelling*, vol. 115, no. 1–2, pp. 119–129.

Bergman, C. M., Schaefer, J. A. and Luttich, S. (2000) 'Caribou movement as a correlated random walk', *Oecologia*, vol. 123, no. 3, pp. 364–374.

Bjørnstad, O. N. and Grenfell, B. T. (2001) 'Noisy clockwork: Time series analysis of population fluctuations in animals', *Science*, vol. 293, pp. 638–643.

Brillinger, D. R., Preisler, H. K., Ager, A. A. and Kie, J. G. (2004) 'An exploratory data analysis (EDA) of the paths of moving animals', *Journal of Statistical Planning and Inference*, vol. 122, no. 2, pp. 43–63.

Brown, D. G., Riolo, R., Robinson, D. T. North, M. and Rand, W. (2005) 'Spatial process and data models: Towards integration of agent-based models and GIS', *Journal of Geographical Systems*, vol. 7, pp. 25–47.

Chrisman, N. R. (1998) 'Beyond the snapshot: Changing the approach to change, error, and process', in Egenhofer, M. and Golledge, R. (eds.) *Spatial and Temporal Reasoning in Geographic Information Systems*, pp. 85–93, Oxford: Oxford University Press.

Dobson, J. E. and Fisher, P. F. (2003) 'Geoslavery', *IEEE Technology and Society Magazine*, vol. 22, no. 1, pp. 47–52.

Dykes, J. A. and Mountain, D. M. (2003) 'Seeking structure in records of spatio-temporal behaviour: Visualization issues, efforts and application', *Computational Statistics and Data Analysis*, vol. 43, no. 4, pp. 581–603.

Erwig, M., Güting, R. H., Schneider, M. and Vazirgiannis, M. (1999) 'Spatio-temporal data types: An approach to modeling and querying moving objects in databases', *GeoInformatica*, vol. 3, no. 3, pp. 269–296.

Fancy, S. G. and Whitten, K. R. (1991) 'Selection of calving sites by Porcupine herd caribou', *Canadian Journal of Zoology*, vol. 69, pp. 1736–1743.

Fancy, S. G., Pank, L. F., Whitten, K. R. and Regelin, W. L. (1989). 'Seasonal movement of caribou in arctic Alaska as determined by satellite', *Canadian Journal of Zoology*, vol. 67, pp. 644–650.

Fayyad, U., Piatetsky-Shapiro, G. and Smyth, P. (1996) 'From data mining to knowledge discovery in databases', *AI Magazine*, vol. 17, no. 3, pp. 37–54.

Forer, P. (1998) 'Geometric approaches to the nexus of time, space, and microprocess: Implementing a practical model for mundane socio-spatial systems', in Egenhofer, M. J. and Colledge, R. G. (eds.) *Spatial and Temporal Reasoning in Geographic Information Systems*, pp. 171–190, Oxford: Oxford University Press.

Forer, P. (2002) 'Timelines, environments, and issues of risk in health: The practical algebra of x, y, t, a', in Briggs, D., Forer, P. C., Järup, L. and Stern, R. (eds.) *GIS for Emergency Preparedness and Health Risk Reduction,* pp. 35–60, Berlin Heidelberg: Springer.

Forer, P., Chen, H. F. and Zhao, J. F. (2004) 'Building, unpacking and visualising human flows with GIS', *Proceedings of the GIS Research UK 12th Annual Conference,* pp. 334–336, Norwich: University of East Anglia.

Frihida, A., Marceau, D. J. and Thériault, M. (2004) 'Extracting and visualizing individual space-time paths: An integration of GIS and KDD in transport demand modelling', *Cartography and Geographic Information Science*, vol. 31, no. 1, pp. 19–29.

Golledge, R. G. (2002) 'The nature of geographic knowledge', *Annals of the Association of American Geographers*, vol. 92, no. 1, pp. 1–14.

Griffith, B., Douglas, D. C., Walsh, N. E., Yound, D. D., McCabe, T. R., Russell, D. E., White, R. G., Cameron, R. D. and Whitten, K. R. (2002). 'The Porcupine caribou herd', in Douglas, D. C., Reynolds, P. E. and Rhode, E. B. (eds.) *Arctic Refuge Coastal Plain Terrestrial Wildlife Research Summaries*, pp. 8–37, U. S. Geological Survey, Biological Resources Division, Biological Science Report USGS/BRD/BSR-2002-0001.

Grumbach, S., Koubarakis, M., Rigaux, P., Scholl, M. and Skiadopoulos, S. (2003) 'Spatio-temporal models and languages: An approach based on constraints', in Koubarakis, M., Sellis, T., Frank, A. U., Grumbach, S., Güting, R. H., Jensen, C. S., Lorentzos, N., Manolopoulos, Y., Nardelli, E., Pernici, B., Schek, H.-J., Scholl, M., Theodoulidis, B. and Tryfona, N. (eds.) *Spatio-Temporal Databases: The CHOROCHRONOS Approach*, Lecture Notes in Computer Science, vol. 2520, pp.177–201, Berlin Heidelberg: Springer.

Güting, R. H., Boehlen, M. H., Erwig, M., Jensen, C. S., Lorentzos, N., Nardelli, E., Schneider M. and Viqueira, J. R. R. (2003) 'Spatio-temporal models and languages: An approach based on data types', in Koubarakis, M., Sellis, T., Frank, A. U., Grumbach, S., Güting, R. H., Jensen, C. S., Lorentzos, N., Manolopoulos, Y., Nardelli, E., Pernici, B., Schek, H.-J., Scholl, M., Theodoulidis, B. and Tryfona, N. (eds.) *Spatio-Temporal Databases: The CHOROCHRONOS Approach*, Lecture Notes in Computer Science, vol. 2520, pp.117–176, Berlin Heidelberg: Springer.

Hägerstrand, T. (1970) 'What about people in regional science', *Papers of the Regional Science Association*, vol. 24, pp. 7–21.

Hand, D., Manilla, H. and Smyth, P. (2001) *Principles of Data Mining,* Cambridge: MIT Press.

Hermann, M. and Leuthold, H. (2001) 'Mentalities and their social basis in the mirror of Swiss popular referendums', in German, *Swiss Political Science Review*, vol. 7, no. 4, pp. 39–63.

Hermann, M. and Leuthold, H. (2003) *Atlas of the Political Landscape of Switzerland*, in German, Zurich: vdf-Verlag.

Hill, N. and Häder, D.-P. (1997) 'A biased random walk model for the trajectories of swimming micro-organisms', *Journal of Theoretical Biology*, vol. 186, no. 4, pp. 503–526.

Hornsby, K. and Egenhofer, M. J. (2002) 'Modeling moving objects over multiple granularities', *Annals of Mathematics and Artificial Intelligence*, vol. 36, no. 1–2, pp. 177–194.

Huisman, O. and Forer, P. (1998), 'Computational agents and urban life spaces: a preliminary realisation of the time-geography of student lifestyles', *Proceedings of the 3rd International Conference of GeoComputation* [Electronic], GeoComputation CD-ROM.

Imfeld, S. (2000) *Time, Points and Space: Towards a better analysis of wildlife data in GIS*, unpublished PhD thesis, Zurich: University of Zurich.

Iwase, S. and Saito, H. (2003) 'Tracking soccer players based on homography among multiple views', in Ebrahimi, T and Sikora, T. (eds.) *Visual Communications and Image Processing, Proceedings of SPIE*, vol. 5150, pp. 288–292.

Jain, A. K., Duin, R. P. W. and Mao, J. (2000) 'Statistical pattern recognition: A review', *IEEE Transactions on Pattern Recognition and Machine Intelligence*, vol. 22, no. 1, pp. 4–37.

Jones, P. W. and Smith, P. (2001) *Stochastic Processes: An Introduction*, London: Arnold Publishers.

Jonsen, I. D., Myers, R. A. and Flemming, J. M. (2003) 'Meta analysis of animal movement using state-space models', *Ecology*, vol. 85, no. 11, pp. 3055–3063.

Kraak, M.-J. and Koussoulakou, A. (2004) 'A visualization environment for the space-time-cube', in Fisher, P. F. (ed.) *Developments in Spatial Data Handling, Proceedings of the 11th International Symposium on Spatial Data Handling*, pp. 189–200, Berlin Heidelberg: Springer.

Kwan, M.-P. (2000) 'Interactive geovisualization of activity-travel patterns using three dimensional geographical information systems: a methodological exploration with a large data set', *Transportation Research Part C*, vol. 8, no. 1–6, pp. 185–203.

Laube, P. and Imfeld, S. (2002) 'Analyzing relative motion within groups of trackable moving point objects', in Egenhofer, M. J. and Mark, D. M. (eds.) *Geographic Information Science, Second International Conference, GIScience 2002*, Lecture Notes in Computer Science, vol. 2478, pp. 132–144, Berlin Heidelberg: Springer.

Laube, P. and Purves, R. S. (2005) 'Evaluation of a geographic knowledge discovery approach using random walk models', in Billen, R., Drummond, J., Forrest, D. and João, E. (eds.) *Proceedings of the GIS Research UK: GISRUK 2005*, pp. 130–135, Glasgow: University of Glasgow.

Laube, P., Imfeld, S. and Weibel, R. (2005) 'Discovering relative motion patterns in groups of moving point objects', *International Journal of Geographical Information Science*, vol. 19, no. 6, pp. 639–668.

Laube, P., Van Kreveld, M. and Imfeld, S. (2004) 'Finding REMO - detecting relative motion patterns in geospatial lifelines', in Fisher, P. F. (ed.) *Developments in Spatial Data Handling, Proceedings of the 11th International Symposium on Spatial Data Handling*, pp. 201–214, Berlin Heidelberg: Springer.

Mark, D. M. (1998) 'Geospatial lifelines', *Integrating Spatial and Temporal Databases, Dagstuhl Seminars*, no. 98471.

Massey, D. (1999) 'Space-time, "science" and the relationship between physical geography and human geography', *Transactions of the Institute of British Geographers*, vol. 24, no. 3, pp. 261–276.

Miller, H. J. and Han, J. (2001) 'Geographic data mining and knowledge discovery: An overview', in Miller, H. J. and Han, J. (ed.) *Geographic Data Mining and Knowledge Discovery*, pp. 3–32, London: Taylor and Francis.

Miller, H. J. and Wentz, E. A. (2003) 'Representation and spatial analysis in geographic information systems', *Annals of the Association of American Geographers*, vol. 93, no. 3, pp. 574–594.

Moreira, J., Ribeiro, C. and Saglio, J.-M. (1999) 'Representing and manipulation of moving points: An extended data model for location estimation', *Cartography and Geographic Information Science*, vol. 26, no. 2, pp. 109–123.

O'Sullivan, D. and Unwin, D. J. (2003) *Geographic Information Analysis*, Hoboken: John Wiley & Sons.

Openshaw, S. (1984) *The Modifiable Areal Unit Problem*, Norwich: Geo Books.

Padmanabhan, B. (2004) 'The interestingness paradox in pattern discovery', *Journal of Applied Statistics*, vol. 31, no. 8, pp. 1019–1035.

Pfoser, D. and Jensen, C. S. (1999) 'Capturing the uncertainty of moving-object representations', in Güting, R. H., Papadias, D. and Lochowsky, F. (eds.) *Advances in Spatial Databases, 6th International Symposium, SSD'99*, Lecture Notes in Computer Science, vol. 1651, pp. 111–131, Berlin Heidelberg: Springer

Pfoser, D. and Theodoridis, Y. (2003) 'Generating semantics-based trajectories of moving objects', *Computers, Environment and Urban Systems*, vol. 27, no. 3, pp. 243–263.

Ramos-Fernandez, G., Mateos, J. L., Miramontes, O., Cocho, G., Larralde, H. and Ayala-Orozco, B. (2004) 'Lévy walk patterns in the foraging movement of spider monkeys (*Ateles geoffroyi*)', *Behavioral Ecology and Sociobiology*, vol. 55, no. 3, pp. 223–230.

Schmitt, F. G. and Seuront, L. (2001) 'Multifractal random walk in copepod behaviour', *Physica A*, vol. 301, no. 1–4, pp. 375–396.

Silberschatz, A. and Tuzhilin, A. (1996) 'What makes patterns interesting in knowledge discovery systems', *IEEE Transactions on Knowledge and Data Engineering*, vol. 8, no. 6, pp. 970–974.

Sistla, A. P., Wolfson, O., Chamberlain, S. and Dao, S. (1998) 'Querying the uncertain position of moving objects', in Etzion, O., Jajodia, S. and Sripada, S. (eds.) *Temporal Databases – Research and Practice*, Lecture Notes in Computer Science, vol. 1399, pp. 310–337, Berlin Heidelberg: Springer.

Smyth, C. S. (2001) 'Mining mobile trajectories', in Miller, H. J. and Han, J. (eds.) *Geographic Data Mining and Knowledge Discovery*, pp. 337–361, London: Taylor and Francis.

Trajcevski, G., Wolfson, O., Hinrichs, K. and Chamberlain, S. (2004) 'Managing uncertainty in moving objects databases', *ACM Transactions on Database Systems (TODS)*, vol. 29, no. 3, pp. 463–507.

Turchin, P. (1998) *Quantitative Analysis of Movement: Measuring and Modelling Population Redistribution in Animals and Plants*, Sunderland, Massachusetts: Sinauer Publishers.

Wentz, E. A., Campell, A. F. and Houston, R. (2003) 'A comparison of two methods to create tracks of moving objects: Linear weighted distance and constrained random walk', *International Journal of Geographical Information Science*, vol. 17, no. 7, pp. 623–645.

Westervelt, J. D. and Hopkins, L. D. (1999) 'Modeling mobile individuals in dynamic landscapes', *International Journal of Geographical Information Science*, vol. 13, no. 3, pp. 191–208.

Part V

Epilogue

Chapter 15

Current and Future Trends in Dynamic and Mobile GIS

Jane Drummond [1], Elsa João [2] and Roland Billen [3]

[1]Department of Geographical and Earth Sciences, University of Glasgow, UK
[2]Graduate School of Environmental Studies, University of Strathclyde, UK
[3]Department of Geography, University of Liège, Belgium

The *terms* 'dynamic GIS' and 'mobile GIS' have been around for some time. For example, back in 1990 Perez-Trejo suggested that a *dynamic GIS* could help analyse the impacts of climatic change on complex ecosystems. According to the author, climatic changes cannot be assessed by studying one aspect of the system alone, but a dynamic GIS might contribute to the understanding of the dynamic interactions of physical and ecological subsystems within an integrated framework (Perez-Trejo, 1990). Or in 1995, when Olsen described the use of an enhanced version of the Highways Works Order Costing System (HiWOCS) by the UK Gloucester City Council's highways department; the system was integrated into a pen-based, *mobile GIS* for the management of roads, paving, etc. The mobile GIS allowed the location of faults on-site and was linked directly into the council's financial system. Additional elements allowed cyclic inspections providing a link from initial fault detection and an issued works order through to final inspection (Olsen, 1995). However, despite this early start, current research is generating particularly exciting results both in terms of dynamic and mobile GIS as can be seen in the different chapters of this book. This last chapter aims to summarise the main key findings, recent advances and opportunities (Section 15.1) and identify key problems, threats or constraints (Section 15.2). The chapter concludes with suggestions for future research (Section 15.3) and recommendations for future practice (Section 15.4).

15.1 Key findings, recent advances and opportunities

15.1.1 Dynamic processes

The real world is dynamic. Consequently, it should be self-evident that characterising and simulating real-world processes implies modelling their dynamic nature. To date, GIS have provided useful tools for investigating spatial patterns but have suffered from an inability to explore the dynamic aspects of geographic phenomena. Therefore, new models dealing with these dynamic aspects are needed. This implies a dramatic evolution in GI systems: the mixing of space and time. One

Dynamic and Mobile GIS: Investigating Changes in Space and Time. Edited by Jane Drummond, Roland Billen, Elsa João and David Forrest. © 2006 Taylor & Francis

has to move from static geographic feature (object) representations inherited from traditional cartography to new space-time representations addressing the very nature of change. The result should be a new generation of GIS tools incorporating multi-dimensional space-time modelling as proposed by Maguire (Chapter 1), and leading to so-called spatiotemporal information systems (STIS).

Recent advances in the field consider occurrent entities, such as *events* (Beard, Chapter 4) and *processes* (Reitsma and Albrecht, Chapter 5), instead of static objects. Events with associated attributes of change such as rate of change or rate constancy provide key units for the exploration and analysis of mechanisms of change. Furthermore, events provide a basis for the integration of information from heterogeneous spatiotemporal data streams. Such streams are currently quite challenging to integrate, due to the diversity of spatial and temporal regimes that one can expect to encounter. In a process-based simulation, information about a whole process is represented—not only its state at a precise moment of time, which has been the case, to date, for most existing models dealing with dynamic phenomena. This represents a real improvement but is still at an early stage of development.

As we can see, current research and advances in dynamic modelling are based on a redefinition of core entities. However, other aspects should be taken into account when thinking about dynamic processes. The management of spatial constraints through time is one of them (Oosterom, Chapter 7). This shows the complexity of handling space and time in a coherent way; a constraint, for example, can be true at time t and false at time t+1.

Considering research and business opportunities is both straightforward and challenging. The research potential is tremendous. The applications' potential almost infinite. However, commercial GIS are currently far removed from functioning as STIS and modelling dynamic processes is still in its research infancy.

15.1.2 Mobility

Mobility is unquestionably a fundamental aspect of contemporary life. This has been recognised for some time. For example, as quoted by Mateos and Fisher in Chapter 11, 20 years ago Prato and Trivero (1985) suggested that mobility was the primary activity of contemporary societies. What is particularly relevant to Geographical Information Science is that those movements (e.g. of people) are increasingly leaving 'digital trails' that can be tracked, collected in large databases and then analysed. In the past, wearable tracking devices to collect motion data were mainly used by small populations under study. This was usually for ecological studies, for example for tracking endangered species like the Amur leopard in Siberia. However, nowadays most people (some unknowingly) wear tracking devices in the form of mobile phones; thus greatly increasing the volume of tracking data (see Chapter 11).

Laube et al., in Chapter 14, consider that Geographical Information Science can contribute to finding out about patterns made by individuals and groups while at the same time coping with the large volume of tracking data. For this reason, the authors argue that the study of *motion* (i.e. exploring the dynamic processes of such digital trails) is an emerging research area in Geographical Information Science. Laube et al. in their chapter advocate *quantitatively* analysing motion, as opposed to just *visualising* motion. Laube et al. argue that one effective way to analyse motion quantitatively is through a geographic knowledge discovery technique called 'mining motion patterns' that allows the integration of space and time

A major technological development relevant to motion, and a key tool for a mobile society, is the advent of mobile GIS and other mobile devices such as cellular phones. The next section evaluates key findings, recent advances and opportunities related to mobile devices such as mobile phones and mobile GIS.

15.1.3 Mobile Devices

Developments currently underway in mobile technology will inevitably increase the automated gathering of individual route data. Loyalty cards, cash cards and other ID cards can automatically add attributes to these location data. Projected data volumes are even predicted, by Laube et al. in Chapter 14, to outstrip GIS analytical capabilities in the near future. But ignoring this gloomy prognosis, we have, through mobile phones, a technology representing a wearable computing device accepted by about 80% of the adult population. In terms of a location system, mobile phone technology is cheaper, more acceptable and functioning more effectively within buildings and in urban canyons, than GPS. Through the analysis of each phone's 'spatiotemporal' signature the mobility patterns of large groups of people can be characterised and analysed, to form 'New Cellular Geographies' which will allow data sets from different 'timespaces' to be linked, according to Mateos and Fisher, in Chapter 11.

Because mobile GIS, through its portability, usability and flexibility extends the functionality of GIS it will greatly strengthen disaster management (see Chapter 12), other GIS applications that benefit from rapid data gathering and data gathering where communal discussion of issues, such as in participatory GIS (see Chapter 13), is beneficial. The extension of participatory GIS into developing societies has been hampered by the expense of the hardware. But mobile GIS may provide an achievable entry level.

Of course, there are problems associated with mobile devices. In Part III of this book those associated with visualisation have been raised. The small low-resolution screens offer quite a challenge to good visualisation, obliging us to think about what the user really needs to be able to see.

However, regardless of developments within the GI sector itself, the explosive evolution of mobile devices does mean that opportunities to extend the sphere of GI's influence are likely to explode, too!

15.2 Problems, threats or constraints

15.2.1 Systems and technology

It has been some years since GIS has been constrained by screen resolution and the number of available display colours, but certainly these are, once more, currently issues with mobile GIS. If visualisation is a problem, just applying the rules that have worked for paper maps is unlikely to be effective. According to Plesa and Cartwright (Chapter 8) new approaches are needed.

Another perceived 'threat', or at least constraint, associated with mobile GIS technology raised by this book's authors relates to locational privacy. Duckham and Kulik (Chapter 3) offer 'obfuscation' as a solution. There has been popular privacy invasion concern over phone cameras, with suggestions, for example, that they either emit a flash or a loud noise when used to take a photograph. Tracking individuals through the signals emanating from their mobile phones has been increasingly resorted to by law enforcement agencies. Records of these movements can be kept, and in the EU these must be, for at least 12 months. Beneficial use of such archives has been well publicised, but their unscrupulous sale and subsequent exploitation has not, yet, become a public issue. When society does debate this issue, and if the conclusion is that locational privacy is a right, then the technology must be available to protect it.

At the moment there is a huge range of hardware and systems available: rather like in the early days of personal computers. This offers major barriers to the creation of a collaborative environment in which effective mobile GIS can flourish. However, as with personal computers, standards must, and will, emerge.

15.2.2 Data, accuracy and scale

Another important source of possible problems, threats or constraints that can be detrimental to the development of location-aware devices and mobility studies is associated with data, accuracy and scale issues. First there is the issue of *data availability* that was mentioned in several of the chapters in the book. Reitsma and Albrecht, in Chapter 5, for example suggest that there is a lack of appropriate data for validating process definitions and the results of process-oriented data models. While Laube et al., in Chapter 14, point out that there is a lack of tracking data for large (i.e. more than 200) groups of individuals. Cost—this increases with the number of individuals being tracked, and the extent of spatial and temporal coverage—is a major contributor to this lack. It is therefore not surprising that many animal tracking studies focus on a small number of individuals (e.g. Curtis, 2000). In the case of humans, Mateos and Fisher, in Chapter 11, suggest that the need for user consent can also limit the size of the population sample than can be surveyed.

More fundamentally, the underlying data model can also affect the availability and quality of tracking data. How the data model can constrain data collection can be illustrated by the fact that tracks of mobile phones give cell information but do not disclose more accurate x,y coordinate observations. Mateos and Fisher, in Chapter 11, observe that the measurement of the mobility patterns of large groups of people through the analysis of the 'spatiotemporal signature' of their mobile phone is limited by the spatiotemporal *accuracy* imposed by the technology. They suggest that the current limited spatiotemporal accuracy of mobile phones makes it only appropriate to measure inter-urban mobility. Laube et al., in Chapter 14, also suggest that data originating from certain moving object database applications (e.g. taxi management systems – see for example Yeh et al., 2004) feature long static periods and rare updates and therefore might not be appropriate for some mobility studies. Finally, Laube et al., in Chapter 14, point out that investigating objects that move on a network, for example vehicles moving on a street network, may reveal more about the structure of the traffic network than about the behaviour of the drivers.

The other major issue that may possibly cause problems is *scale*. There are two key aspects to be considered here: the appropriateness of scale choice *and* scale effects (i.e. how the choice of scale may affect the results). First, in relation to scale choice there is a fine balance between collecting too much data and not collecting enough. For example, in relation to the spatial scale, O'Neill et al. (1996, p. 169) recommended that 'in reporting landscape pattern, grain should be 2 to 5 times smaller than the spatial features of interest'. In relation to the temporal scale, the 'granularity of time' and its importance for incorporating the temporal dimension in a GIS has also been studied (e.g. see Kemp and Kowalczyk, 1994, p. 91). Laube et al., in Chapter 14, warn that in order to avoid semantic mismatches, the knowledge discovery process must be performed at an 'adequate granularity': 'undersampling a lifeline causes information loss, while oversampling may drown out the track's signal and may even feign autocorrelation between successive moves'. For example, in their caribou case study Laube et al. suggest that in order to search for seasonal migration patterns an analysis granularity of hours would not be adequate because it might introduce noise caused by daily movement patterns.

Regarding scale effects it is well known that patterns of objects will change according to the spatial or temporal scale (e.g. see Fernandes et al., 1999 in the case of ecology; João, 2002, in the case of environmental assessment; Meentemeyer and Box, 1987, in the case of landscape studies; Osterkamp, 1995, in the case of water quality; Sposito, 1998, in the case of hydrology; and Stein and Linse, 1993, in the case of archaeology). Gray (1999, p. 330), for example, found that 'one's conclusions about whether land is degraded are influenced by the scope and scale of the analysis. For example, if we examined changes at the local or regional scale using aerial photographs, we would most likely arrive at a different conclusion than if we examined soils at the farm scale. The scale at which studies are undertaken affects the conclusion because processes and parameters important at one scale may not be important or predictive at another scale'. Openshaw (1984) discussed the Modifiable Areal Unit Problem (MAUP) in the case of the spatial scale. Laube et al., in Chapter 14, propose something equivalent but for the temporal scale. Laube et al. suggest that if in their study the temporal units were differently specified, different patterns and relationships would have been observed—i.e. a 'modifiable temporal unit problem' or MTUP (cf. MAUP mentioned above).

It is crucial to have accurate and up-to-date information (see Hummel, in Chapter 10) as no clever algorithm can compensate for poor data. However, it is also important to consider the human and legal aspects that may, for example, oblige a dilution of accuracy and this is discussed in the next section.

15.2.3 Human and legal aspects

We quite literally broadcast our location while using mobile GIS. This may prompt actions, which are life-saving, life-threatening or invade our privacy. Thus we are obliged to question the human and legal implications of dynamic and mobile GIS; such implications have been addressed significantly by Matt Duckham and Lars Kulik in Chapter 3 and are alluded to by Qingquan Li in Chapter 2. Further, Patrick Laube and his co-authors in Chapter 14, comment that 'in a globalised world, people, goods, data and ideas move in increasing volumes at increasing speeds over increasing distances, and more and more leave a digital trail behind them'. Data representing such a digital trail can be automatically collected, either overtly or covertly, in databases, implying not only active surveillance but the possibility of misrepresentation and adverse decision-making through the (mis)matching and analysis of spatially referenced and other data that identifies individuals.

To consider these data and their databases, are the implications (with regard to the expectations that individuals remain unidentified, private) understood by policy-makers, systems developers and the public? Bennet and Raab (2003) remind us that Government proposals for the electronic delivery of services and information; the rationalisation of information processes; and, open government, depend, for their effectiveness and acceptability, on controlling the potential misuse of personal data. Are controls to accessing these data in place, or being adequately thought about at government level? Electronic identity theft and fraud are now publicly discussed (Gowen and Hernadez, 2005) and obviously of concern to the financial sector. Perhaps if research for the prevention of financial fraud can be aligned with that for the protection of privacy, then technical solutions will emerge. The benefits of such an alignment can be understood from figures proposed by Ingrian Networks (2005), which claim that each security breach costs a financial firm, on average, $1.65 billion in market capitalisation. Without this alignment, there is a good chance that those developing techniques to abuse data privacy will 'win out'.

The current situation has prompted the technical response, outlined in this book's Chapter 3, namely obfuscation. Duckham and Kulik propose that an individual's location is protected by broadcasting a set of locations (an obfuscation set), only one of which is the individual's true location. For this, or any technical solution, to be effective not only does the proposed technology have to be thoroughly researched, but also the techniques employed, now, or having future potential, to invade a person's privacy (circumventing location privacy protection and attempting to discover an individual's exact location) must be understood. Other extant technical solutions include authentication of all access; audit trials of all access; identification of security breaches and suspicious attempted access; data masking; encryption hardware; and above all internal security. It is claimed that 50% of all security breaches arise after being internally (Ingrian Networks, 2005) initiated.

We need to answer some questions, such as what level of protection we actually want and how ethical concerns should constrain the availability of geo-spatial, especially lifeline, data in the years ahead. Not alluded to in this book, despite its international authorship, are the very different levels of privacy incursion found acceptable by different societies. Given the global nature of the problem, awareness of these varying attitudes should inform any discussion.

15.3 Future research

15.3.1 Spatiotemporal information theory and spatiotemporal analysis

Considering dynamic or mobile GIS without accordingly extending the available spatial analyses tools would be meaningless. But as well as technical advances, a deep reflection on core spatiotemporal information concepts must be undertaken. In this respect, the work still to be done is tremendous.

Clearly considering events, processes and dynamic objects (instead of static objects) involves a huge conceptual evolution impacting every aspect of information capture, maintenance, analysis and visualisation. In a way, an upper-level ontology of dynamic geographic phenomena has still to be defined. New sets of spatiotemporal relationships should be described which will have the same impact on modelling strategies as topological relationships had on 2D GIS in the last quarter of the 20th century.

From the subjects tackled in this book, we can detect some important future directions. First of all, getting true 3D (2D + time) and 4D (3D + time) models remains a challenge. Multi-dimensional motion patterns (i.e. encompassing two of more motion properties such as speed, change of speed, motion azimuth and sinuosity) indicate another promising direction research could take (Chapter 14). Likewise work concerning the analytical and statistical methods needed to test the significance and similarities among event patterns is needed (Chapter 4).

One needs, at least, to define and implement new indexing strategies, new query languages, new visualisation methods, new analytical tools, advanced spatial analysis, statistical functions and new spatiotemporal constraints. Concrete examples are the definition of optimisation algorithms, for the rapid processing of spatial information accommodated in a small-capacity memory, fast extraction and compression of spatial information in the context of large user groups and concurrent data manipulation (see Chapter 1).

15.3.2 Equipment and devices

Considering research directions relating to hardware, several issues emerge. There are no user interfaces designed specifically for mobile GIS. Most current mobile GIS software still follows the traditions of desktop GIS interfaces, but a tiny stylus and on-screen keyboard do not support these nor are they right for mobile GIS, at least in the emergency context. Direct voice commands or a touch screen simply used by human fingers are both more appropriate for emergency responders and field workers, according to Tsou and Sun (Chapter 12).

Currently, a GIS professional has to manually convert the data submitted from field workers to a Web-based GIS framework. Some predicted advances in Web Services technologies and improvement in distributed database functions might automate these tasks in the future. However, it is always dangerous to rely on automatic data conversion without verifying the data accuracy and data quality. Quality control procedures have to be established to verify the accuracy of submitted geo-spatial data from the field.

Mobile GIS also allows geo-spatial analysis to take place in the field, at the site of interest. Many emergency and disaster management tasks need advanced GIS analytical functions requiring significant computing power. Most mobile GIS devices are tiny and only have very limited computing capability; thus the processing time for spatial analysis and image processing might prevent the adoption of mobile GIS for real-time response. One possible solution is to execute the power-hungry GIS functions via the Internet at remote GIS engine services. Then, the results can be sent back to the mobile GIS devices, also via the Internet.

Since most mobile GIS devices are small and fragile, emergency responders and managers might be reluctant to use them to share their maps with others. One possible alternative is to print out paper maps directly from mobile GIS devices via wirelessly portable printers or from an in-built printer inside a Pocket PC or a notebook computer. So far we have considered mainly the human eye as the sensor. How can mobile phones be equipped to make them environmental monitoring kits? Gouveia et al. (in Chapter 13) ask which other devices can be integrated or coupled with mobile GIS?

15.3.3 Data and accuracy

It seems quite strange to still be talking about data quality and accuracy as issues to be placed on the GIS research agenda, given that there are at least two international conference series, namely: International Seminar on Spatial Data Quality (ISSDQ, 2005) and Spatial Accuracy Assessment in Natural Resources and Environmental Sciences (ACCURACY2006, 2005), devoted to the topic and that it has been sessioned in every GISRUK conference. Nevertheless Maguire reminds us, in this book's first chapter that 'we need to develop techniques for reducing, quantifying, and visually representing uncertainty in geographic data and for analysing and predicting the propagation of this uncertainty through GIS-based data analyses'. But we are now talking about a different set of practices. Perhaps the advent of mobile GIS, with its limited visualisation and processing capabilities and less GIS-attuned users will really focus our minds on these issues of data and accuracy? Certainly it would be irresponsible not to be, at least, considering ways of transmitting the quality of geo-spatial information to this potentially huge group of GIS novice users.

Mobile GIS is not just about information display. It can also be about data capture. Experienced GIS users are aware of several primary data capture methods, and their relative qualities. A danger with mobile GIS is that low-resolution GPS positioning techniques will be the only primary data capture methods implemented. Are there ways of prompting the user to achieve appropriate data capture standards?

Information generation not only uses data, it also needs processing models. If they are of low quality so is the information. Rietsma and Albrecht, in Chapter 5, suggest that they do not know of any measurement approach that quantitatively records process information. This probably ignores the long history of error estimation supported by Least Squares Adjustment which will be familiar territory to those GIS workers with a Geomatics background (e.g. Mikhail, 1976), and more recent developments in crisp and fuzzy set theory where the probability or certainty of a rule holding (Drummond, 1991) can be determined from observing the outcome of information generation procedures. But certainly these procedures need more of an airing; a wider consideration by the dynamic and mobile GIS community.

The tenor of this section, so far, has been that this book has not augmented the *data, accuracy and scale* research agenda in any way. But this is far from the case if we turn to the dynamic aspect of this book's title. As Laube and his co-authors say in Chapter 14 'tracking data are in many cases not perfect'. One should never expect them to be perfect. How can the imperfection be quantified? How can the effect of imperfect tracking data on generated information be known? This must be a research item.

A cautionary note. Surprisingly some claim tracking data are in short supply. Are there the sources to carry out research into the quality of these data? Again Chapter 14's authors have a suggestion, 'where real observation motion data are lacking or suffer from poor quality, carefully synthesised artificial motion data offer a feasible alternative to studying some processes [...] artificial life forms are always visible, healthy, don't die, don't get shot, don't lose their GPS receiver, don't need privacy and are willing to report their location at any desired time.'

15.3.4 Behaviour

Do we need to consider behaviour as a component of the Dynamic and Mobile GIS research agenda? We may consider human behaviour, but animal behaviour is an issue too. We may consider the behaviour of the GIS user gathering, transmitting and processing data at a remote location. We may consider the GIS user as an economic being. We may consider the behaviour of the dynamic objects we represent in our databases. We may consider research into behaviour as being something that raises privacy or other ethical issues.

Considering the last of these, it has already been noted by Duckham and Kulik (Chapter 3) that existing approaches to location privacy are static in nature and the development of truly spatiotemporal approaches to location privacy are needed.

Turning to the user, given the level of GIS skill expected amongst the majority of future mobile GIS practitioners, issues related to the nature and orientation of geo-spatial visualisation are of concern. Plesa and Cartwright in Chapter 8 make a case for adding an assessment of realistic visualisation to the research agenda, claiming a 'need to develop some system of classification of images between abstract and photorealistic' as an early step in this research.

Dynamic and Mobile GIS offers several technologies, each with cost implications. Which business models support the use of mobile technologies and which will be acceptable? How will this new pool of GIS users behave economically?

The accurate representation of a tracked object's movement, between recordings, requires research into interpolation methods based on an understanding of the object's behaviour. This involves the integration of the geometric properties of the object's motion with semantic information (such as cultural background, socio-economic status, transport mode) and details of the geographic environment harbouring the motion. As suggested by Laube and coauthors in Chapter 14, any assumption of objects moving through undifferentiated space does not hold for the complex motion of genetically imprinted or intelligent objects, following their chosen corridors, valleys or ridges.

15.4 Recommendations for future practice

15.4.1 Standards

Important work has been done on the formalising of 2D geo-spatial information. This has given birth to several norms and standards. Such work, prompted by the need for interoperability between data and systems, has mainly arisen after the first commercial GIS emerged. Although norms and standards should be used at the early stages of GIS or spatial database implementation, it is not uncommon that organisations using GIS or maintaining spatial databases are still occupied by defining core or domain ontologies, building data dictionaries or conceptual data models, upgrading data models and data structures. This is obviously not good practice, but is sometimes the result of the non-availability, poor understanding or poor definition of norms and standards.

While dynamic and mobile GIS are still in their early stages it is essential not to make the same mistakes as were made with 2D GIS. Norms and standards should be adopted which are based on deep theoretical reflection, particularly taking into account this new representation of real-world processes. Citing Li (Chapter 2):

> 'LBS standards, for spatial information abstraction, mobile services integration, spatial data compression, positioning and data transformation in Mobile GIS should be based on OpenGIS specifications of OGC, wireless application protocol (WAP) forum, mobile location protocol of Open Mobile Alliance, mobile location service, Web services specifications, GSM (including GPRS and EDGE) and W-CDMA specifications […] from third generation partnership project (3GPP), UMTS (Universal Mobile Telecommunications System) technical specifications, and the standards from W3C including Scalable Vector Graphics (SVG), Mobile Web Initiative and Web Services'.

This incomplete, but already impressive list, demonstrates the variety and complexity of norms, specifications and standards that have to be considered.

15.4.2 Institutional aspects

The possible implementation of the suggestion posed by Mateos and Fisher in Chapter 11, that mobile devices might form the basis of a new spatial reference system to analyse population, has institutional consequences. New legislation would be needed to govern the sampling of the location of the population, for example on selected survey days. Mateos and Fisher propose that similar guidelines to the national census of population could require coverage of a large part of the population while at the same time safeguarding anonymity (e.g. individual privacy could be assured by only publishing and visualising information in aggregated ways).

Accuracy and privacy issues are absolutely key and need to be addressed before starting an extensive collection and analysis of mobile data (see Mateos and Fisher, Chapter 11, on spatiotemporal accuracy in mobile phone location, and Matt Duckham and Lars Kulik, Chapter 3, on location privacy and location-aware computing). The likely large-scale systematic storage of location data in the future is a key challenge to Geographical Information Science not only in terms of database storage and processing capacity, but also in terms of GIS data models and the ontology of spatiotemporal representation.

References:

ACCURACY2006 (2005) [Online], Available:
    http://terrestrial.eionet.eu.int/activities/announcements/ann1132141342/announcement.

Bennet, C. J. and Raab, C. D. (2003) *The Governance of Privacy: Policy Instruments in Global Perspective*, Aldershott, UK: Ashgate Publishing.

Curtis, A. R. (2000) [Online], Available:
    http://www.spacetoday.org/Satellites/Tracking/Animals/MalaysiaElephant.html [24/01/2006].

Drummond, J. E. (1991) *Determining and Processing Quality Parameters in Geographic Information Systems,* The Netherlands: ITC.

Fernandes, T. F., Huxham, M. and Piper, S. R. (1999) 'Predator caging experiments: a test of the importance of scale', *Journal of Experimental Marine Biology and Ecology*, vol. 241: pp. 137–154.

Gowen, A. and Hernandez, N. (2005) 'Pickpocketing has changed to identity theft', *The Washington Post*, [Online], Available: http://sltrib.com/business/ci_3253201 [29/11/2005].

Gray, L. C. (1999) 'Is land being degraded? A multi-scale investigation of landscape change in Southwestern Burkina Faso'. *Land Degradation & Development*, vol. 10: pp. 329–343.

ISSDQ, (2005) *Proceedings of the International Symposium on Spatial Data Quality, ISSDQ'05*, Beijing, China, August 25-26, 2005. Inst. Remote Sensing and Geographic Information System, Peking University, China, 100871

Ingrian Networks, (2005) 'Achieving Data Privacy in the Enterprise', [Online], Available: http://www.ingrian.com/resources/whitepapers [November 2005].

João, E. (2002) 'How scale affects environmental impact assessment', *Environmental Impact Assessment Review*, vol. 22 (4): pp. 287–306.

Kemp, Z. and Kowalczyk, A. (1994) 'Incorporating the temporal dimension in a GIS', in: Worboys, M. (ed.), *Innovations in GIS 1: Selected papers from the First National Conference on GIS Research UK*, London: Taylor & Francis.

Meentemeyer, V. and Box, E. (1987) 'Scale effects in landscape studies', in Turner, M. G. (ed.), *Landscape Heterogeneity and Disturbance*, New York: Springer Verlag, pp. 15–34.

Mikhail, E. M. (1976) *Observation and Least Squares*, New York: IEP.

O'Neill, R., Hunsaker, C., Timmins, S., Jackson, B., Jones, K., Riitters, K. and Wickham, J. (1996) 'Scale problems in reporting landscape pattern at the regional scale', *Landscape Ecology*, vol. 11 (3): pp. 169–180.

Olsen, H. (1995) 'Gloucester gets connected', *Surveyor*, 182 (5340): pp. 22–24.

Openshaw, S. (1984) *The Modifiable Areal Unit Problem*, Norwich: Geo Books.

Osterkamp, W. (ed.) (1995), *Effects of Scale on Interpretation And Management of Sediment and Water Quality*. IAHS Publication No. 226 International Association of Hydrological Sciences.

Perez-Trejo, F. (1990) 'Dynamic-GIS: An "intelligent" tool for understanding the impacts of climatic change on complex ecosystems', in: Boer, M. M. and De-Groot, R. S. (eds.), *Landscape-Ecological Impact of Climatic Change*. Proc. conference, Lunteren, 1989. (IOS Press, Amsterdam), pp. 318–324.

Sposito, G. (ed.) (1998) *Scale Dependence and Scale Invariance in Hydrology*, Cambridge: Cambridge University Press.

Stein, J. K. and Linse, A. R. (eds.) (1993) *Effects of Scale on Archaeological and Geoscientific Perspectives*, Boulder, CO.: Geological Society of America.

Yeh, A.G.O., Lai, P.C., Wong, S.C. and Yung, N.H.C. (2004) 'The architecture for a real-time traffic multimedia internet geographic information system', *Environment and Planning B: Planning-and-Design*, 31 (3): 349-366.

Subject Index

T - #0373 - 071024 - C7 - 234/156/15 - PB - 9780367389932 - Gloss Lamination